A Modern Introduction to Probability and Statistics

A Modern Introduction to Probability and Statistics

Understanding Statistical Principles in the Age of the Computer

Graham J. G. Upton

OXFORD UNIVERSITY PRESS

Great Clarendon Street, Oxford, OX2 6DP,
United Kingdom

Oxford University Press is a department of the University of Oxford.
It furthers the University's objective of excellence in research, scholarship,
and education by publishing worldwide. Oxford is a registered trade mark of
Oxford University Press in the UK and in certain other countries

Published in the United States of America by Oxford University Press
198 Madison Avenue, New York, NY 10016, United States of America

British Library Cataloguing in Publication Data

Data available

Library of Congress Control Number: 2025930380

ISBN 9780198943129
ISBN 9780198943136 (pbk.)

DOI: 10.1093/oso/9780198943129.001.0001

Printed and bound by
CPI Group (UK) Ltd, Croydon, CR0 4YY

The manufacturer's authorised representative in the EU for product safety is
Oxford University Press España S.A. of El Parque Empresarial San Fernando
de Henares, Avenida de Castilla, 2 – 28830 Madrid (www.oup.es/en or product.
safety@oup.com). OUP España S.A. also acts as importer into Spain of products
made by the manufacturer.

PREFACE

Audience

This is an introductory course suitable for anyone who is meeting the subject of statistics for the first time. This includes students at school, undergraduates, postgraduates and anyone else who wants to learn about the subject. Since statistics is a fundamental component of data science, this book should also be read by students of that discipline.

Syllabus

My aim has been to cover virtually all of the material that is included in the first-year syllabi of a dozen universities (selected, of course, at random!). However, the same material is just as appropriate at school level and this book is heavily based on my (joint with Ian Cook) earlier book, *Understanding Statistics*, which was aimed at 16-year-olds, but has been a set text for the first year statistics course at one university for more than 30 years.

R

Most universities introduce statistics students to the R programming language as part of their first-year statistics course. However, introductory statistics is also taught as part of non-mathematical degrees at both undergraduate and postgraduate level. These students may already be familiar with statistical packages (for example, SPSS, Stata, SAS) better suited to their own disciplines. For this reason, I have not dwelt on the use of R in the main text, but I do provide information concerning the relevant R commands at the end of each chapter. I strongly encourage those unfamiliar with R to get acquainted! It is free to download for any platform. I also recommend using it in conjunction with the *RStudio* integrated development environment (IDE) for R. *RStudio* is free and can also be used with *Python*. The platform can keep your program, results, program history, and plots all neatly arranged.

Use of statistical tables

With access to an appropriate computer program, it will be possible to obtain exact answers to many standard questions, with just a few lines of computer code. This is obviously preferable. However, I am conscious that there will be occasions where all the reader has to hand is a set of tables (for example, in an examination). For that reason, I have included, within the running text, brief extracts from relevant tables, so as to provide some familiarity with what may be available.

Use of approximations

When I was a student, for many situations, exact calculations were impractical and one was obliged to rely on approximations. Nowadays, with a computer, those approximations are rarely needed. However, for completeness, and as a nod to yesteryear, I have included some that I still find useful.

Computers and formulae

It is no longer necessary to work through complicated formulae in order to analyse data or answer some question of interest. A short computer command will often suffice. However, it seems to me to be wrong to regard the computer simply as a black box: one should surely have an idea of what has been calculated. Therefore, in Part B (Statistics), there are many cases where I derive (or simply present) the underlying formulae, so that the reader knows what is going on in that black box.

The more mathematical sections

Although this book covers the syllabi of the first-year statistics courses of most universities, the intention is that it should also be suitable for any reader who is familiar with the basic methods of calculus. For this reason, the more mathematical sections (or chapters) are indicated with an asterisk, while the more mathematical passages elsewhere in the text are shaded. The idea is that these can be omitted without interrupting the flow of information.

Two parts

The book is divided into two parts, with the first part concentrating on the mathematics of probability. There are only fleeting references to data in this part, which includes some rather technical chapters that can be ignored by those more averse to mathematics. The second part is concerned with methods for interpreting and analysing data. Here, the probability distributions introduced in the first part play a critical role.

Exercises

Small numbers of exercises are sprinkled liberally through the book. There are answers at the end of the book. For the most part, the emphasis is on interpretation of results rather than mundane number-crunching.

Projects and group work

Statistics is about random variation. The best way to understand random variation is to experience it. For that reason I have included a number of projects. Some of these can be done on one's own, but every project will benefit by a comparison of the results obtained by different people.

Textbooks versus real life

Every statistics textbook (and this one is no exception) is prone to give the reader the impression that this is the only way in which the analysis can be carried out. This is partly influenced by the fact that there are sets of questions (with the expected answers given in the back of the book). Real-life problems are usually more complex, with many variables being relevant. Give ten experienced statisticians a real-life problem, and they will very probably tackle it in ten different fashions. However, their final conclusions are likely to be broadly similar. In this book I have tried to provide you with the basic tools of statistical analysis, but exactly how you use them will be up to you.

Graham Upton, Wivenhoe, February 2025

ACKNOWLEDGEMENTS

It is a pleasure to acknowledge the forensic scrutiny applied by my former colleagues, Ian Cook and Dan Brawn, to earlier versions of this text. I also thank the copy editor, C. N. Lau, for greatly improving the presentation. I would also like to thank the commissioning editor, Dan Taber, and the other OUP staff, for their assistance with this publication. Any errors that remain are, of course, my responsibility.

CONTENTS

PART II STATISTICS

PART I
Probability

Probability and statistics are, of course, closely related subjects, since a knowledge of probability is essential to appreciate the random variation that is apparent in observed data. Nevertheless, in the late eighteenth and early nineteenth centuries, probability questions were the subjects of letters between the eminent mathematicians of the time, because they were more interested in the application of probability to gambling than data analysis!

1

Probability

Suppose someone says that it will probably rain today. We understand that that means that it is more likely to rain than not to rain, but we can't be more definite than that. Indeed, how believable the statement is will depend on who makes it: if it were made by my Aunt Matilda, then that would be quite a different matter to a statement that had been made by a meteorologist. In this chapter, we will be more specific: we will assign a numerical value to the occurrence of an event, with 0 meaning that the event cannot occur, and 1 meaning the event is certain to occur. We need rules to determine such values. These rules are the subject of this chapter and Chapter 2.

1.1 Relative frequency

Suppose we roll a six-sided die[1] and are interested in the outcome '6'. To get some idea of how likely the outcome is, we roll the die repeatedly. Here are the first 10 rolls:

2 4 4 1 2   3 2 4 3 1

After 10 rolls we have had no 6s. We might therefore think that getting a 6 was impossible. But, here are the next 20 rolls:

4 5 6 4 3   2 3 6 2 4   3 4 2 2 5   4 6 5 3 3

After 30 rolls, we have had three 6s $\longrightarrow$ a **relative frequency** of $3/30 = 0.1$.

What will happen as we increase the number of rolls? The answer is that the number of 6s will continue to rise, whereas the relative frequency will slowly tend to some limiting value. It is this limiting value that we call the **probability**.

So, if all six sides of the die are equally likely (which is the case for a **fair die**), then the limit of the relative frequency will be $1/6$ and we will say that the probability of a 6 is $1/6$.

1.2 Preliminary definitions

- A **statistical experiment** is one in which there are a number of possible outcomes and we have no way of predicting which outcome will actually occur. Sometimes the experiment may have already taken place, but we remain ignorant of the outcome.

[1] The word 'die' is the singular form of the familiar 'dice'. Unless otherwise stated, all six-sided dice in this book have sides numbered, respectively, 1, 2, 3, 4, 5, and 6.

- The **sample space**, often denoted by S, is the set of all possible outcomes of the experiment.
- An **event** is any set of possible outcomes of the experiment (thus, an event is a subset of S). When rolling a die we might be interested in events such as 'getting a number greater than 3', 'getting a 6' or 'getting an even number'.
- A **simple event** is an event consisting of a single outcome. When rolling a normal six-sided die, the simple events are 1, 2, ..., 6.

1.3 The probability scale

Assigned to the event E is a number, the probability of the event E, which takes a value in the range 0 to 1 (inclusive). The number is denoted by $\mathrm{P}(E)$. In addition to satisfying

$$0 \leq \mathrm{P}(E) \leq 1,$$

the value of $\mathrm{P}(E)$ is such that

$$\begin{aligned} \text{if } E \text{ is impossible} &\longrightarrow \mathrm{P}(E) = 0, \\ \text{if } E \text{ is certain to occur} &\longrightarrow \mathrm{P}(E) = 1. \end{aligned}$$

Intermediate values of $\mathrm{P}(E)$ have natural interpretations:

$$\begin{array}{lcll} \mathrm{P}(E) & = & 0.5 & \longrightarrow & E \text{ is as likely to occur as not to occur,} \\ \mathrm{P}(E) & = & 0.001 & \longrightarrow & E \text{ is very unlikely,} \\ \mathrm{P}(E) & = & 0.999 & \longrightarrow & E \text{ is highly likely.} \end{array}$$

Example 1.1

We toss an ordinary coin. We are interested in the probabilities of the events A, *and* B, *where:*

A : *The coin comes down heads.*
B : *The coin explodes in a flash of green light.*

We can reasonably assume that $\mathrm{P}(A) = 1/2$ and that $\mathrm{P}(B) = 0$.

1.4 Probability with equally likely outcomes

Suppose that the sample space, S, consists of $n(S)$ equally likely possible outcomes, with $n(E)$ being the number of outcomes that correspond to the event E. Then $\mathrm{P}(E)$, the probability that the event E occurs, is given by

$$\mathrm{P}(E) = \frac{n(E)}{n(S)}. \tag{1.1}$$

This clearly satisfies the requirement that $0 \leq \mathrm{P}(E) \leq 1$.

Example 1.2

A fair six-sided die is tossed. Suppose that we are interested in the event A defined as 'the number obtained is a multiple of 3'.

The sample space, S, consists of the equally likely outcomes $\{1, 2, 3, 4, 5, 6\}$, with $n(S) = 6$. The outcomes corresponding to A are $\{3, 6\}$, so $n(A) = 2$ and $\mathrm{P}(A) = n(A)/n(S) = 2/6 = 1/3$.

Example 1.3

Two fair coins are tossed. We require the probability that exactly one head is obtained.

Each coin is equally likely to give a head (H) or a tail (T). Even if the coins are tossed simultaneously we can be sure that one lands a few nanoseconds before the other. The sample space, S, therefore contains four equally likely possible sequences: HH, HT, TH, TT.

The event of interest, A, corresponds to the outcomes HT, TH. Thus $n(A) = 2$, $n(S) = 4$ and hence $\mathrm{P}(A) = n(A)/n(S) = 2/4 = 1/2$.

Hopefully, that last problem seemed very straightforward. However, in the eighteenth century, the noted mathematician d'Alembert[2] incorrectly argued that, since the coins were fair and there were only three possible outcomes, the answer was 1/3. d'Alembert had failed to enumerate the sample space.

Exercises 1a

1. An unbiased die is thrown. Find the probability that:
   - **(a)** the score is even,
   - **(b)** the score is at least 2,
   - **(c)** the score is at most 2,
   - **(d)** the score is divisible by 3.
2. A box contains four red balls, six green balls, and five yellow balls. A ball is drawn at random. Find the probability that:
   - **(a)** the ball is green,
   - **(b)** the ball is red,
   - **(c)** the ball is not yellow.

[2] Jean le Rond d'Alembert (1717–1783) was a French mathematician and physicist whose name is now associated with equations, principles, theorems, and a paradox. However, it seems that probability was not his strong point!

3. A disc carries the numbers 1 and 2 on its faces. It is thrown with a fair die. The score is the sum of the two numbers that show. Find the probability that:
   - **(a)** the score is at least 4,
   - **(b)** the score is at most 6.
4. I have 14 coins in my pocket. There are two 1p coins, three 2p coins, four 5p coins, and five 10p coins. I choose a coin at random. Find the probability that:
   - **(a)** it is a 2p coin,
   - **(b)** it is worth at least 5p.
5. An ordinary pack of 52 playing cards contains, in addition, one joker. A card is drawn at random from the well-shuffled pack. What is the probability that it is the joker?
6. Two unbiased dice, one red and one green, are thrown and their separate scores are noted. Represent the result as (r, g), where r and g are the scores on the red and green dice, respectively. Explain why there are 36 possible simple events. Hence find the probability that:
   - **(a)** a double six is obtained,
   - **(b)** a double (any score) is obtained,
   - **(c)** the sum of the two scores is 4,
   - **(d)** the score on the red die is greater than that on the green die,
   - **(e)** both scores are divisible by 3.

1.5 The complementary event, E'

The complementary event, E', is the event defined as 'the event E does not occur'.[a]

[a] You may also find the complementary event denoted by $\bar{E}$, $C(E)$, or E^c

If $n(E)$ is the number of outcomes for which E occurs, with $n(S)$ denoting the total number of outcomes in the sample space, then $n(S) - n(E)$ is the number of outcomes corresponding to the event E'. Thus,

$$P(E') = \frac{n(S) - n(E)}{n(S)} = 1 - \frac{n(E)}{n(S)} = 1 - P(E).$$

This result

$$P(E') = 1 - P(E), \tag{1.2}$$

or its equivalent

$$P(E) = 1 - P(E'),$$

often enables us to simplify calculations.

$(E')' = E$, since, if E' does not occur, then E occurs, and viceversa.

Example 1.4

> *We toss a red die and a blue die. Both dice are fair with six sides. We wish to find P(A), where A is the event 'the total of the numbers shown by the two dice exceeds 3'.*

We begin by finding $n(S)$, the number of possible outcomes in the sample space. There are six equally likely outcomes for the red die. Whichever of these outcomes arises, there will also be six equally likely outcomes for the blue die. In all, therefore, there are thirty-six equally likely outcomes: $n(S) = 36$. We can see this easily on a diagram that shows the possible totals of the two dice:

		Red die					
		1	2	3	4	5	6
	1	2	3	4	5	6	7
	2	3	4	5	6	7	8
Blue	3	4	5	6	7	8	9
die	4	5	6	7	8	9	10
	5	6	7	8	9	10	11
	6	7	8	9	10	11	12

The complementary event, A', is the event that 'the total of the two dice does not exceed 3'. Now, whereas there are lots of outcomes for which A occurs, there are very few for which A' occurs and it is easy to count them: the (red die, blue die) possibilities are $\{(1,1), (1,2), (2,1)\}$. Hence $n(A') = 3$. The 33 remaining outcomes in the diagram correspond to the event A.

Since all the outcomes are equally likely, $P(A) = 33/36 = 11/12$.

In the last example, the fact that the dice were coloured was scarcely relevant. It simply made it easier to describe what is happening. All that is required is some method of distinguishing the dice, and this is *always* possible, even if the dice are described as being identical! We could imagine the dice as being rolled one after the other, or being rolled by different people, or being rolled at the same time from different starting points in the die shaker.

1.6 Venn diagrams

A Venn[3] diagram is a simple representation of the sample space and any relevant events. It is often helpful in seeing 'what is going on'. Usually the sample space is represented by a rectangle, with individual regions within the rectangle representing events.

[3] John Venn (1834–1923) was a Cambridge lecturer whose speciality was logic. His major work, *The Logic of Chance*, was published in 1866. It was in this work that he introduced the diagrams that now bear his name.

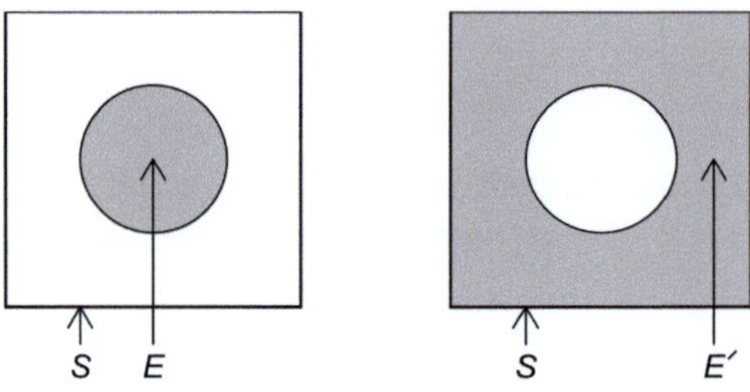

Figure 1.1 Venn diagrams illustrating the sample space S and the events E and E'.

1.7 Unions and intersections of events

Suppose A and B are two events associated with a particular statistical experiment. We now consider the events denoted by $A \cup B$ and $A \cap B$, which are defined as follows:

$A \cup B$	A **OR** B	At least one of A and B occurs.
$A \cap B$	A **AND** B	Both A and B occur.

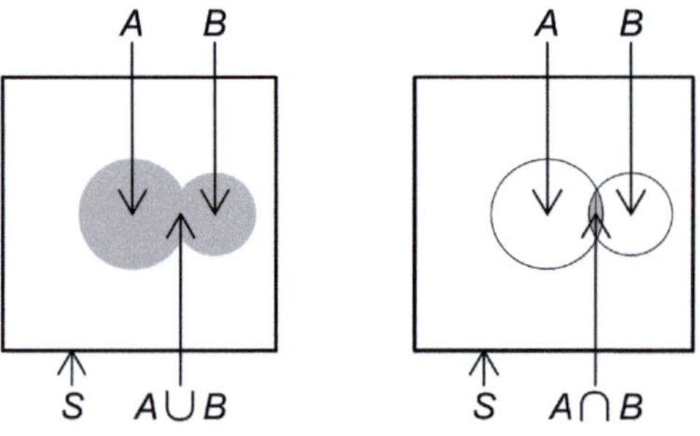

Figure 1.2 Venn diagrams illustrating the events $A \cup B$ and $A \cap B$.

- $A \cup B$ is called the **union** of A and B,
- $A \cap B$ is called the **intersection** of A and B.

Suppose that the numbers of outcomes in A, B, and $A \cap B$ are, respectively, $n(A)$, $n(B)$, and $n(A \cap B)$. Those outcomes in $n(A \cap B)$ are included in both $n(A)$ and $n(B)$. The outcomes in $A \cup B$ include all those in A and all those in B but no others. If we simply add together $n(A)$ and $n(B)$, we will overstate the number in $A \cup B$, because we will have counted those in $A \cap B$ twice.

Hence,

$$n(A \cup B) = n(A) + n(B) - n(A \cap B).$$

Dividing throughout by $n(S)$ we get

$$\mathrm{P}(A \cup B) = \mathrm{P}(A) + \mathrm{P}(B) - \mathrm{P}(A \cap B). \tag{1.3}$$

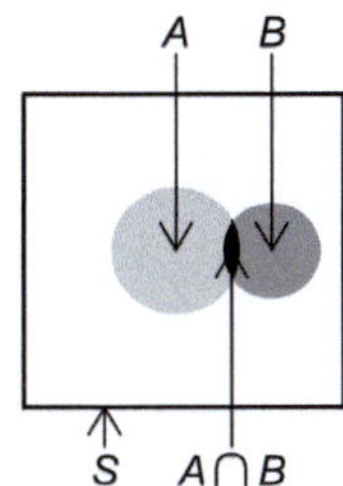

Figure 1.3 Venn diagram illustrating the relation
$P(A \cup B) = P(A) + P(B) - P(A \cap B)$.

This is sometimes referred to as the **inclusion–exclusion formula**. It generalizes easily (but messily) to more than two events. For example, for three events A, B, and C, we have

$$\begin{aligned} P(A \cup B \cup C) &= \{P(A) + P(B) + P(C)\} \\ &\quad -\{P(A \cap B) + P(A \cap C) + P(B \cap C)\} \\ &\quad +P(A \cap B \cap C). \end{aligned}$$

Example 1.5

Each month a mail order firm awards a 'Star Prize' to a randomly chosen shopper. The firm uses the following procedure. It first chooses eight shoppers at random. The names of these eight shoppers are put into a hat. A guest celebrity then draws the name of the lucky winner of the 'Star Prize' from the hat, with the other seven shoppers being awarded consolation prizes.

One month, the first of the eight shoppers was a male living in the south of the country. The complete list of those chosen was

Shopper number	1	2	3	4	5	6	7	8
North (N) or South (S)	S	S	N	S	N	S	S	N
Male (M) or female (F)	M	F	F	F	M	F	F	F

We define the events A and B as follows:

A: The winner of the 'Star Prize' lives in the south.
B: The winner of the 'Star Prize' is male.

- The event $A \cap B$ is the event: 'the winner of the "Star Prize" is a male living in the south'.
- The event $A \cup B$ is the event: 'the winner of the "Star Prize" is either a male, or lives in the south (or both)'.

The situation is illustrated in the Venn diagram, with the eight simple events (the shoppers), which are all equally likely, being identified by their numbers.

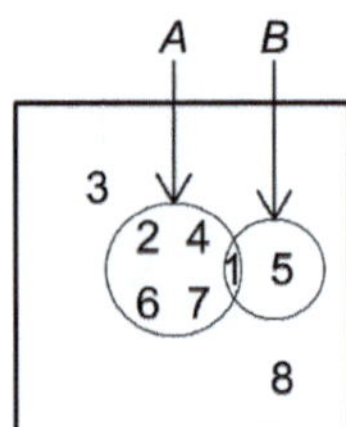

It can be seen that only the first of the eight simple events corresponds to $A \cap B$. The following table provides a comprehensive list of the various events:

Event (E)	Simple events in E	$n(E)$	P(E)
Sample space, S	1, 2, 3, 4, 5, 6, 7, 8	8	1
A	1, 2, 4, 6, 7	5	5/8
B	1, 5	2	$2/8 = 1/4$
$A \cap B$	1	1	1/8
$A \cup B$	1, 2, 4, 5, 6, 7	6	$6/8 = 3/4$

As a check, note that

$$\mathrm{P}(A \cup B) = \mathrm{P}(A) + \mathrm{P}(B) - \mathrm{P}(A \cap B) = 5/8 + 2/8 - 1/8 = 6/8.$$

1.8 Mutually exclusive events

Events A, B, ..., M, are said to be **mutually exclusive** if the occurrence of one of them implies that none of the others can have occurred. If D and E are two mutually exclusive events, then $\mathrm{P}(D \cap E) = 0$.

All simple events are mutually exclusive.

1.8.1 The addition rule

If the events A and B are mutually exclusive, then $\mathrm{P}(A \cap B) = 0$. This simplifies Equation (1.3) to give

$$\mathrm{P}(A \cup B) \quad = \quad \mathrm{P}(A) + \mathrm{P}(B), \tag{1.4}$$

which is known as the **addition rule**.

The addition rule *only* applies to mutually exclusive events.

Example 1.6

An Irish rugby club contains 40 players, of whom 7 are called O'Brien, 6 are called O'Connell, 4 are called O'Hara, 8 are called O'Neill, and there are 15 others. The 40 players draw lots to decide who should be captain of the first team. Determine the probability that the captain of the first team is

(a) *called either O'Brien or O'Connell,*

(b) *is not called either O'Hara or O'Neill.*

The sample space consists of the 40 players, each of whom is equally likely to be selected as captain. Denote the event that 'the captain is an O'Brien' by the symbol B, with C, H, and N denoting the similar events. The events B, C, H, and N are mutually exclusive, since a player cannot have two surnames.

$$\textbf{(a)}\ \mathrm{P}(B \text{ or } C) = \mathrm{P}(B \cup C) = \mathrm{P}(B) + \mathrm{P}(C) = \frac{7}{40} + \frac{6}{40} = \frac{13}{40}.$$

The probability that the captain is called O'Brien or O'Connell is 13/40.

$$\begin{aligned}\textbf{(b)}\ \mathrm{P}(\text{Neither } H \text{ nor } N) &= 1 - \mathrm{P}(H \text{ or } N)\\ &= 1 - \{\mathrm{P}(H) + \mathrm{P}(N)\}\\ &= 1 - \left(\frac{4}{40} + \frac{8}{40}\right) = \frac{28}{40} = \frac{7}{10}.\end{aligned}$$

The probability that the captain is not called O'Hara or O'Neill is 7/10.

1.9 Sigma notation

Suppose that we are interested in the probability that one of the mutually exclusive events $A_1, A_2, ..., A_5$ occurs. Using the addition rule, this probability is $\mathrm{P}(A_1) + \mathrm{P}(A_2) + \cdots + \mathrm{P}(A_5)$. We now introduce some simplifying notation that will be frequently encountered later in the book. In place of $x_1 + x_2 + \cdots + x_n$ we write

$$\sum_{i=1}^{n} x_i,$$

so that $\mathrm{P}(A_1) + \mathrm{P}(A_2) + \cdots + \mathrm{P}(A_5)$ becomes:

$$\sum_{i=1}^{5} \mathrm{P}(A_i).$$

The $\sum$ sign is the Greek equivalent of the upper-case letter S and is pronounced 'sigma'. This sigma should not be confused with σ, which is the Greek equivalent of the lower-case letter s.

In the shorthand formula the letter i is simply an index. Any letter could be used, but it must replace i everywhere it appears. Thus:

$$\sum_{j=1}^{4} y_j = \sum_{i=1}^{4} y_i = \sum_{r=1}^{4} y_r = y_1 + y_2 + y_3 + y_4.$$

Changing the value of n results in a change in the number of terms being summed. For example:

$$\sum_{j=1}^{3} y_j = y_1 + y_2 + y_3, \text{ but } \sum_{j=1}^{2} y_j = y_1 + y_2.$$

1.9.1 Applications of sigma notation

Here are some further examples of the use of the Σ sign:

$$\sum_{r=1}^{3} r = 1 + 2 + 3 = 6,$$

$$\sum_{s=2}^{4} s^2 = 2^2 + 3^2 + 4^2 = 29,$$

$$\sum_{j=1}^{2} (2j + 5) = \{(2 \times 1) + 5\} + \{(2 \times 2) + 5\} = 16,$$

$$\sum_{k=2}^{3} (k^2 + 6k) = \{2^2 + (6 \times 2)\} + \{3^2 + (6 \times 3)\} = 43.$$

In running text we write $\sum_{i=1}^{n} x_i$ instead of

$$\sum_{i=1}^{n} x_i.$$

There are four particularly useful results that involve manipulation of the $\sum$ sign:

1. $\sum_{i=1}^{n}(x_i + y_i) = \sum_{i=1}^{n} x_i + \sum_{i=1}^{n} y_i$
2. $\sum_{i=1}^{n} cx_i = c\sum_{i=1}^{n} x_i$
3. $\sum_{i=1}^{n} c = nc$
4. $\sum_{i=1}^{n} x_i = \sum_{i=1}^{m} x_i + \sum_{i=m+1}^{n} x_i$

In the above, c is a constant, and m is an integer such that $1 \leq m < n$. Result 3 should be particularly noted. It follows immediately from result 2 by putting all the x-values equal to 1. All four results are easily proved by writing out the various summations in full.

Often the limits of the summation are obvious, in which case they may be dropped from the formula. The suffix i can also be omitted, so that, for example, if, from the context, was obvious that we were discussing the probabilities of the events $A_1, A_2, \dots, A_n$ we might simply write $\sum \mathrm{P}(A)$.

1.10 Exhaustive events

Two events are said to be **exhaustive** if it is certain that at least one of them occurs. For example, when rolling a die it is certain that at least one of the events A, 'the number obtained is 1, 2, 3 or 5' and B, 'the number obtained is even' will occur. In this example, if a 2 is obtained, then both A and B occur. If the events A and B are exhaustive then

$$\mathrm{P}(A \cup B) = 1. \tag{1.5}$$

Any event E and its complement, E', are both exhaustive and mutually exclusive:

$$\mathrm{P}(E \cup E') = 1, \qquad \mathrm{P}(E \cap E') = 0.$$

The events $A, B, \dots, N$, are said to be exhaustive if it is certain that at least one of them occurs:

$$\mathrm{P}(A \text{ or } B \text{ or } \cdots \text{ or } N) = \mathrm{P}(A \cup B \cup \cdots \cup N) = 1.$$

Thus the simple events that make up the sample space, S, are mutually exclusive and exhaustive.

Exercises 1b

1. A fair die is thrown. Events A, B, C, D are defined as follows:

   A: The score is even.
   B: The score is divisible by 3.
   C: The score is not more than 2.
   D: The score exceeds 3.

   Verify that $P(A) + P(B) = P(A \cup B) + P(A \cap B)$.
   Find:

   (a) $P(C')$.

   (b) $P(A \cup B \cup C)$.

   (c) $P(C \cap D)$.

2. A survey of 1000 people revealed the following voting intentions:

	Women	Men	Total
Conservative	153	130	283
Labour	220	194	414
Liberal Democrat	157	146	303
Total	530	470	1000

   A person is chosen at random from those in the sample. Find the probability that the person chosen:

   (a) intends to vote Conservative,

   (b) is a woman intending to vote Labour,

   (c) is either a woman or intends to vote Conservative,

   (d) is neither a man nor intends to vote Liberal Democrat,

   (e) is a man and intends to vote either Labour or Liberal Democrat.

3. A man tosses two fair dice. One is numbered 1 to 6 in the usual way. The other is numbered 1, 3, 5, 7, 9, 11. The events A to E are as follows:
   A: Both dice show odd numbers.
   B: The number shown by the normal die is the greater.
   C: The total of the two numbers shown is greater than 10.
   D: The total is less than or equal to 4.
   E: The total is odd.

   (a) Determine the probability of each event.

   (b) State which pairs of events (if any) are exclusive.

   (c) State which pairs of events (if any) are exhaustive.

1.11 Probability trees

Probability trees are diagrams that help us to see what is happening. Consider the following problem:

> *A fair coin is tossed three times.*
> *Determine P(exactly two heads are obtained).*

Each time we toss the coin the number of distinguishable outcomes increases:

After first toss	Either H or T
After second toss	The sequence of outcomes must be HH, HT, TH or TT
After third toss	Either HHH, HHT, HTH, HTT, THH, THT, TTH or TTT

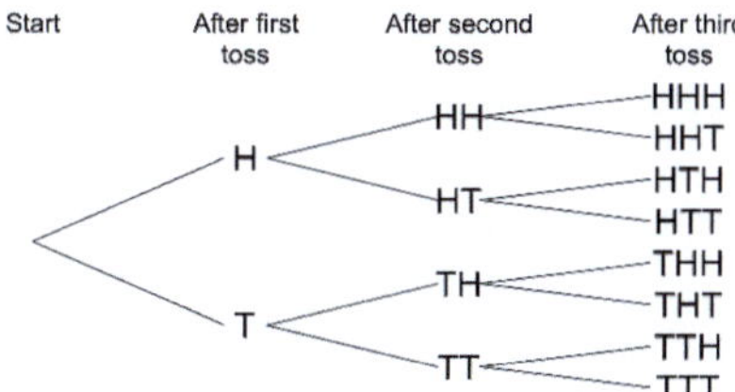

The same possibilities are represented more simply in a tree diagram in which the final column lists the entire sample space.

Each of the eight sequences is equally likely to occur. Three sequences include exactly two heads (HHT, HTH, THH) and so the probability of obtaining exactly two heads is 3/8.

Consider the new problem:

> *A man tosses a 5p coin, then he tosses a 10p coin, and finally he tosses a 20p coin.*
> *Determine P(exactly two heads are obtained).*

Essentially the same tree diagram does the trick:

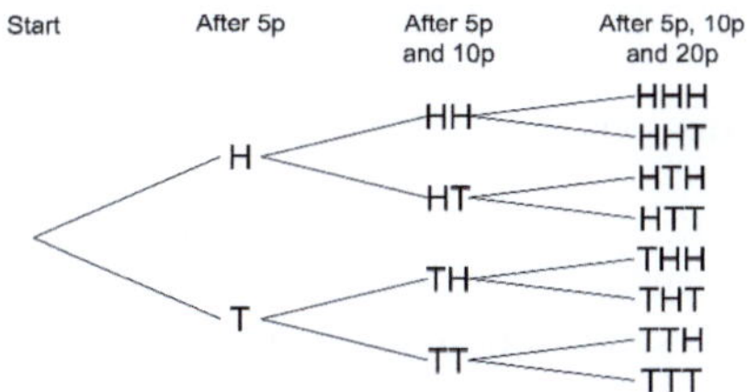

Assuming that all three coins are fair, the probability is again equal to 3/8.

Consider one final problem:

A man tosses three fair coins. Determine P(exactly two heads are obtained).

Once again we can use the same tree diagram: although all three coins are tossed simultaneously, we can be sure that they will not all fall at the same time. If we had a slow-motion camera, then we could observe the situation after one coin has landed, and again after two coins. Since they are all fair coins, we don't need to worry about which coin came down first. The required probability is again 3/8.

Although all three problems refer to coin tosses, they describe different physical situations that are all equivalent in terms of their probability structure. This is an example of a general principle: most probability problems can be translated into problems concerning either the tossing of (possibly bent) coins, the rolling of (possibly biased) dice, or the drawing of coloured balls from boxes!

1.12 Sample proportions and probability

So far, the probability to be associated with an event, has been expressed in terms of the number of simple events in a sample space, in which all the possible outcomes are equally likely. An alternative view of probability is a consequence of the general idea that a sample of observations gives information about the population from which it is derived. The bigger the sample, the more reliable is the information.

We have to adapt this approach when the outcomes in the sample space are no longer equally likely. For example, if we are interested in the probability that a bent penny comes down heads, then we could simply toss the penny a number of times (our sample) and see what proportion of the time a head is obtained:

Determine the sample proportion	→	Estimate the population probability

As the number of observations increases, so the observed sample proportion of occasions on which the event E occurs will vary. However, the variations will generally decrease in magnitude, and we expect that the observed sample proportion will approach a value $\mathrm{P}(E)$, that we will take to be the probability of E.

Consider the following two situations:

Experiment	*Event*
A fair die is tossed	A : A 6 is obtained;
A car is chosen at random	B : The car is white.

For event A it seems reasonable that if we were to roll a fair die a huge number of times then 'obviously' the event A would occur on about one-sixth of occasions: $\mathrm{P}(A) = 1/6$. There is no need to do any real sampling — we need only think about it.

For event B, however, there is no alternative to real sampling. To have any idea of the value of $\mathrm{P}(B)$, we need to examine a large sample of cars to find out what (roughly) is the proportion of cars that are white.

Put theory into practice: So, what *is* the probability of the event, B, that a randomly chosen car is white? To answer this, all we can do is to count cars, keeping a tally of the number of cars and the number that are white. Complete the following table:

Number of cars n	Number of white cars w	Sample Proportion $p = w/n$
2		
5		
10		
50		
100		
200		

You may wish to stop before seeing 200 cars, if the road is not a busy one!

Your best estimate of $P(B)$ is simply your final value for p.

1.13 Unequally likely possibilities

The results so far have been obtained while considering equally likely simple events. However, the relations hold equally well for unequally likely events.

Example 1.7

The events A *and* B *are such that* $P(A) = 0.4$, $P(B') = 0.3$ *and* $P(A \cap B) = 0.2$. *We wish to determine:*

(a) $P(A \cup B)$,

(b) $P(A' \cap B')$.

When there are just two or three events a Venn diagram is very useful. We can assign a probability to each section in turn:

(1) We start with $P(A \cap B) = 0.2$.

(2) Next, since $P(A \cap B)$ is part of $P(A)$, we can insert the balance of $P(A)$ into the following diagram. This is $P(A \cap B') = 0.4 - 0.2 = 0.2$.

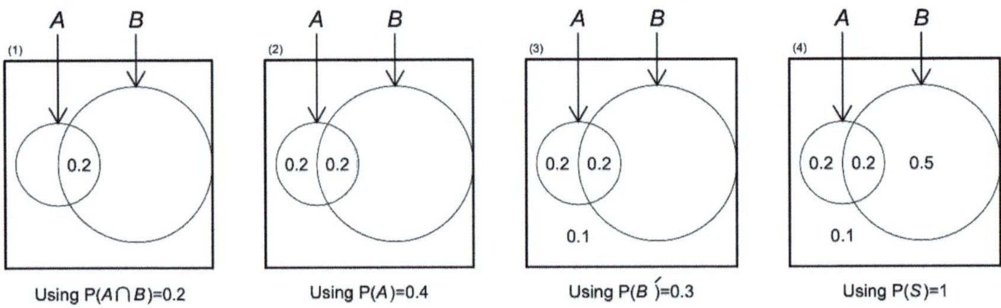

(3) We are told that $P(B') = 0.3$. Part of that is $P(A \cap B')$. The remainder is $P(A' \cap B') = 0.3 - 0.2 = 0.1$.

(4) The probabilities associated with three of the four sections are now identified. Since the four probabilities sum to 1, we obtain the balance, which is $P(B \cap A')$ by subtraction as 0.5.

By inspection we see that:

(a) $P(A \cup B) = 0.2 + 0.2 + 0.5 = 0.9$
and we have already found that:

(b) $P(A' \cap B') = 0.1$.

1.14 Physical independence

The coin-tossing examples of previous sections were examples of situations in which the separate components (e.g. toss of one coin and toss of another coin) were physically independent events. By **physical independence**, we mean that the outcome of one component (e.g. the first toss) can have no possible influence on the outcome of any other component (e.g. the second toss).

1.14.1 The multiplication rule

If A and B are two events *relating to physically independent situations*, then:

$$P(A \cap B) = P(A) \times P(B). \tag{1.6}$$

More generally, if $A, B, \ldots, N$ *all relate to physically independent situations* (for example, N separate tosses of a coin), then

$$P(A \cap B \cap \cdots \cap N) \quad = \quad P(A) \times P(B) \times \cdots \times P(N). \tag{1.7}$$

This very useful result is known as the **multiplication rule**.

Example 1.8

A bent penny has probability 0.8 of coming down heads when it is tossed. The penny is tossed six times. What is the probability that it shows heads on every occasion?

The six tosses are physically independent — there is no way that the outcome of one of the tosses can affect the outcomes of the other tosses. Therefore:

$$\begin{aligned}\text{P(6 heads)} &= \text{P('Head on first toss' } \textit{and} \text{ 'Head on second toss'} \\ &\quad \cdots \textit{ and} \text{ 'Head on sixth toss')} \\ &= \text{P('Head on first toss')} \times \text{P('Head on second toss')} \times \\ &\quad \cdots \times \text{P('Head on sixth toss')} \\ &= 0.8 \times 0.8 \times \cdots \times 0.8 \\ &= 0.8^6 \\ &= 0.262 \text{ (to 3 d.p.)}\end{aligned}$$

The probability of getting six heads with the bent penny is just over a quarter.

Example 1.9

A computer system consists of a mouse, a keyboard, and the computer itself. From past experience it is known that, on delivery, the probability that the mouse works correctly is 0.99, the probability that the keyboard works correctly is 0.98, and the probability that the computer works correctly is 0.95. We wish to determine the probability that only two of the components work correctly.

Define the events M, K, and C as follows: M = 'the mouse works correctly', K = 'the keyboard works correctly', and C ='the computer works correctly'. We need the probability that two components work correctly and the third does not. Using the union/intersection notation, the event of interest, E, is given by:

$$E = (M' \cap K \cap C) \cup (M \cap K' \cap C) \cup (M \cap K \cap C').$$

The situations are exclusive, so, using the addition rule,

$$\text{P}(E) = \text{P}(M' \cap K \cap C) + \text{P}(M \cap K' \cap C) + \text{P}(M \cap K \cap C').$$

Assuming that, for each component, the probability of its working is independent of the condition of the other components, we can use the multiplication rule so that, for example, $\text{P}(M' \cap K \cap C) = \{1 - \text{P}(M)\} \times \text{P}(K) \times \text{P}(C)$. Thus,

$$\begin{aligned}\text{P}(E) &= (0.01 \times 0.98 \times 0.95) + (0.99 \times 0.02 \times 0.95) + (0.99 \times 0.98 \times 0.05) \\ &= 0.077 \quad \text{(to 3 d.p.)}.\end{aligned}$$

An alternative approach uses a probability tree:

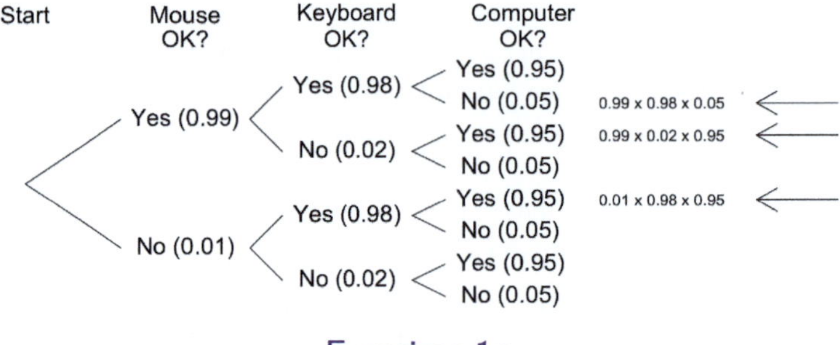

Exercises 1c

1. Two four-sided dice have sides numbered 1, 2, 3, 4. The dice are tossed. Assuming that they are both fair dice, determine the probability of a total score of 7.
2. A box contains three red balls, two yellow balls, and four green balls. Two balls are chosen at random and removed from the box. Determine the probability that the two balls are of different colours.
3. A red die is equally likely to show any of its six sides, which are numbered 1 to 6 in the usual way. A green die is also fair, but it has two sides numbered 3, two numbered 4, and two numbered 5. Both dice are rolled. Determine the probability that the value shown by the green die is greater than that shown by the red die.
4. An oxcart is delivering sacks of rice from a farm to the rice depot. There are three junctions between the farm and the depot. At each junction the oxcart may need to wait for more than five minutes in order to join the main traffic stream. The probabilities of these occurrences (which are independent of one another) are, respectively, 0.1, 0.2, and 0.3.
   - **(a)** Determine the probability that the cart never needs to wait for more than five minutes.
   - **(b)** Determine the probability that the cart has to wait for more than five minutes at precisely two junctions.
5. On Mondays, Tuesdays, and Wednesdays, a man travels to work by train. Each day, the probability that the train is late is 0.1 (irrespective of its performance on other days). Determine the probability that, on a randomly chosen week, the train is late on at most two occasions.
6. The probability that a biased coin comes down heads is 0.4. It is tossed three times. Find the probability of obtaining:
   - **(a)** exactly two heads,
   - **(b)** at least two heads.
7. Two chess grand masters, Xerxes and Yorick, play a tournament of three games. Past experience of their games suggests that, for each game, P(Xerxes wins)=1/4, P(Yorick wins)=1/5, and P(draw) = 11/20. Assuming that the outcomes of the games are independent of one another, determine the probabilities of each of the following:
   - **(a)** Xerxes wins all three games.
   - **(b)** Exactly two games are drawn.
   - **(c)** Yorick wins at least one game.
   - **(d)** Xerxes wins more games than Yorick wins.

1.15 Orderings

Consider the following problem:

> *Four markers are arranged in a line. The markers are labelled A, B, C, and D. Assuming that all possible arrangements are equally likely, determine the probability that the markers are in the order ABCD.*

The possible arrangements that might have occurred are the following:

ABCD	ABDC	ACBD	ACDB	ADBC	ADCB
BACD	BADC	BCAD	BCDA	BDAC	BDCA
CABD	CADB	CBAD	CBDA	CDAB	CDBA
DABC	DACB	DBAC	DBCA	DCAB	DCBA

In all there are 24 possible orderings of the markers. Since each ordering is equally likely, the required probability is 1/24.

To see how to generalize this approach, note that there are four possibilities for the first marker. Suppose that this is A (the possible orderings are those in the first row of the table). There are then three possible candidates for the second marker (B, C, or D). Suppose that B is second. Then there are two possibilities for third place (C or D), with whichever is left being in last place. We see that the number 24 occurred because $4 \times 3 \times 2 \times 1 = 24$.

In general, therefore, if there were n objects, the number of possible orderings would be $n \times (n - 1) \times (n - 2) \times \cdots \times 3 \times 2 \times 1$. This is tedious to write out, so we use the notation:

$$n! = n \times (n - 1) \times (n - 2) \times \cdots \times 3 \times 2 \times 1.$$

The quantity $n!$ is read as 'n **factorial**'.

> For convenience, $0!$ is defined to be equal to 1.

Example 1.10

A supermarket uses a code to identify each product that it stocks. The code consists of an ordering (without repetition) of the letters A to E, followed by an ordering (without repetition) of the digits 1 to 6.

The number of orderings of the 5 letters is $5! = 120$.

The number of orderings of the 6 digits is $6! = 720$.

Since each ordering of the letters can be associated with any one of the 720 orderings of the digits, the number of possible codes is $120 \times 720 = 86,400$.

1.15.1 Orderings of similar objects

Consider the following problem:

Four markers are arranged in a line. The markers are labelled A, B, A ,and B. Assuming that all possible arrangements are equally likely, determine the number of distinguishable orderings.

The only change from the previous situation is that the marker labelled C is now labelled A, while D has become B. Making the appropriate adjustments to the previous table, we get

ABAB	ABBA	AABB	AABB	ABBA	ABAB
BAAB	BABA	BAAB	BABA	BBAA	BBAA
AABB	AABB	ABAB	ABBA	ABAB	ABBA
BABA	BAAB	BBAA	BBAA	BAAB	BABA

There is now a lot of repetition! The only distinguishable arrangements are AABB, ABBA, ABAB, BAAB, BABA and BBAA. The reduction comes about because A followed by C and C followed by A now give an identical result (A followed by A). This halves the number of distinguishable orderings. A similar halving results from the replacement of D by B.

The general rule is as follows:

If there are n objects, consisting of a of one type, b of a second type, and so on, then the number of *distinct* arrangements of the objects in a line is

$$\frac{n!}{a!b!\ldots}.$$

Example 1.11

The four letters of the word COOK are arranged in a line.

(a) *How many distinct arrangements are there?*

(b) *If an arrangement is chosen at random, what is the probability that the two O's are consecutive?*

(a) There are four letters, consisting of 1 C, 2 O's and 1 K. The number of arrangements is therefore

$$\frac{4!}{1!1!2!} = \frac{24}{1 \times 1 \times 2} = 12.$$

There are 12 possible arrangements of the letters in the word COOK.

(b) We require the two O's to be consecutive. Imagine that they are glued together. We then have only three items to arrange in order: C, OO, and K. The number of possible orderings is

$3! = 6$. Thus 6 of the 12 possible arrangements of the letters in the word COOK involve a double O: the required probability is $6/12 = 1/2$.

In the last question the number of orderings is small enough that we could write them all out, but life is not always that easy.

Example 1.12

Five chairs are arranged in a line. Five boys are to be seated on the chairs. If Alfred and Bruce sit next to each other, then a fight is sure to start.

(a) *How many possible arrangements are there if there are no restrictions on the seating arrangements?*

(b) *If the boys are assigned seats at random, what is the probability that Alfred and Bruce are not sitting next to one another?*

(a) There are $5! = 120$ possible arrangements, all equally likely.

(b) An easy way to answer this is to consider the complementary event 'a fight starts'! Imagine that Alfred and Bruce are 'glued' together in the order AB. There are now 4 'objects' (boys or doubleboys) to be arranged in order.

There are then $4! = 24$ possible arrangements of the objects. There are a further 24 possible arrangements with Albert and Bruce 'glued' in the order BA. In all, therefore, there are 48 unsatisfactory arrangements and therefore $120 - 48 = 72$ satisfactory arrangements. The probability that Albert and Bruce are not sitting next to each other is therefore $72/120 = 3/5$.

Arrangements of n objects in a circle are more restrictive because there are n possible 'starting points' for the circle. Denoting the directions north, south, east and west by the letters N, S, E and W, the familiar clockwise ordering NESW could also be represented as ESWN, SWNE, or WNES, depending upon one's starting point.

The number of arrangements of n objects arranged in a circle is equal to the corresponding number of arrangements on a line, divided by n.

If the circle can be 'turned over', so that clockwise and anti-clockwise arrangements are indistinguishable, then the number of arrangements is equal to the corresponding number of arrangements on a line, divided by $2n$ rather than n.

Example 1.13

If the five chairs of the previous example are now arranged in a circle, what is the probability that Albert and Bruce are not sitting next to each other?

The number of equally likely distinct arrangements is now $120/5 = 24$. The number of AB arrangements is now $24/4 = 6$, and the number of BA arrangements is also 6, so the total number of unsatisfactory arrangements is 12. The probability that Albert and Bruce do not sit next to each other is therefore $1/2$, somewhat smaller than before.

Exercises 1d

1. Six children, Alice, Brenda, Caroline, David, Edward, and Frank stand in line in a random order. Find the probability that:

- **(a)** The three girls (Alice, Brenda, and Caroline) are next to each other,
- **(b)** Brenda and Frank are next to each other,
- **(c)** Caroline and David are not next to each other.

2. The ten letters of the word STATISTICS are arranged in a random order. Find the probability that:

- **(a)** The two Is are next to each other,
- **(b)** The first and last letters are both S.

3. Six novels, labelled A, B, C, D, E, F, have to be arranged in order of merit for a literary prize. Find the total number of different ways in which this can be done.
Suppose that the novels are arranged in a random order. Find the probability that:

- **(a)** F is first,
- **(b)** A is last,
- **(c)** C is first and D is second,
- **(d)** D comes immediately after C,
- **(e)** either B or E (or both) appear in the first two places.

4. The six children, Alice, Brenda, Caroline, David, Edward, and Frank, now stand in a circle. Distinguishing between clockwise order and anticlockwise order, find the number of different orders.
Find the probability that:

- **(a)** The three girls are next to each other.
- **(b)** Brenda and Frank are next to each other .
- **(c)** Caroline and David are not next to each other.

1.16 Permutations and combinations

Consider the following problem:

A pack of 52 playing cards (all different) is shuffled. Determine the probability that the top card in the pack is the ace of spades, the next is the ace of hearts, and the next is the ace of diamonds.

Now any one of the 52 cards could have been at the top of the pack. This leaves 51 cards, any one of which might have been next. Similarly, there are 50 possibilities for the third card. There are therefore a total of $52 \times 51 \times 50 = 132,600$ possibilities for the first three cards in order. Only one of these corresponds to the event described, so the probability of that event is 1/132,600.

The number of *ordered* arrangements of r objects chosen from a collection of n objects, is denoted by ${}^{n}\mathrm{P}_{r}$ (read as **'n p r'** or **'n perm r'**) and each ordering is called a **permutation** of the selected objects.

The value of ${}^{n}\mathrm{P}_{r}$ is given by

$$ {}^{n}\mathrm{P}_{r} = n \times (n-1) \times \cdots \times (n-r+1). \tag{1.8} $$

Note that there are r terms in the expression on the right of this equation. An equivalent expression, using factorials, is:

Permutations $$ {}^{n}\mathrm{P}_{r} = \frac{n!}{(n-r)!}. $$

Using the factorial formula for the previous problem, we have $n = 52$ and $r = 3$ so that

$$ {}^{52}\mathrm{P}_{3} = \frac{52!}{(52-3)!} = \frac{52!}{49!} = \frac{52 \times 51 \times 50 \times (49!)}{49!}. $$

Cancelling out the 49!, we get $52 \times 51 \times 50 = 132,600$, as before.

Consider now the slightly different problem:

A pack of 52 playing cards (all different) is shuffled. Determine the probability that the top three cards in the pack are the ace of spades, the ace of hearts, and the ace of diamonds.

This problem differs from the previous one in that *the order in which the cards appear is irrelevant*. There are $3! = 6$ possible orders for three cards, so the number of *distinguishable* groups of three cards, chosen from 52, is the number of ordered possibilities (132,600) divided by 6 giving the answer 22,100. The probability that the first three cards are the three aces is therefore 1/22,100.

The number of *unordered* arrangements of r objects selected from a collection of n objects, is denoted in this book[4] by $\binom{n}{r}$ which is read as either **'n c r'** or **'n choose r'**. Each collection of selected objects is a **combination**.

$$\text{Combinations} \qquad \binom{n}{r} = \binom{n}{n-r} = \frac{n!}{(n-r)!r!}. \qquad (1.9)$$

Using Equation (1.9) with $n = 52$ and $r = 3$, we get

$$\begin{aligned}\binom{52}{3} &= \frac{52!}{49! \times 3!} \\ &= \frac{52 \times 51 \times 50}{3 \times 2 \times 1} \\ &= 22,100.\end{aligned}$$

Example 1.14

A man is planting rose bushes. He has eight different bushes, each with a different colour flower, and he will plant five bushes in a row in his back garden. How many different possibilities does he have?

Assuming that order matters here, the number of possible arrangements is

$$^{8}\mathrm{P}_5 = \frac{8!}{5!} = 8 \times 7 \times 6 = 336.$$

Example 1.15

A pack of 52 cards is shuffled and a 'hand' of 13 randomly chosen cards is dealt to one card player. How many possible hands can that player receive?

In this case the order in which the player receives the cards is irrelevant. The number of possible hands is therefore:

$$\begin{aligned}\binom{52}{13} &= \frac{52!}{39!13!} = \frac{52 \times 51 \times 50 \times 49 \times \cdots \times 41 \times 40}{13 \times 12 \times 11 \times \cdots \times 2 \times 1} \\ &\simeq 6.35 \times 10^{11}.\end{aligned}$$

There are about 635 thousand million possible hands!

[4] An alternative is $^{n}\mathrm{C}_r$.

Example 1.16

At the beginning of a game show, a contestant is allowed a five-second glimpse of a table on which is placed a fluffy toy and four other objects (all different). At the end of the show the contestant is asked to name as many of the objects as possible.

(a) *How many different combinations of objects might be named?*

(b) *What proportion include the fluffy toy?*

(a) The contestant may name 0, 1, 2, 3, 4, or 5 of the objects. The total number of combinations is therefore

$$\binom{5}{0}+\binom{5}{1}+\binom{5}{2}+\binom{5}{3}+\binom{5}{4}+\binom{5}{5}=1+5+10+10+5+1=32.$$

An alternative approach is to argue that each of the five objects can either be 'chosen' or 'not chosen'. There are therefore two possibilities for each of five objects, so the total number of combinations is $2^5 = 32$.

(b) Given that the fluffy toy is named, the contestant may name up to four of the remaining objects. The total number of combinations including the fluffy toy is therefore

$$\binom{4}{0}+\binom{4}{1}+\binom{4}{2}+\binom{4}{3}+\binom{4}{4}=1+4+6+4+1=16(=2^4).$$

The proportion of the combinations that include the fluffy toy is therefore $16/32 = 1/2$.

Exercises 1e

1. Determine how many different groups of three students can be chosen from a class of 15. The class contains 5 males and 10 females.

   (a) The chosen group must contain one male and two females. Determine the number of possible groups.

   (b) If the group must contain at least one male and at least one female, how many different groups are possible?

2. For this year's examination the professor chooses 3 questions at random from the bank of 10 questions that was previously prepared. Assuming that the question order is irrelevant, how many different examinations might be set?

3. A bowl of fruit contains four bananas, three apples, and two oranges. These are to be placed in nine differently coloured bowls, one item of fruit per bowl. Find the number of different allocations.

1.17 Sampling with replacement

This is easy! The situation is one of physical independence and we can use the addition and multiplication rules and probability trees. Here is a typical problem.

Example 1.17

A pack of seven cards consists of the queens of spades, hearts, diamonds, and clubs together with the ace, king and jack of Spades. The pack is shuffled and a card is chosen at random. After its identity has been noted, the card is replaced in the pack, which is again shuffled. This is repeated on two further occasions. Determine the probability that a queen is chosen on only one occasion.

On each occasion the probability that a queen is chosen is 4/7. Using Q to denote a queen and R to denote one of the other cards, the possibilities that include exactly one queen are RRQ, RQR, and QRR. For each of these possibilities, the probability is the product of 3/7, 3/7, and 4/7, so the overall probability is

$$3 \times \left(\frac{3}{7}\right)^2 \times \frac{4}{7} = \frac{108}{343},$$

which is about 0.315 (to 3 d.p.).

1.18 Sampling without replacement

We start with an example that is similar to the previous problem, but requires the cards selected to be different.

Example 1.18

A pack of seven cards consists of the queens of spades, hearts, diamonds, and clubs together with the ace, king, and jack of spades. The pack is shuffled and three cards are chosen at random. Determine the probability that just one of the three cards is a queen.

In our new problem the order of selection is again unimportant and we are therefore concerned with combinations rather than permutations. The number of distinct combinations of three cards chosen from seven cards is $\binom{7}{3} = \frac{7!}{4! \times 3!} = \frac{7 \times 6 \times 5}{3!} = 35$. These are listed systematically in the following table using the shorthand of A, K and J for the ace, king and jack and with S, H, D and C representing the four queens.

When making lists it is important to work systematically (or we will get hopelessly lost!). In this case we work alphabetically:

ACD	ACH	$\underline{\text{ACJ}}$	$\underline{\text{ACK}}$	ACS	ADH	$\underline{\text{ADJ}}$	$\underline{\text{ADK}}$	ADS
$\underline{\text{AHJ}}$	$\underline{\text{AHK}}$	AHS	AJK	$\underline{\text{AJS}}$	$\underline{\text{AKS}}$	CDH	CDJ	CDK
CDS	CHJ	CHK	CHS	$\underline{\text{CJK}}$	CJS	CKS	DHJ	DHK
DHS	$\underline{\text{DJK}}$	DJS	DKS	$\underline{\text{HJK}}$	HJS	HKS	$\underline{\text{JKS}}$	

The 12 outcomes corresponding to the event of interest are underlined.

For each of the $\binom{4}{1} = 4$ possible selections of a queen there are $\binom{3}{2} = 3$ possible selections of two other cards from the three available. The total number of possibilities is the product $\binom{4}{1} \times \binom{3}{2} = 4 \times 3 = 12$. The probability of the event of interest is

$$\frac{\binom{4}{1} \times \binom{3}{2}}{\binom{7}{3}} = 12/35.$$

The general result is as follows:

The probability of simultaneously choosing n_1 from N_1, n_2 from N_2, and so on, is:

$$\frac{\binom{N_1}{n_1} \times \binom{N_2}{n_2} \times \cdots \times \binom{N_k}{n_k}}{\binom{N}{n}}.$$

Example 1.19

A committee of five is chosen by drawing lots from a group of eight men and four women. Determine the probability that the committee contains exactly three men.

Since nobody can be chosen more than once, selection is without replacement. An outcome consists of an unordered group of three people. We now suspend thought and simply identify the values of the parts of N and n. We have $N = 12$, $n = 5$, $N_1 = 8$, $N_2 = 4$, $n_1 = 3$, and $n_2 = 2$. Hence,

$$\begin{aligned}\frac{\binom{N_1}{n_1} \times \binom{N_2}{n_2}}{\binom{N}{n}} &= \frac{\binom{8}{3} \times \binom{4}{2}}{\binom{12}{5}} \\ &= \frac{8!}{3!5!} \times \frac{4!}{2!2!} \times \frac{5!7!}{12!} \\ &= \frac{8 \times 7 \times 6}{3 \times 2 \times 1} \times \frac{4 \times 3}{2 \times 1} \times \frac{5 \times 4 \times 3 \times 2 \times 1}{12 \times 11 \times 10 \times 9 \times 8} \\ &= 56 \times 6 \times \frac{1}{792} \\ &= 14/33.\end{aligned}$$

The probability that the committee contains exactly three men is 14/33, which is 0.424 to 3 decimal places.

Example 1.20

A notorious gang of outlaws contains five gunfighters called Smith, four called Jones, and one called Cassidy. In a gunfight, three of the gang are killed. Assuming that each gunfighter had the same probability of being killed, what is the probability that the three killed in the gunfight all had different names?

This time the outcomes are unordered groups of three outlaws. There are three types of outlaw: Smith, Jones, and Cassidy. The numbers of these are 5, 4, and 1 (total 10), while the numbers required from each group are 1, 1, and 1 (total 3). Hence the required probability is

$$\begin{aligned}\frac{\binom{5}{1} \times \binom{4}{1} \times \binom{1}{1}}{\binom{10}{3}} &= \frac{5 \times 4 \times 1}{120} \\ &= 1/6.\end{aligned}$$

The probability that the three gunfighters had different names is 1/6.

Example 1.21

> *The **birthday problem**. How many people does there need to be in a room before it is more likely than not that two will have the same birthday?*

You might like to guess what the answer will be!

Expressing the question more formally:

> A room contains N people. The event of interest, E, is that at least two people in the room have the same day as their birthday. The question is 'What is the smallest value of N for which the probability of E exceeds 0.5?'.

For simplicity, assume that every year has 365 days, and assume that equal numbers are born on each day of the year. Clearly, for $N = 1$ the event is impossible, while for $N = 366$, the event is certain. You might guess that the answer would be around 180, but this is a long way from the truth.

Imagine that we stand the N people in line and ask each in turn for their birthday. The probability that the second person does not have the same birthday as the first is 364/365. Similarly, the probability that the third person does not have the same birthday as either of the first two, is 363/365. So, using the multiplication rule, the probability that the first three people have different birthdays to one another is

$$\frac{364}{365} \times \frac{363}{365} \approx 0.992.$$

In the same way, the probability that the first four all have different birthdays is

$$\frac{364}{365} \times \frac{363}{365} \times \frac{362}{365} \approx 0.984.$$

The probability then drops rapidly: after 10 people it is 0.883 and after 20 people it is 0.589. After 23 people it is 0.493 which means that the probability that a birthday is shared is 0.507. So the answer is that in a room containing just 23 people there is a better than 50% chance that two have the same birthday. [a]

[a] If every day were the same, then their common probability would be 1/365 = 0.274%. In reality the probability of a day being a birthday varies, with, in the UK, the least frequent birthday (ignoring February 29th) being Boxing Day (0.205% of all birthdays) and the most frequent being September 26th (0.297%). Two other rare days are Christmas Day (0.214%) and New Year's Day (0.237%). In the United States it is Christmas Day that is the least frequent birthday, followed by New Year's Day, with September 9th being the most common.

Exercises 1f

1. There are 10 bottles arranged in a random order on a shelf. Five are green, three are blue, and two are yellow. Two bottles are knocked off the shelf. Determine the probability that:

   (a) both bottles are green,

   (b) both bottles are the same colour,

   (c) the bottles are of different colours.

2. A Scottish cricket team consists of seven batsmen and four bowlers. Six of the team are members of the McTavish family and the remainder are members of the McDonald family. Assuming that family membership and type of cricketer are independent of one another, determine the probability that exactly two batsmen are members of the McTavish family.

3. A man is taking 12 shirts with him on a flight. He takes 4 formal shirts and 8 casual shirts. Of the latter 3 are long-sleeved and 5 are short-sleeved. The man has two cases. He puts 6 shirts, chosen at random, in each case. One of his cases is lost. Determine the probability that he has lost:

   (a) exactly three formal shirts,

   (b) more than two formal shirts,

   (c) all his long-sleeved casual shirts.

4. A committee consists of three women (Anne, Bridget, Christine) and two men (David and Edward). Two members are chosen to be the Chair and Vice-chair. In how many different ways can these offices be filled?
   Find the probability that:

   (a) both the members chosen are men,

   (b) both are women,

   (c) the Chair is a woman and the Vice-chair is a man,

   (d) the Chair is a man and the Vice-chair is a woman,

   (e) the two are of opposite sex.

5. Manjula has the following coins in her purse: eight 1p coins, three 2p coins, four 5p coins, two 10p coins, and four 20p coins. In the dark she drops three coins. Find the probability that:

   (a) each of the coins dropped is worth at least 5p,

   (b) the total value of the three coins is 3p,

   (c) all three coins have the same value.

Key facts

- An **event**, E, is an outcome, or set of possible outcomes, of an experiment or trial.
- The limiting value of a relative frequency is called the **probability**, with $0 \leq P(E) \leq 1$.
- The **complementary event**, E', is the event that E does not occur.
-

Union	$A \cup B$	A **OR** B	*At least one of* A *and* B *occurs*
Intersection	$A \cap B$	A **AND** B	*Both* A *and* B *occur*

- The entire set of possible outcomes is denoted by S (the **sample space**).
- **Generalized addition rule**:

$$P(A \cup B) = P(A) + P(B) - P(A \cap B).$$

- Events $A, B, \ldots, M$, are mutually **exclusive** if the occurrence of one of them implies that none of the others can have occurred.
- If $A, B, \ldots, M$ are **mutually exclusive** events, then

$$P(A \text{ or } B \text{ or } \cdots \text{ or } M) = P(A) + P(B) + \cdots + P(M)$$

- If A and B are two **independent** events, then the **multiplication rule** applies:

$$P(A \text{ and } B) = P(A \cap B) = P(A) \times P(B).$$

- Events are said to be (collectively) **exhaustive** if it is certain that at least one of them occurs.

R

- The value of $x!$ is provided by `factorial(x)`.
- The value of $\binom{n}{r}$ is provided by `choose(n,r)`.

2

Conditional probability

The probability that we assign to an event is always influenced by the information that we have available. For example, suppose that I see a man lying motionless on the grass in a nearby park and I am interested in the probability of the event 'the man is dead'. In the absence of other information a reasonable guess might be that the probability is one in a million. However, if I have just heard a shot ring out, and a suspicious-looking man with a smoking revolver is standing nearby, then the probability would be rather higher!

2.1 Notation

We write

$$\mathrm{P}(B|A)$$

to mean the probability that the event B occurs (or has occurred) given the information that the event A occurs (or has occurred).

The quantity $B|A$ is read as 'B *given* A' and $\mathrm{P}(B|A)$ is described as a **conditional probability** since it refers to the probability that B occurs (or has occurred) *conditional* on the event that A occurs (or has occurred).

Example 2.1

A statistician has two coins, one of which is fair, while the other is double-headed. She chooses one coin at random and tosses it. The events A_1, A_2 *and* B *are defined as follows:*

A_1	:	*The fair coin is chosen.*
A_2	:	*The double-headed coin is chosen.*
B	:	*A head is obtained.*

Determine the values of $P(\mathrm{B}|\mathrm{A}_1)$ *and* $P(\mathrm{B}|\mathrm{A}_2)$.

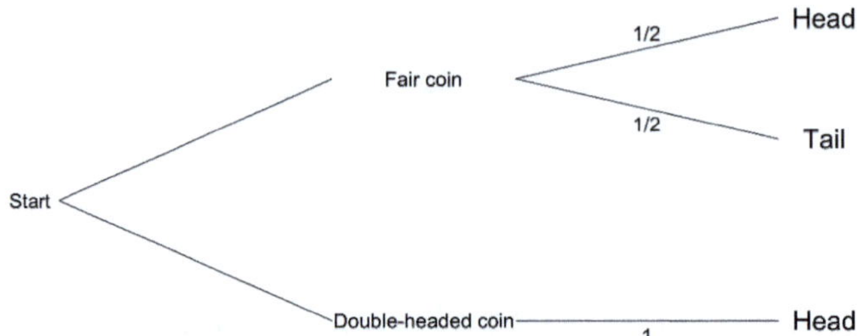

Figure 2.1 Probability tree illustrating the outcomes with one fair coin and one double-headed coin.

The possibilities are illustrated in Figure 2.1.

- If the fair coin is tossed, then the probability of a head is 1/2: $P(B|A_1) = 1/2$.
- If the double-headed coin is tossed, then the probability is 1: $P(B|A_2) = 1$.

The next example involves equally likely simple events.

Example 2.2

An electronic display is equally likely to show any of the digits $1, 2, \ldots, 9$. *Determine the probability that it shows a prime number, (i.e. one of 2, 3, 5, and 7):*

(a) *given no knowledge about the number,*

(b) *given the information that the number is odd.*

Let B be the event 'a prime number' and A be the event 'an odd number'. Thus $A \cap B$ is the event 'an odd prime number'.

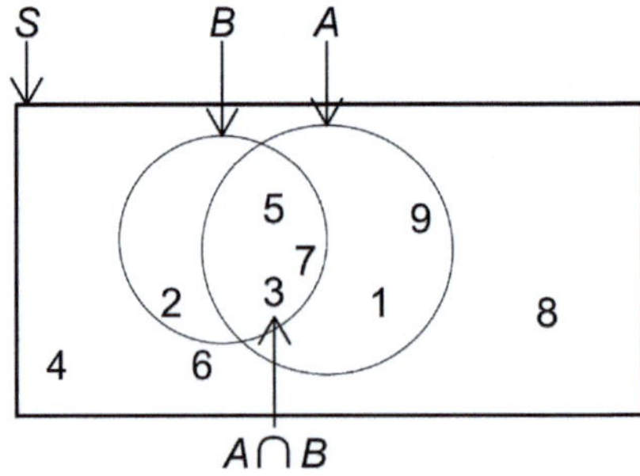

Figure 2.2 ***B*** **is the event 'a prime number' and** ***A*** **is the event 'an odd number'.**

(a) Since there are nine possible outcomes, $n(S) = 9$. Since there are four outcomes corresponding to the event of interest, $n(B) = 4$. Since the outcomes are all equally likely, $\mathrm{P}(B) = n(B)/n(S) = 4/9$.

(b) Given the information that the number is odd, we know that it must be one of the $n(A)$ numbers: 1, 3, 5, 7, or 9. Initially, each of these outcomes was equally likely. The knowledge that one of them has occurred does not make their chances of occurrence unequal. Of these five possible outcomes, three (3, 5, and 7) are prime. These outcomes are the simple events corresponding to the event $A \cap B$. Thus, $\mathrm{P}(B|A) = n(A \cap B)/n(A) = 3/5$.

The previous example illustrated, for a particular case, the result that, for equally likely simple events

$$\mathrm{P}(B|A) = \frac{n(A \cap B)}{n(A)}.$$

If we divide both the numerator and the denominator of the right-hand side of this equation by $n(S)$, we obtain

$$\frac{n(A \cap B)}{n(S)} \bigg/ \frac{n(A)}{n(S)} = \frac{\mathrm{P}(A \cap B)}{\mathrm{P}(A)}.$$

The result

$$\mathrm{P}(B|A) = \frac{\mathrm{P}(A \cap B)}{\mathrm{P}(A)} \tag{2.1}$$

is always true (provided A is a possible event!) and is not confined to equally likely events.

Knowing that A has occurred means that we can ignore all of the sample space except for that part occupied by the event A. The part of A in which B also occurs is the part denoted by $A \cap B$, and Equation (2.1) is seen to be a simple statement about proportions.

Rearranging Equation (2.1) gives

$$\mathrm{P}(A \cap B) = \mathrm{P}(A) \times \mathrm{P}(B|A). \tag{2.2}$$

Reversing the roles of A and B gives

$$\mathrm{P}(B \cap A) = \mathrm{P}(B) \times \mathrm{P}(A|B).$$

Since $A \cap B$ and $B \cap A$ are descriptions of the same event, namely the intersection of A and B, we have

$$\mathrm{P}(A \cap B) = \mathrm{P}(B \cap A).$$

Hence,

$$\mathrm{P}(A \cap B) = \mathrm{P}(A) \times \mathrm{P}(B|A) = \mathrm{P}(B) \times \mathrm{P}(A|B). \tag{2.3}$$

2.2 Statistical independence

Two events A and B are said to be **statistically independent** if knowledge that one occurs does *not* alter the probability that the other occurs. Formally, if A and B are two statistically independent events with nonzero probabilities, then

- $\mathrm{P}(A|B) = \mathrm{P}(A)$,
- $\mathrm{P}(B|A) = \mathrm{P}(B)$,
- $\mathrm{P}(A \cap B) = \mathrm{P}(A) \times \mathrm{P}(B)$.

Any one of the above equations is enough to guarantee independence of A and B (assuming that both events have nonzero probability of occurrence).

Physically independent events are always statistically independent.

The words 'statistically' and 'physically' are often omitted and events are simply referred to as being 'independent'.

Exclusive events with positive probability cannot be independent.

Example 2.3

Two events A *and* B *are such that* $\mathrm{P}(\mathrm{A}) = 0.5$, $\mathrm{P}(\mathrm{B}) = 0.4$, *and* $\mathrm{P}(\mathrm{A}|\mathrm{B}) = 0.3$.

(a) *State whether the events are independent.*

(b) *Find the value of* $\mathrm{P}(\mathrm{A} \cap \mathrm{B})$.

(a) The events A and B are *not* independent since $\mathrm{P}(A) \neq \mathrm{P}(A|B)$.

(b) $\mathrm{P}(A \cap B) = \mathrm{P}(B) \times \mathrm{P}(A|B) = 0.4 \times 0.3 = 0.12$.

Example 2.4

Two events A *and* B *are such that* $\mathrm{P}(A) = 0.7$, $\mathrm{P}(B) = 0.4$, *and* $\mathrm{P}(\mathrm{A}|\mathrm{B}) = 0.3$. *Determine the probability that neither* A *nor* B *occur.*

It is not obvious how to answer this! One way is to 'doodle', by writing down the probabilities of things we do know! So, from Equation (2.3),

$$\mathrm{P}(A \cap B) = \mathrm{P}(B) \times \mathrm{P}(A|B) = 0.4 \times 0.3 = 0.12.$$

From Equation (1.3) we can now obtain $P(A \cup B)$:

$$P(A \cup B) = P(A) + P(B) - P(A \cap B) = 0.7 + 0.4 - 0.12 = 0.98.$$

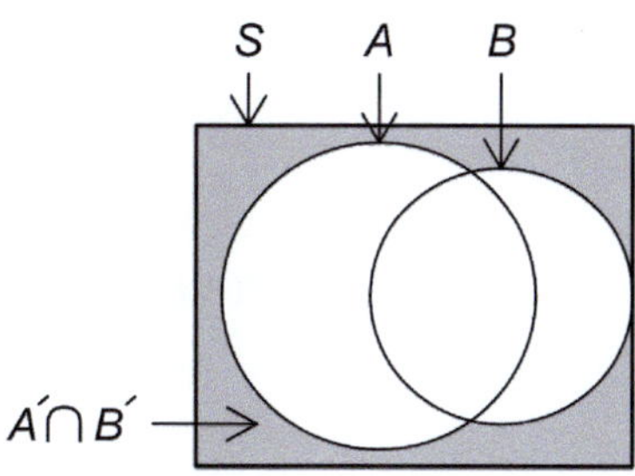

Looking at a Venn diagram, P(neither A nor B) $= P(A' \cap B') = 1 - P(A \cup B)$. The required probability is therefore $1 - 0.98 = 0.02$.

Example 2.5

A person is chosen at random from the population. Let A *be the event 'the person is female' and let* B *be the event 'the person is aged at least 80'. Suppose that* $P(A) = 0.5$, $P(B) = 0.1$ *and* $P(A|B) = 0.7$. *Let the event* C *be defined by* $C = A \cap B'$.

(a) *Describe the event* C *in real terms.*

(b) *Determine* $P(A|B)'$.

This question is much easier to answer when written in plain English!

In a certain population, 50% are female, 10% are aged at least 80, and 70% of these aged people are female.

(a) The event C is 'a female aged less than 80'.

(b) We need to find the probability that someone aged under 80 is female. A simple approach is to form a table. It may also help to give the population a definite size, N, say. The number of females aged 80 or over is therefore $0.7 \times 0.1N = 0.07N$. The remainder of the table is filled by subtraction.

		B'	B	
		< 80	≥ 80	
A'	Males	$0.47N$	$0.03N$	$0.50N$
A	Females	$0.43N$	$0.07N$	$0.50N$
		$0.90N$	$0.10N$	N

The proportion of females amongst those aged under 80 is therefore $0.43N/0.90N = \frac{43}{90}$. So $P(A|B')$ is just less than 1/2.

Exercises 2a

1. The events A and B are such that $P(A) = 0.4$, $P(B) = 0.7$, $P(A \cap B) = 0.2$. Find:

   (a) $PA|B)$,

   (b) $P(A'|B)$,

   (c) $P(A'|B')$.

2. The events A and B are such that $P(A) = 0.8$, $P(A|B) = 0.8$, $P(A \cap B) = 0.5$. Using a Venn diagram, or otherwise, find:

   (a) $P(B)$,

   (b) $P(B|A)$,

   (c) $P(A \cup B)$,

   (d) $P(A|A \cup B)$,

   (e) $P(A \cap B|A \cup B)$.

3. A box contains 5 red balls and 3 white balls. A second box contains 4 red balls and 4 white balls. Two balls are drawn at random from the first box and placed in the second box. One ball is then drawn from the 10 balls currently in the second box. Determine the probability that this ball is red.

4. Three ordinary unbiased dice, one red, one green, and one blue, are thrown simultaneously. Events R, G, S, and T, are defined as follows:
   R: The score on the red die is 3.
   G: The score on the green die is 2.
   S: The sum of the scores on the red and green dice is 4.
   T: The total score for the three dice is 5.
   Find the following probabilities:

   (a) $P(R \cap G)$,

   (b) $P(S|R)$,

   (c) $P(R|S)$,

   (d) $P(R \cup G)$,

   (e) $P(T)$,

   (f) $P(S|T)$.

5. Assume that children are equally likely to be born a boy or a girl. A family has two children. Determine the probability that both are girls:

   (a) if we know that the youngest is a girl,

   (b) if we know that at least one of them is a girl.

6. On the sunny tropical island of Utopia, one quarter of the large number of adult inhabitants are male and the remainder are female. The island's tourist welcoming committee consists of six individuals drawn at random from the adult inhabitants of the island. Determine the probability that:

   (a) exactly one committee member is male,

   (b) all the committee members are female,

(c) at least five committee members are female,

(d) all the committee members are of the same sex,

(e) all the committee members are female, given that it is known that they are all of the same sex.

7. The Green Hand gang used to consist of 12 individuals, of whom 8 were called Smith and 4 were called Jones. One bad year, they fell foul of a rival gang and every month one member of the Green Hand gang was 'eliminated' at random. Determine the probability of each of the following events:
A: Exactly three of the first five eliminated were named Jones.
B: The last two eliminated were both named Smith.
Determine also $P(A|B)$ and $P(B|A)$.

2.3 Mutual and pairwise independence

If the events $A, B, C, \ldots, M$, each having nonzero probability, are **mutually independent**, then their probabilities and the probabilities of their intersections satisfy all possible equations of the general form

$$P(E \cap F \cap \cdots \cap K) = P(E) \times P(F) \times \cdots \times P(K),$$

including

$$P(A \cap B \cap C \cap \cdots \cap M) = P(A) \times P(B) \times P(C) \times \cdots \times P(M). \tag{2.4}$$

If the events $A, B, C, \ldots, M$, each having nonzero probability, are **pairwise independent**, then their probabilities and the probabilities of their intersections satisfy all possible equations of the type

$$P(E \cap F) = P(E) \times P(F), \tag{2.5}$$

where E and F are any pair of the events.

Mutual independence clearly implies pairwise independence, but the reverse is untrue, as the example that follows illustrates.

Example 2.6

Two fair coins are tossed and the events A, B, *and* C *are defined as follows:*

A: *The first coin shows a head.*
B: *The second coin shows a head.*
C: *The two coins show different faces.*

Demonstrate that A, B, *and* C *are pairwise independent but not mutually independent.*

The outcomes corresponding to the various events of interest are summarized in the following table:

Event	Outcomes	Probability	Event	Outcomes	Probability
A	(H,H), (H,T)	1/2	B	(H,H), (T,H)	1/2
C	(H,T), (T,H)	1/2	$A \cap B$	(H,H)	1/4
$A \cap C$	(H,T)	1/4	$B \cap C$	(T,H)	1/4
$A \cap B \cap C$		0			

Thus,

$$\mathrm{P}(A \cap B) = 1/4 = \mathrm{P}(A) \times \mathrm{P}(B),$$

$$\mathrm{P}(A \cap C) = 1/4 = \mathrm{P}(A) \times \mathrm{P}(C),$$

$$\mathrm{P}(B \cap C) = 1/4 = \mathrm{P}(B) \times \mathrm{P}(C).$$

Thus the events A, B, and C display pairwise independence. However,

$$\mathrm{P}(A \cap B \cap C) = 0 \neq \mathrm{P}(A) \times \mathrm{P}(B) \times \mathrm{P}(C),$$

so the three events are *not* mutually independent.

A set of physically independent events will be both mutually independent and pairwise independent.

2.4 The total probability theorem (the partition theorem)

A simple example is provided by the following situation:

A statistician has three fair coins and a double-headed coin. She chooses one of the coins at random and tosses it. Determine the probability that she obtains a head.

We can illustrate this situation with a probability tree:

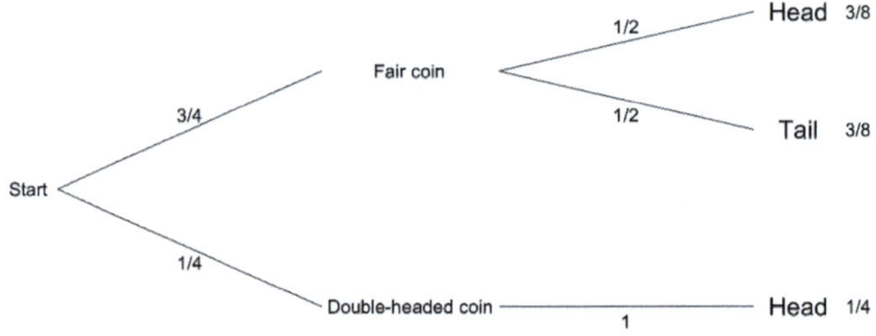

The total probability that she obtains a head is 5/8, the sum of the two branches of the tree that end with the outcome 'Head'. (This probability could also be deduced, by noting that the four coins have eight sides between them, with five of these equally likely sides being heads.)

The total probability theorem therefore amounts to the statement that

the whole is the sum of its parts.

A simple illustration of the general idea is provided by the following diagram.

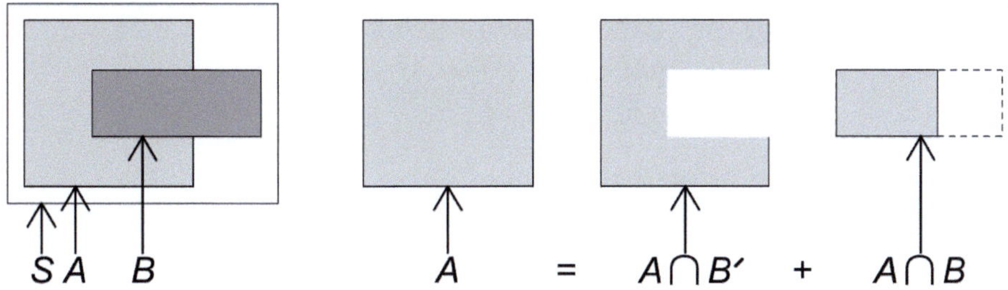

Translating the diagram into probability statements that use the fact that $A \cap B$ and $A \cap B'$ are mutually exclusive, we have:

$$\begin{aligned} P(A) &= P(A \cap B) + P(A \cap B') \\ &= \{P(B) \times P(A|B)\} + \{P(B') \times P(A|B')\}. \end{aligned}$$

In this case, A consists of just two 'slices': $A \cap B$, and $A \cap B'$. The result generalizes easily to m 'slices' as follows. Suppose that $B_1, B_2, \ldots, B_m$ are m *mutually exclusive and exhaustive* events in the sample space S. Let A be some other event. Then, using sigma notation (see Section 1.9),

$$P(A) = \sum_{i=1}^{m} P(A \cap B_i), \tag{2.6}$$

and hence, using Equation (2.3),

$$P(A) = \sum_{i=1}^{m} P(B_i) \times P(A|B_i). \tag{2.7}$$

Example 2.7

Of those students who do well in physics, 80% also do well in mathematics. Of those who do not do well in physics, only 30% do well in mathematics. If 40% do well in physics, what proportion do well in mathematics?

Define the events A, B_1, and B_2 as follows:

A: Does well in mathematics,
B_1: Does well in physics,
B_2: Does not do well in physics.

The information given tells us that $P(A|B_1) = 0.8$, $P(A|B_2) = 0.3$, and $P(B_1) = 0.4$. From the latter we can deduce that $P(B_2) = 0.6$. The events B_1 and B_2 are mutually exclusive and exhaustive, so, using Equation (2.7),

$$\begin{aligned} P(A) &= \{P(B_1) \times P(A|B_1) + \{P(B_2) \times P(A|B_2)\} \\ &= (0.4 \times 0.8) + (0.6 \times 0.3) \\ &= 0.50. \end{aligned}$$

Thus half the students do well in mathematics.

Example 2.8

Suppose that a box contains three balls, numbered, respectively, 0, 1 and 2. A ball is drawn at random from the box and is found to have the number n. We now toss n coins. What is the probability that we get exactly one head?

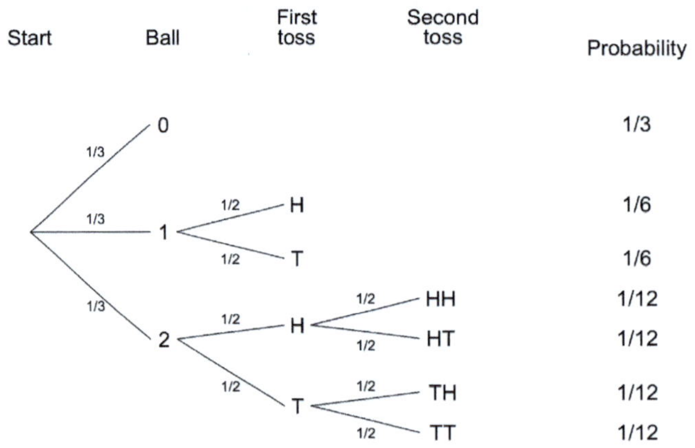

Start | Ball | First toss | Second toss | Probability

0: 1/3
1 → H: 1/6
1 → T: 1/6
2 → H → HH: 1/12
2 → H → HT: 1/12
2 → T → TH: 1/12
2 → T → TT: 1/12

We begin by drawing a probability tree, with the strands of the tree bearing the conditional probabilities of occurrence of that branch rather than its alternatives.

Formally, we can define events:

A Exactly one head is obtained.
B_i The ball chosen is numbered i, where $i = 0, 1$, or 2.

As the diagram shows, $P(B_0) = P(B_1) = P(B_2) = 1/3$, while $P(A|B_0) = 0$, $P(A|B_1) = 1/2$, and $P(A|B_2) = (1/2 \times 1/2) + (1/2 \times 1/2) = 1/2$.

The total probability of the event A is given by:

$$\begin{aligned} P(A) &= \{P(B_0) \times P(A|B_0)\} + \{P(B_1) \times P(A|B_1)\} + \{P(B_2) \times P(A|B_2)\} \\ &= (1/3 \times 0) + (1/3 \times 1/2) + (1/3 \times 1/2) \\ &= 0 + 1/6 + 1/6 \\ &= 1/3. \end{aligned}$$

The probability that we get exactly one head is 1/3.

Example 2.9

A car is made in three versions: petrol, hybrid, and electric. The proportions of the three types are 25, 40, and 35%, respectively. Each version of the car has either a normal trim or an executive trim. Of the petrol version, 70% have normal trim. The proportions for the hybrid and electric versions are 40 and 35%, respectively. In a publicity stunt the car-makers choose an owner at random to receive the prize of a free car service for the lifetime of the car. Determine the probability that the winner's car has an executive trim.

Define the events A, B_1, B_2, and B_3 as follows:

A	Owner's car has an executive trim,
B_1	Owner's car is the petrol version,
B_2	Owner's car is the hybrid version,
B_3	Owner's car is the electric version.

The events B_1, B_2, and B_3 are mutually exclusive and exhaustive, so, using Equation (2.7),

$$\begin{aligned} P(A) &= \{P(B_1) \times P(A|B_1)\} + \{P(B_2) \times P(A|B_2)\} + \{P(B_3) \times P(A|B_3)\} \\ &= (0.25 \times 0.3) + (0.4 \times 0.6) + (0.35 \times 0.65) \\ &= 0.5425. \end{aligned}$$

The probability that the owner's car has an executive trim is approximately 54%.

Exercises 2b

1. A bag contains red balls and blue balls. The probability that a red ball weighs more than 10 g is 0.6, whereas for a blue ball the probability is just 0.2. In the bag there are twice as many red balls as there are blue balls. A ball is chosen at random from those in the bag. Determine the probability that it weighs more than 10 g.
2. A fair die with faces numbered from 1 to 6 is rolled. If the side uppermost is a 1 or a 2, then the die is rolled for a second time. Determine the probability that the side uppermost on the final roll is a 6.

3. There are two groups of students. One group consists of 7 boys and 8 girls. The other group contains 5 boys and 15 girls. A group is chosen at random, and a student is chosen at random from that group. Determine the probability that a boy has been chosen.
4. It is known that 5% of the population have a particular disease. A test is used to detect this disease. The test gives a positive result for 95% of people with the disease, but it is also shows positive for 1% of those without the disease. If a person from this population is randomly selected and tested, what is the probability that the test is positive?

2.5 Bayes' theorem

In introducing the idea of conditional probability we effectively asked the question

> 'Given that event B has occurred in the past, what is the probability that event A will occur?'.

We now consider the 'reverse' question

> 'Given that the event A has just occurred, what is the probability that it was preceded by the event B?'.

We now develop a general result, beginning with a restatement of Equation (2.3):

$$\mathrm{P}(A) \times \mathrm{P}(B|A) = \mathrm{P}(B) \times \mathrm{P}(A|B).$$

Dividing through by $\mathrm{P}(A)$ we get

$$\mathrm{P}(B|A) = \frac{\mathrm{P}(B) \times \mathrm{P}(A|B)}{\mathrm{P}(A)}. \tag{2.8}$$

Suppose that, instead of a single event, B, there were m alternative previous events that could have happened, namely, $B_1, B_2, \ldots, B_m$. Assume that, as was the case with the total probability theorem, these events are mutually exclusive and exhaustive. From Equation (2.8),

$$\mathrm{P}(B_j|A) = \frac{\mathrm{P}(B_j) \times \mathrm{P}(A|B_j)}{\mathrm{P}(A)},$$

and, on substituting for $\mathrm{P}(A)$ using Equation (2.7), we get[1]

Bayes' theorem

$$\mathrm{P}(B_j|A) = \frac{\mathrm{P}(B_j) \times \mathrm{P}(A|B_j)}{\sum_{i=1}^{m}\{\mathrm{P}(B_i) \times \mathrm{P}(A|B_i)\}}. \tag{2.9}$$

Note that, in Equation (2.9), the numerator is one of the terms in the sum in the denominator.

[1] The Reverend Thomas Bayes (1701–61) was a Nonconformist minister in Tunbridge Wells, Kent. The theorem that bears his name has led to the development of a distinctive approach to statistics referred to as Bayesian statistics, with its advocates being referred to as 'Bayesians'. A brief introduction is provided in Chapter 20.

Example 2.10

A statistician has a fair coin and a double-headed coin. She chooses one of the coins at random and tosses it. She obtains a head. Determine the probability that the coin that she tossed was double-headed.

Define the events A, B_1, and B_2 as follows:

A:	A head is obtained.
B_1:	The fair coin is chosen.
B_2:	The double-headed coin is chosen.

We want $\text{P}(B_2|A)$ and we know the following probabilities: $\text{P}(B_1) = 1/2$, $\text{P}(B_2) = 1/2$, $\text{P}(A|B_1) = 1/2$, $\text{P}(A|B_2) = 1$. Using Bayes' theorem we have

$$\begin{aligned} \text{P}(B_2|A) &= \frac{\text{P}(B_2) \times \text{P}(A|B_2)}{\text{P}(B_1) \times \text{P}(A|B_1) + \text{P}(B_2) \times \text{P}(A|B_2)} \\ &= \frac{1/2 \times 1}{(1/2 \times 1/2) + (1/2 \times 1)} \\ &= \frac{1/2}{1/4 + 1/2} \\ &= 2/3. \end{aligned}$$

Example 2.11

According to a firm's internal survey, of those employees living more than 2 miles from work, 90% travel to work by car. Of the remaining employees, only 50% travel to work by car. It is known that 75% of employees live more than 2 miles from work. Determine:

(a) *the overall proportion of employees who travel to work by car,*

(b) *the probability that an employee who travels to work by car lives more than 2 miles from work.*

Define the events A, B_1 and B_2 as follows:

A:	Travels to work by car,
B_1:	Lives more than 2 miles from work,
B_2:	Lives not more than 2 miles from work.

The events B_1 and B_2 are mutually exclusive and exhaustive, with $\text{P}(B_1) = 0.75$, $\text{P}(B_2) = 0.25$, $\text{P}(A|B_1) = 0.9$, and $\text{P}(A|B_2) = 0.5$.

(a) From the total probability theorem,

$$\begin{aligned} \mathrm{P}(A) &= \{\mathrm{P}(B_1) \times \mathrm{P}(A|B_1)\} + \{\mathrm{P}(B_2) \times \mathrm{P}(A|B_2)\} \\ &= (0.75 \times 0.9) + (0.25 \times 0.5) = 0.675 + 0.125 = 0.8, \end{aligned}$$

so 80% of employees travel to work by car.

(b) From Bayes' theorem,

$$\mathrm{P}(B_1|A) = \{\mathrm{P}(B_1) \times \mathrm{P}(A|B_1)\}/\mathrm{P}(A) = (0.75 \times 0.9)/0.8 = 0.84375,$$

so the probability that an employee, who travels to work by car, lives more than 2 miles from work, is about 0.84.

An alternative approach involves constructing the following table from the information in the question:

	More than 2 miles	Not more than 2 miles	Total
Travels by car	67.5	12.5	80.0
Does not travel by car	7.5	12.5	20.0
Total	75.0	25.0	100.0

The entries are percentages of the workforce. The first entry, 67.5%, is obtained by calculating the value corresponding to 90% of the 75% who live more than 2 miles from work (using $0.90 \times 0.75 = 0.675$).

(a) The answer is the first row total, 80%.

(b) The answer is the proportion of the first row that are contained in the top left cell of the table, namely $67.5/80 = 0.84$ (to 2 d.p.).

Example 2.12

The test for a particular rare disease is very effective. If a patient does not have the disease, then in 99.95% of cases this is confirmed by the test. If a patient does have the disease, then in 99.9% of cases this is confirmed by the test. Given that only 0.01% of patients actually have the disease, determine the probability that a patient who tests positive for the disease does indeed have the disease.

Given the accuracy of the test, we might expect the answer to be in excess of 99%, but this is not the case. We will set out the calculations in a table:

	Proportion of population	Test positive
No disease	0.9999	$0.9999 \times 0.0005 = 0.00049995$
Disease	0.0001	$0.0001 \times 0.999 = 0.00009990$
Total		0.00059985

Thus the proportion of those who test positive that actually have the test is $0.0000999/0.00059985 \approx 0.167$. In other words, only about 1 in 6 of those testing positive will actually have the disease. Because of the rarity of the disease, most of those testing positive are, in fact, **false positives**.

Exercises 2c

1. Suppose that on one-third of the days of the year some rain falls on my garden. Suppose also that when it rains there is a probability of 0.7 that my barometer will be indicating rain, but when it does not rain there is a probability 0.1 that my barometer will nevertheless indicate rain.

(a) Determine the probability that, on a randomly chosen day of the year, my barometer indicates rain.

(b) Given that my barometer is indicating rain, determine the probability that it is actually raining.

2. Four machines A, B, C, and D produce respectively 30, 30, 15, and 25% of the total number of items from a factory. The percentages of defective output of these machines are 1, 1.5, 3 and 2% respectively. One item is randomly selected from the combined output of the four machines.

(a) Find the probability that the item is defective.

(b) Given that the item is defective, determine the probability that it was produced by machine A.

3. A red die is equally likely to show any of its six sides, which are numbered 1 to 6 in the usual way. A green die is also fair, but it has two sides numbered 3, two numbered 4, and two numbered 5. Both dice are rolled.

(a) Determine the probability that the value shown by the green die is greater than that shown by the red die.

(b) Given the information that the values shown by the two dice are different, determine the probability that it is the green die that shows the larger number.

4. A bag contains 7 white balls and 3 black balls. A white box contains 5 green balls and 2 red balls. A black box contains 3 green balls and 1 red ball. A ball is taken at random from the bag. If this ball is white, a ball is taken at random from the white box. Otherwise, a ball is taken at random from the black box. Given that the final ball is red, determine the probability that it has been taken from the white box.

5. In an examination, the probabilities of three candidates, Aloysius, Bertie, and Claude, solving a certain problem are $\frac{4}{5}$, $\frac{3}{4}$, and $\frac{2}{3}$, respectively. Determine the probability that the examiner will receive from these candidates:

(a) one, and only one, correct solution,

(b) not more than one correct solution,

(c) at least one correct solution.

Given that the examiner receives just one correct solution, determine the probability that the solution was provided by Bertie.

2.6 *The Monty Hall problem

The problem is named after Monty Hall, who was the original host of a television game show called *Let's Make a Deal*. The problem is as follows:

> *There are three doors: Behind one door is a car; behind the other two doors are goats. You pick a door, and will win whatever it conceals. The host, who knows what is behind each door, opens another door to reveal a goat. The host then asks, 'Do you want to change your choice of door?. The question is should you stick or should you switch?*

At the start of the game all three doors were equally likely to conceal the car, so the probability of a win was 1/3. After the host has revealed a goat, the probability of having chosen the car appears to have risen to 1/2. However, this is not the case, because the host did not choose a door at random. He knew where the car was and deliberately did not reveal it. To see how this makes a difference, let us suppose that originally you have chosen door 1. Now consider all the possibilities, each having probability 1/3:

Prob.	Door 1	Door 2	Door 3	Door opened	Stick	Switch
1/3	Goat	Goat	Car	2	Wins Goat	Wins Car
1/3	Goat	Car	Goat	3	Wins Goat	Wins Car
1/3	Car	Goat	Goat	2 or 3	Wins Car	Wins Goat

So you improve your chance of winning from 1/3 to 2/3 by switching. This only becomes apparent after a careful examination of all the possibilities.

Key facts

- $\mathrm{P}(B|A)$ denotes the probability that the event B occurs (or has occurred) given the information that the event A occurs (or has occurred):

$$\mathrm{P}(B|A) = \mathrm{P}(A \cap B)\big/\mathrm{P}(A)$$

- Two useful results:

$$\mathrm{P}(A) \times \mathrm{P}(B|A) = \mathrm{P}(B) \times \mathrm{P}(A|B)$$

$$\mathrm{P}(A) = \mathrm{P}(A \cap B) + \mathrm{P}(A \cap B')$$

- A and B are said to be independent if any of the following occur:

$$\mathrm{P}(A|B) = \mathrm{P}(A|B') = \mathrm{P}(A)$$

$$\mathrm{P}(B|A) = \mathrm{P}(B|A') = \mathrm{P}(B)$$

$$\mathrm{P}(A \cap B) = \mathrm{P}(A) \times \mathrm{P}(B)$$

  If one is true, then all are true.

- The **total probability theorem**:

$$\mathrm{P}(A) = \sum_{i=1}^{m} \mathrm{P}(A \cap B_i) = \sum_{i=1}^{m} \mathrm{P}(B_i) \times \mathrm{P}(A|B_i)$$

- **Bayes' theorem**:

$$\mathrm{P}(B_j|A) = \mathrm{P}(B_j) \times \mathrm{P}(A|B_j)\Big/\sum_{i=1}^{m}\{\mathrm{P}(B_i) \times \mathrm{P}(A|B_i)\}$$

3

Probability distributions

This chapter is concerned with discrete random variables. A **variable** is the characteristic, measured or observed, when an experiment is carried out, or an observation is made. The variable is described as **random** if its value is not predictable. The adjective **discrete** implies that a list of its possible numerical values could be made. Here are some examples:

Discrete random variable	*Possible values*
The number obtained when rolling a fair six-sided die	1, 2, 3, 4, 5, 6
The number of heads obtained when four fair coins are tossed.	0, 1, 2, 3, 4
The amount (in £) won in a lottery having prizes of 50p, £5, and £50	0, 0.5, 5, 50
The net gain (in £) from buying a 25p ticket in the above lottery	-0.25, 0.25, 4.75, 49.75
The number of rainy days in May	0, 1, ..., 31
The number of heads obtained when a single fair coin is tossed once	0, 1
The number of tosses of a fair coin until a head is obtained	1, 2, 3, ... (no limit!)

In each case, the possible outcomes can be thought of as a list of numerical values. These values do not have to be positive, nor do they have to be integers. Usually, but not always, the list is limited to just a few values.

3.1 Notation

We write

RANDOM VARIABLES	as e.g.	X, Y, Z;
observed values	as e.g.	x, y, z.

This leads to a statement such as

$$P(X = x) = 1/4,$$

which should be read as

'The probability that the random variable X takes the value x is 1/4.'

We can link this statement to the probability of an event, by defining the event A as 'the random variable X takes the particular value x'. Thus $P(A) = 1/4$.

To simplify formulae, we will often replace the lengthy $P(X = x)$, by the simpler P_x.

3.2 Probability distributions

Suppose we roll a biased die which has sides numbered 1 to 6. Define the random variable X to be 'the number showing on the top of the die'. We know two things:

1. The observed value of X must be 1, 2, 3, 4, 5, or 6.
2. On a given roll the random variable X can only take *one* of those values.

These correspond to statements that the six outcomes are both exhaustive and exclusive; hence,

$$P_1 + P_2 + \cdots + P_6 = 1.$$

Generalizing, for a discrete random variable X, that can take only the distinct values $x_1, x_2, ..., x_m$:

$$\sum_{i=1}^{m} P_{x_i} = 1. \tag{3.1}$$

The sizes of $P_{x_1}, P_{x_2}, ...,$ show how the total probability of 1 is *distributed* amongst the possible values of X—the most likely value for X will be the one with the highest probability. This is analogous to a frequency distribution, and, since the quantities are probabilities, the values $P_{x_1}, P_{x_2}, ...,$ are said to define a **probability distribution**.

Example 3.1

Tabulate the probability distribution of the number of heads obtained when a fair coin is tossed twice.

Let X be the random variable 'the number of heads obtained'. The possible values are 0, 1, and 2.

The simplest way of finding the required probabilities is to use a probability tree:

Start	After first toss	After second toss	Number of heads
	H	HH	2
		HT	1
	T	TH	1
		TT	0

The summary table is

Number of heads, x	0	1	2
P_x	1/4	1/2	1/4

Put theory into practice: Hopefully you found the last example trivially easy. However, the distinguished French mathematician Jean-le-Rond d'Alembert (1717–83) gave the incorrect answer 1/3 because he failed to notice that there were two ways of obtaining the outcome 1.

D'Alembert's error was to assume that the three possibilities were equally likely. To verify that they are not, toss two coins a total of twenty times. Draw up a tally chart of the number of heads (0, 1, or 2) obtained on each occasion. If d'Alembert had seen your results, he would probably have spotted his error!

3.2.1 The probability function

For many situations, it will not be necessary to make a list of all m probabilities in order to specify the probability distribution, because some convenient function (called the **probability function** (pf) or **probability mass function** (pmf)) can be found.

Example 3.2

Obtain a formula for the probability distribution of the random variable X defined as 'the result of rolling a fair six-sided die'.

Each of the six possible values for X has probability 1/6, so we can write:

$$P_x = 1/6, \qquad (x = 1, 2, \ldots, 6).$$

3.2.2 *Illustrating probability distributions*

As always with the subject statistics, it is a good idea to draw pictures. Since a discrete random variable can only take discrete values, a bar chart is appropriate, with the y-axis measuring probability.

Example 3.3

> *The random variable X is defined as 'the sum of the scores shown by two fair six-sided dice'. Draw an appropriate diagram of the probability distribution of X. Tabulate the distribution.*

We begin by drawing up a table showing the 36 possible outcomes, all of which (since the dice are fair) are equally likely:

		First die					
		1	2	3	4	5	6
	1	2	3	4	5	6	7
	2	3	4	5	6	7	8
Second	3	4	5	6	7	8	9
die	4	5	6	7	8	9	10
	5	6	7	8	9	10	11
	6	7	8	9	10	11	12

Entries in table are the sums of the numbers shown by the two dice

By inspection of the table we can see that, of the 36 equally likely possibilities, there is just 1 possibility leading to the outcome $X = 2$, so $P_2 = 1/36$. The most likely value for X is 7, which has probability $6/36 = 1/6$.

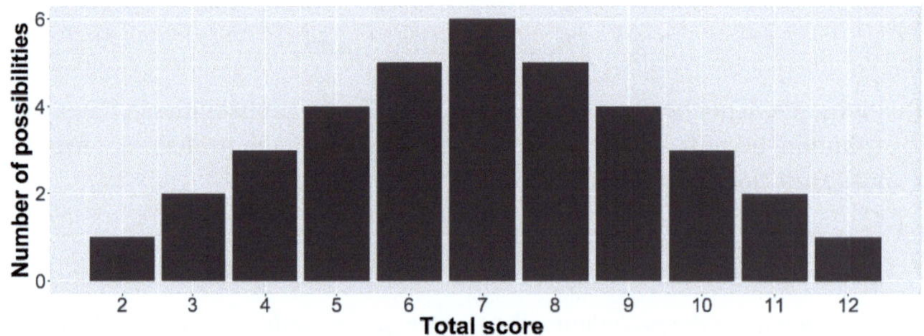

The full distribution is tabulated below:

Value of x	2	3	4	5	6	7	8	9	10	11	12
P_x	$\frac{1}{36}$	$\frac{2}{36}$	$\frac{3}{36}$	$\frac{4}{36}$	$\frac{5}{36}$	$\frac{6}{36}$	$\frac{5}{36}$	$\frac{4}{36}$	$\frac{3}{36}$	$\frac{2}{36}$	$\frac{1}{36}$

Exercises 3a

1. A box contains three red marbles and five green marbles. Two marbles are taken at random (without replacement) from the box. Let X be the number of green marbles taken from the box. Determine the probability that X takes each of the possible values.
2. A box contains three red marbles and five green marbles. A marble is taken at random from the box and its colour is noted. The marble is replaced in the box. For a second time a marble is taken at random from the box. Let X be the number of green marbles taken from the box. Determine the probability that X takes each of the possible values.
3. A fair six-sided die is thrown along with a fair coin whose sides are numbered 1 and 2. The total of the numbers shown by the coin and the die is denoted by T. Determine the probability distribution of T.
4. Which of the following experiments refers to a discrete random variable?
   - **(a)** A book is chosen at random from a shelf with 50 books and its author is noted.
   - **(b)** A book is chosen at random from a shelf with 50 books and the number of pages is noted.
   - **(c)** A book is chosen at random from a shelf with 50 books and the fifth letter on the tenth page is noted.
   - **(d)** A pupil is chosen at random from a particular class and the pupil's name is noted.
   - **(e)** A pupil is chosen at random from a particular class and the pupil's height is recorded in inches, to the nearest inch.
   - **(f)** The number of cars passing a given point on the road between 09:00 and 10:00 tomorrow.
   - **(g)** The colour of the first car to pass a given point on the road after 09:00 tomorrow.

3.2.3 Estimating probability distributions

Just as probabilities can be thought of as being the limiting values of relative frequencies as the sample size increases, so probability distributions describe the limiting proportions of the various possible values. If we concentrate on a single outcome, such as a six, and plot relative frequency against number of tosses, then we get a graph such as Figure 3.1, which was obtained using a computer to simulate the tossing of a fair die.

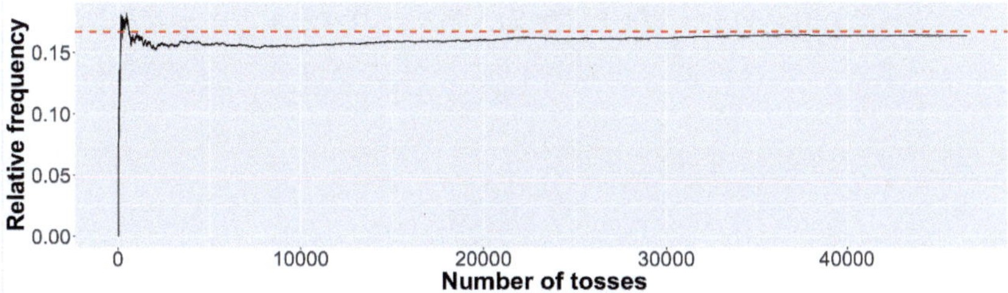

Figure 3.1 Relative frequency tends to its probability limit as the number of trials increases.

Note how the 'wiggles' die away as the number of tosses increases, while the relative frequency becomes generally closer to its limiting value of 1/6. A summary of the results for all six outcomes follows.

Number of rolls	*Relative frequency of* 1	2	3	4	5	6
36	0.222	0.250	0.111	0.194	0.139	0.083
216	0.194	0.199	0.111	0.167	0.162	0.167
1296	0.173	0.193	0.167	0.152	0.153	0.162
7776	0.166	0.175	0.168	0.170	0.168	0.153
46656	0.166	0.166	0.167	0.166	0.171	0.163
Target	1/6	1/6	1/6	1/6	1/6	1/6

As the sample size (the number of rolls) increases, so the relative frequencies slowly converge on the theoretical probabilities, and the observed distribution of the possible outcomes converges on the theoretical probability distribution.

3.2.4 The cumulative distribution function (cdf)

Often simply referred to as the **distribution function**, this is an alternative function for summarizing a probability distribution. It provides a formula for $\mathrm{P}(X \leq x)$ in place of that for $\mathrm{P}(X = x)$.

Example 3.4

Obtain the cumulative distribution function for the random variable X defined as 'the result of rolling a fair six-sided die'.

The following formula does the trick:

$$\mathrm{P}(X \leq x) = \begin{cases} 0 & x < 1, \\ x/6 & x = 1, 2, \dots, 5, \\ 1 & x \geq 6, \end{cases}$$

since, for example,

$$\mathrm{P}(X \leq 3) = \mathrm{P}(X = 1) + \mathrm{P}(X = 2) + \mathrm{P}(X = 3) = \frac{1}{6} + \frac{1}{6} + \frac{1}{6} = \frac{3}{6}.$$

3.3 The discrete uniform distribution

Here the random variable X is equally likely to take any of k values $x_1, x_2, \dots, x_k$, so that the distribution is summarized by:

$$P_{x_i} = \frac{1}{k}. \qquad (i = 1, 2, \dots, k).$$

Example 3.5

> *A familiar example occurs when X is defined as 'the score obtained when a fair six-sided die is rolled'.*

In this case $k = 6$. The distribution is tabulated as

Value of X	1	2	3	4	5	6
Probability	1/6	1/6	1/6	1/6	1/6	1/6

3.4 The Bernoulli distribution

The Bernoulli[1] distribution is very simple, since it refers to a random variable X that can take only the values 0 and 1:

$$P_0 = 1 - p, \qquad P_1 = p.$$

An example of the random variable X is 'the number of heads obtained on a single toss of a bent coin', where the probability of a head is p.

3.5 The binomial distribution

This distribution is concerned with the situation where:

- There is a fixed number, n, of independent trials,
- Each trial results in either a 'success' or a 'failure',
- The probability of success, p, is the same for each trial,
- The question of interest is the probability of obtaining exactly r successes.

The case where $n = 1$ corresponds to the Bernoulli distribution.

[1] Named after James Bernoulli (1654–1705), who was a member of an extremely talented Swiss family: there were seven Bernoullis who deserve a mention in a mathematician's *Who's Who*. James's principal work, *Ars Conjectandi* (The Art of Conjecture), was a treatise on probability.

Example 3.6

Determine the probability of getting two heads in three tosses of a bent coin which has P(Head) = 1/5.

The eight possible outcomes are as follows:

$$\begin{array}{lclclcl}
\text{HHH} & : & \frac{1}{5}\times\frac{1}{5}\times\frac{1}{5}=\frac{1}{125} & \quad & \text{HHT} & : & \frac{1}{5}\times\frac{1}{5}\times\frac{4}{5}=\frac{4}{125}* \\
\text{HTH} & : & \frac{1}{5}\times\frac{4}{5}\times\frac{1}{5}=\frac{4}{125}* & & \text{HTT} & : & \frac{1}{5}\times\frac{4}{5}\times\frac{4}{5}=\frac{16}{125} \\
\text{THH} & : & \frac{4}{5}\times\frac{1}{5}\times\frac{1}{5}=\frac{4}{125}* & & \text{THT} & : & \frac{4}{5}\times\frac{1}{5}\times\frac{4}{5}=\frac{16}{125} \\
\text{TTH} & : & \frac{4}{5}\times\frac{4}{5}\times\frac{1}{5}=\frac{16}{125} & & \text{TTT} & : & \frac{4}{5}\times\frac{4}{5}\times\frac{4}{5}=\frac{64}{125}
\end{array}$$

There are three possible sequences (indicated by a *) that lead to the outcome 'exactly two heads'. Each sequence has probability 4/125. Hence the total probability of obtaining exactly two heads is 12/125.

Listing all the possibilities is only feasible when the number of trials, n, is small. For other cases we need a formula!

In the last example, there were three possible sequences leading to the desired outcome, and each sequence had the same probability. Denoting the probability of a head by p, the answer we calculated was, in effect,

$$\begin{array}{rcccccc}
\text{P(exactly 2 heads)} & = & \text{(Number of sequences)} & \times & p^2 & \times & (1-p)^1 \\
 & = & 3 & \times & (1/5)^2 & \times & (4/5)^1 \\
 & = & 12/125. & & & &
\end{array}$$

For the general case of n independent trials, we need to know how many sequences in a probability tree lead to exactly r successes. The answer was given in Section 1.16 where we introduced the notation:

$$\binom{n}{r} = \frac{n}{1} \times \frac{n-1}{2} \times \frac{n-2}{3} \times \cdots \times \frac{n-r+1}{r}$$

to represent the number of ways of choosing r out of n.

The probability of obtaining r successes out of n independent trials, when, for each trial, the probability of a success is p, is given by:

$$P_r = \binom{n}{r} p^r (1-p)^{n-r}. \qquad (3.2)$$

This result, which provides the definition of the **binomial distribution**, makes no assumptions about the size of r and is therefore valid for all values of r from 0 to n inclusive.

Remember $\binom{n}{r} = \binom{n}{n-r}$, with $\binom{n}{0} = 1$ and $p^0 = 1$.

The quantity $1 - p$ is often written as q.

Writing q for $(1 - p)$, consider the **binomial expansion** of $(q + p)^n$, which is

$$(q+p)^n = q^n + \binom{n}{1}q^{n-1}p^1 + \binom{n}{2}q^{n-2}p^2 + \cdots + \binom{n}{n-1}q^1p^{n-1} + p^n.$$

The probabilities $P_0, P_1, ..., P_n$ are the successive terms in this expansion. Since $q + p = 1$, this confirms that the sum of the binomial probabilities is 1.

The most usual error in calculating a binomial probability is to forget that, in order for there to be exactly r successes, there must also be $n - r$ failures. For this reason the $(1 - p)^{n-r}$ factor must not be omitted from the formula.

Example 3.7

In 2021 in the UK, according to an official website, the most popular car brand was Volkswagen, with 9% of the market. Whilst held up in a traffic jam I occupy my time by examining the cars racing past on the other side of the road. Assuming that the website is correct, determine the probability that, of the first 50 cars that pass me, exactly 5 are Volkswagens.

Each car is either a Volkswagen (a 'success') or not a Volkswagen. Assuming that the traffic jam is not immediately outside a Volkswagen manufacturing plant, the $n = 50$ cars can be assumed to be a random sample of the cars on the road. For each car the probability of a 'success', p, is 0.09.

With $r = 5$ the required probability is

$$\binom{50}{5}(0.09)^5(0.91)^{45} = 0.180 \text{ (to 3 d.p.).}$$

Exercises 3b

1. A bent coin has probability 0.6 of showing a head when tossed. The coin is tossed three times. Determine the probability that exactly two heads are obtained.
2. Five per cent of bluebells actually have white flowers. Determine the probability that a random sample of 10 bluebells include exactly one having white flowers.
3. A batsman is practising his strokes against a fast bowler. Suppose that, for each ball bowled, the probability of the batsman making a good hit of the ball is 0.7, independently of all other balls bowled. Determine the probability that, in a given sequence of 12 balls bowled, the batsman makes exactly eight good hits.
4. There are 15 students in a class. Assuming that each student is equally likely to have been born on any day of the week, find the probability that:
   - **(a)** three or fewer were born on a Monday,
   - **(b)** four or more were born on a Tuesday.
5. Two parents each have the gene for cystic fibrosis. For each of their four children, the probability of developing cystic fibrosis is $\frac{1}{4}$. Find the probability that exactly two of the children develop the disease.
6. The characters in a film are classed as being either 'Good', 'Bad', or 'Ugly'. The proportions in these classes are 0.4, 0.4, and 0.2, respectively. Seven of the characters have red hair. Assuming that class is independent of hair colour, determine the probability that exactly two of the red-headed characters are classed as 'Ugly'.
7. When the Romans decimated a population they lined up the men and executed every tenth man. Six brothers stand at random places in the line. Find the probability that:
   - **(a)** none were executed,
   - **(b)** four or more escaped execution.

3.6 Notation

To save having to write 'The random variable X has a binomial distribution. There are n independent trials. The probability of a "success" is p for each trial', we write

$$X \sim \mathrm{B}(n, p).$$

Here, the symbol $\sim$ means 'has distribution' and 'B' is used as a shorthand for binomial.

The quantities n and p are called the **parameters** of the distribution; they are the quantities whose values are required in order to specify the distribution completely.

3.7 Successes and failures

Some good news: it does not matter which of the two possible outcomes we think of as being a success–the calculations will be the same. For example, suppose I play a game of chance with an opponent and suppose my probability of winning is p. My successes are my opponent's failures, and vice-versa.

Example 3.7 (cont.)

In the previous example, when we required the probability of observing five Volkswagens in a random sample of 50 cars, we defined a success to be 'a Volkswagen'. Suppose instead we define a success to be 'not a Volkswagen'. Thus n is again 50, but the probability of a success, p, is now 0.91 and the value of r is now 45. The required probability is

$$P_{45} = \binom{50}{45}(0.91)^{45}(0.09)^5 = 0.180 \text{ (to 3 d.p.)}.$$

This is (of course) the same value that we obtained previously.

3.8 The shape of the binomial distribution

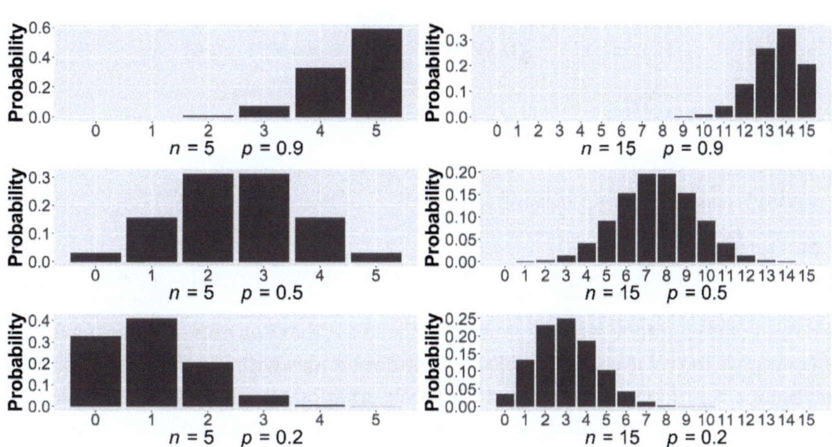

Figure 3.2 Examples of binomial distributions

The shape of the binomial distribution depends upon the value of p. When $p = 1/2$, the distribution is symmetric, since, for all values of r, the probability of obtaining r successes is the same as the probability of obtaining r failures.

For all values of p, the distribution has a mode near np (see the examples in Figure 3.2).

Group project: Take out one suit from a pack of cards. Shuffle these cards and choose one card at random. Replace the card and repeat a further four times. Record the number of times (out of the five) that you obtain a court card (a jack, queen or king). For example, suppose the original card is a 7, and the next four are, respectively, 9, 3, king and 7. A court card has occurred on just one of the five occasions, so the outcome of this experiment is a '1'.

Repeat the entire process so as to obtain a total of 20 observations, each with a value between 0 and 5, inclusive. Combine your results with those of your neighbours in your group and calculate the relative frequencies of the outcomes in your combined sample of results. Calculate the theoretical probabilities for this situation and compare them with your relative frequencies.

In the above experiment the cards were replaced after their values had been noted. Repeat the experiment *without* replacing the cards. This is most easily done by simply choosing five cards from the collection of 13 and noting the number of court cards. Compare your results with those obtained previously.

Group project:

We have seen that coin-tossing provides a simple example of a binomial situation. If a fair coin is tossed six times, and the random variable X denotes the number of heads obtained, then

$$\begin{aligned} P_r &= \binom{6}{r}(0.5)^r(0.5)^{6-r} \qquad (r = 0, 1, \ldots, 6) \\ &= \binom{6}{r}(0.5)^6. \end{aligned}$$

The probabilities of the various values of r are

Outcome, r	0	1	2	3	4	5	6
Probability (to 3 d.p.)	0.016	0.094	0.234	0.313	0.234	0.094	0.016

Toss a coin six times, and record the number of heads, r, that you obtain. Repeat a further nine times and compare the relative frequencies for your outcomes with the theoretical probabilities. The resemblance is unlikely to be close since 10 observations is a very small sample. Combine your results with those of the rest of your group so as to get a larger amount of data. You should find the overall results closely resemble those predicted by the binomial distribution.

3.9 The geometric distribution

We can again use coin-tossing as an illustration. Suppose that we have a bent penny with P(head) $= p$ and P(tail) $= 1 - p$, with $0 < p < 1$. This time we embark on a succession of tosses. We define the random variable X to be the number of tosses up to, and including, the first head (a success).

Evidently $P_1 = p$, since this is the probability of an immediate head. For X to be equal to 2, we must obtain a tail on the first toss and a head on the second toss. Thus,

$$\begin{aligned} P_2 &= \text{P(tail then head)} \\ &= \text{P(tail)} \times \text{P(head)} \qquad \textit{[physically independent events]} \\ &= (1-p)p. \end{aligned}$$

Similarly, for X to be equal to x, we must obtain a sequence of $(x-1)$ tails followed by a head. Each tail occurs with probability $(1 - p)$, so that we get the general result

$$P_x = (1-p)^{x-1}p \qquad (x = 1, 2, ...). \tag{3.3}$$

This general result, which holds for all positive integer values of x, defines a **geometric distribution.**

For a fair penny, $p = 1/2$. In cricket, the match starts with a toss of a coin to see which side plays first. In a recent five-match series, the English captain lost the first four tosses, but won the fifth. Using Equation (3.3), we see that the probability of the captain having to wait this long for a win during the next sequence of tosses is

$$\left(1 - \frac{1}{2}\right)^4 \frac{1}{2} = \frac{1}{32}.$$

The distribution is called *geometric* because the successive probabilities,

$$p, \quad (1-p)p, \quad (1-p)^2p, \quad ...,$$

form a **geometric progression** with first term p and common ratio $(1 - p)$.

Writing q for $(1 - p)$, and noting that $0 < q < 1$,

$$\begin{aligned} \sum_{x=1}^{\infty} P_x &= p(1 + q + q^2 + q^3 + \cdots) \\ &= p\frac{1}{1-q} \qquad \textit{[sum to infinity of a geometric progression]} \\ &= 1 \qquad \textit{[since } q = 1 - p\textit{]}. \end{aligned}$$

This shows that the total probability being distributed is equal to 1, as required. It also proves that, providing $0 < p < 1$, a success will occur eventually.

The geometric distribution can also be written as

$$P_y = (1-p)^y p \qquad (y = 0, 1, 2, \ldots),$$

where y is the number of failures before the first success.

Put theory into practice: Toss a coin 100 times, recording the outcome of each throw as either a head or a tail.

1. Count the numbers of heads and tails. Does the coin appear to be a fair one?
2. Record the length of each run. For example, H TTT HH T H T HHH would be recorded as 1, 3, 2, 1, 1, 1, 3. Draw up the resulting frequency distribution and compare with that to be expected from a geometric distribution.
3. Record which of heads and tails 'is in the lead' as the tosses continue. Thus H TTT HH T H T HHH results in the following:
   H -TT T - T - T - HH (with - indicating equal numbers)
   You probably expect that this will work out to be roughly equal numbers of heads and tails. However, that rarely happens, since once one side 'gets ahead' it can take a long time for the 'other side' to get a look in!
4. If you are good at programming, then you could simulate coin tossing on the computer and see what happens with more tosses.

Group project: Roll a normal six-sided die repeatedly until a 6 is obtained. Record the number of tosses required. Repeat a further nine times. Pool your results with those for the entire group. You should find, as predicted, that the mode is at 1—though some people may have had to roll as many as 20 times in order to get a 6!

3.9.1 Cumulative probabilities

To calculate $\mathrm{P}(X \le x)$, we note that this means that at least one of the first x trials must have been a success. The complement to this event is that all x were failures. If the probability of a failure is $(1-p)$, then the probability of x failures is $(1-p)^x$. Writing q for $(1-p)$ we have

$$\mathrm{P}(X \le x) = 1 - q^x. \tag{3.4}$$

Similarly,

$$\begin{aligned} \mathrm{P}(X < x) &= 1 - q^{x-1} \\ \mathrm{P}(X > x) &= q^x \\ \mathrm{P}(X \ge x) &= q^{x-1}. \end{aligned}$$

We can prove the result in Equation (3.4) as follows:

$$\begin{aligned} P(X \leq x) &= P(X = 1) + P(X = 2) + \cdots + P(X = x) \\ &= p + pq + \cdots + pq^{x-1} \\ &= p(1 + q + \cdots + q^{x-1}). \end{aligned}$$

The bracketed terms are a geometric series with sum $(1 - q^x)/(1 - q)$. Since $p = (1 - q)$, this is the given result.

Example 3.8

Only 1% of the vehicles leaving a motorway are prepared to give lifts to hitch-hikers. George Nerdowell arrives at a motorway exit and sticks out his thumb. Determine the probability that at least four vehicles fail to stop for him (i.e. that he doesn't get a lift until at least vehicle 5).

George will keep his thumb stuck out until he obtains a lift. We must assume that the decisions to stop (or not) made by the drivers are mutually independent. So each vehicle is either a success (with probability, p, equal to 0.01), or a failure (with probability, q, equal to 0.99). Let X be the number of vehicles up to and including the vehicle that gives George a lift. The question requires us to calculate $P(X > 4)$:

$$P(X > 4) = q^4 = 0.99^4 = 0.961 \text{ (to 3 d.p.)}.$$

3.9.2 A paradox!

Assuming that $0 < p < 1$, all geometric distributions have a similar shape: an infinite sequence of ever smaller probabilities. The rate of decline in the size of the probabilities depends upon the value of p, but the mode (the most probable value) is at $x = 1$ in each case.

The practical consequences of this result are, to say the least, surprising! Suppose, for example, that I decide that I will stand outside my house until I see a red sports car. Clearly, I may have to stand there for a long time, since red sports cars are not all that common. Consider therefore the following question: 'What is the most probable number of cars that pass my house up to and including the red sports car?'. The situation is geometric, with the value of p being rather small. Nevertheless, the previous result still holds and the answer to the question is that the most probable number of cars is just 1!

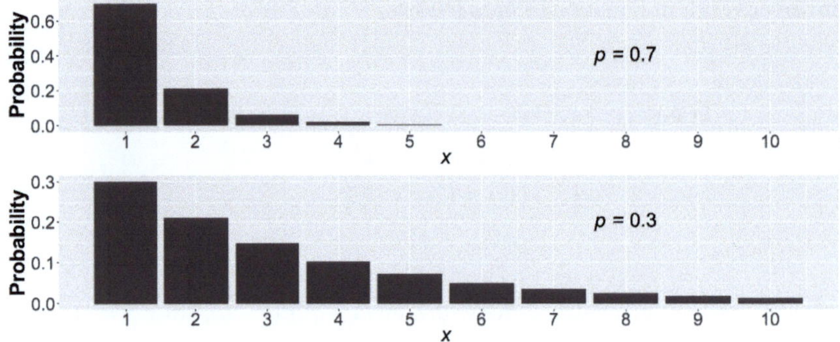

Figure 3.3 All geometric distributions have mode 1.

This result can easily be misinterpreted. The probability of the specific outcome 1 is $P_1 = p$, so $P(X > 1) = 1 - p$. The rarer the event of interest is (i.e the smaller the value of p is), the more likely is the observed value to be greater than 1. Nevertheless, 1 remains the most probable single value.

Exercises 3c

1. In a sales promotion, a random 10% of items sold contain a gift. A customer buys items one at a time until a gift is obtained. Find an expression for the probability distribution of the number of items bought.
2. Every day, a siamese cat attempts to open a kitchen cupboard. The probability of the cat being successful at any attempt is 0.1. Assuming that the outcomes of its attempts are independent of one another, determine the probability that the cat requires more than three attempts before being successful.
3. In the game of ludo, it is necessary to throw a six (with a single fair die) in order to start. The number of throws needed to obtain the first six is N. Find an expression for the probability distribution of N. Hence find:

   (a) $P(N \leq 3)$,

   (b) $P(N > 4)$.

3.10 The Poisson distribution and the Poisson process

In a Poisson[2] distribution the random variable is a count of events occurring *at random* in regions of time or space. 'At random' here has a very particular and strict definition: the occurrences of the events are required to be distributed through time or space so as to satisfy the following:

- Whether or not an event occurs at a particular point in time or space is independent of what happens elsewhere.
- At all points in time (or space) the probability of an event occurring within a small fixed interval of time (or region of space) is the same.
- There is no chance of two events occurring at precisely the same point in time or space.

Events that obey these requirements are said to be described by a **Poisson process**. For a Poisson process the probabilities of the counts in regions of space or intervals of time are observations from a Poisson distribution.

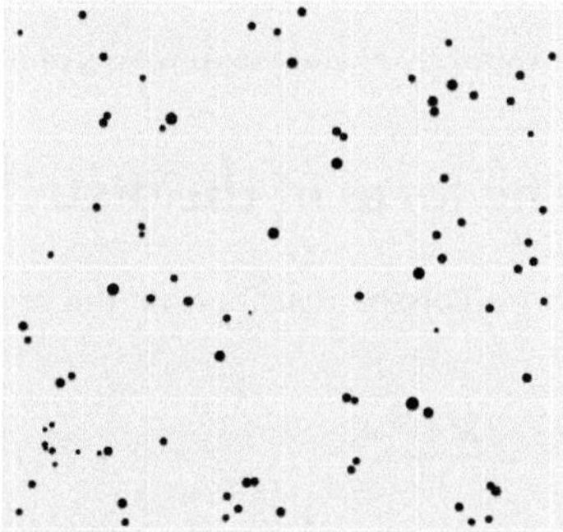

Figure 3.4 Locations of mangroves (of varying size) on a study plot in Kenya.

The relative locations of 86 mangroves growing in a square region of side 20m in Kenya are illustrated in Figure 3.4. The mangroves, which are trees that grow with their roots in water, were not deliberately planted in this way and nature's pattern appears entirely haphazard. Looking at the figure, we cannot deduce where the mangroves in the neighbouring plots will be found. This is a real-life example of an approximate spatial Poisson process. However, it would be wrong to imagine that all spatial patterns can be attributed to a Poisson process! An orchard provides one obvious alternative, with trees approximately regularly positioned.

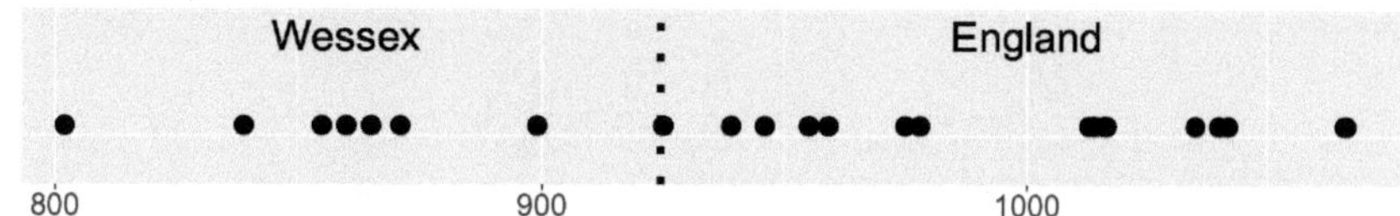

Figure 3.5 The years of accession of the kings of Wessex and England (till 1066).

An example of a time series of events that is an approximate realization of a Poisson process is provided by considering the dates of accession of new rulers in England. Although the overall rate

[2] Siméon Denis Poisson (1781–1840) was a French mathematician. His major work on probability was entitled (in French) *Researches on the Probability of Criminal and Civil Verdicts*. In this long book (over 400 pages) only about one page is devoted to the derivation of the distribution that bears his name!

of occurrence of these 'events' appears to be constant, an outsider, with no knowledge of history, would be hard pressed to find a pattern. Figure 3.5 illustrates the accession points from Egbert, King of Wessex from 802 to 839, via Athelstan, the first King of England in 925, and up to William the Conqueror in 1066.

Here are some other examples of situations that might be modelled by a Poisson process:

- The number of phone calls received on a randomly chosen day.
- The number of cars passing in a randomly chosen five-minute period on a road with no traffic lights or long queues (assuming such a road exists!).
- The number of currants in a randomly chosen currant bun.
- The number of accidents in a factory during a randomly chosen week.
- The number of typing errors on a randomly chosen page of a manuscript.
- The number of daisies in a randomly chosen square metre of playing field.

3.11 The form of the distribution

When X is a random variable having a Poisson distribution, the probability that it takes the value r is given by

$$P_r = \frac{\lambda^r e^{-\lambda}}{r!} \qquad r = 0, 1, 2, \ldots, \tag{3.5}$$

where λ, pronounced 'lambda', is a positive number and $e = 2.71828 \ldots$.

For $r = 0$ this simplifies to

$$P_0 = e^{-\lambda},$$

since $0! = 1$, and $\lambda^0 = 1$ for all values of λ.

One way of defining the value of e is via the expression

$$e^c = 1 + \frac{c}{1!} + \frac{c^2}{2!} + \frac{c^3}{3!} + \frac{c^4}{4!} + \cdots,$$

so that

$$e = e^1 = 1 + \frac{1}{1} + \frac{1}{2} + \frac{1}{6} + \frac{1}{24} + \cdots.$$

We can use the definition of e^c to verify that the probabilities of the Poisson distribution do indeed sum to 1:

$$\begin{aligned} P_0 + P_1 + P_2 + \cdots &= \left(1 + \frac{\lambda}{1!} + \frac{\lambda^2}{2!} + \cdots\right) e^{-\lambda} \\ &= e^{\lambda} e^{-\lambda} \\ &= 1. \end{aligned}$$

Example 3.9

Between 6 p.m. and 7 p.m., a help desk receives calls at the rate of two per minute. Assuming that the calls arrive at random points in time, determine the probability that:

(a) *Four calls arrive in a randomly chosen minute,*

(b) *Six calls arrive in a randomly chosen two-minute period.*

Since calls arrive at random points in time, a Poisson process is being described.

(a) Let X be the number of calls that arrive in a randomly chosen minute. Since the mean number of calls per one-minute period is $\lambda = 2$,

$$P(X = 4) = \frac{2^4 e^{-2}}{4!} = 0.090 \text{ (to 3 d.p.)}.$$

(b) Let Y be the number of calls that arrive in a randomly chosen two-minute period. The mean number of calls per two-minute period is $\lambda = 4$. Hence,

$$P(Y = 6) = \frac{4^6 e^{-4}}{6!} = 0.104 \text{ (to 3 d.p.)}.$$

Example 3.10

Tadpoles are scattered randomly through a pond at the rate of 14 a litre. A random sample of 0.1 litre is examined. What is the probability that it will contain more than 3 tadpoles?

Assuming a Poisson distribution (since the tadpoles are distributed at random in space) with mean 1.4 per 0.1 litre, we require

$$1 - (P_0 + P_1 + P_2 + P_3),$$

which is

$$1 - P(X \leq 3) = 1 - 0.9463.$$

The probability that the sample contains more than 3 tadpoles is 0.054 (to 3 d.p.).

Exercises 3d

1. Currants are randomly distributed in the baker's mixture with an average of 5.6 per bun. Determine the probability that a randomly chosen bun contains

   (a) fewer than four currants,

   (b) more than four currants.

2. The number of emergency calls received by a plumber can be modelled by a Poisson distribution with mean 3.4. Determine the probability that, on a randomly chosen day. the plumber receives more than four calls.

3. In a Suffolk bog, marsh orchids are distributed at random at a density of one per 100 metres^2. Determine the probability that a randomly chosen 500 metre^2 region of the bog holds more than two plants.

4. On a playing field, buttercups are distributed at random at a density of one per 10 metres^2. Determine the probability that a randomly chosen 50 metre^2 region of the field contains at most one plant.

3.11.1 The shape of a Poisson distribution

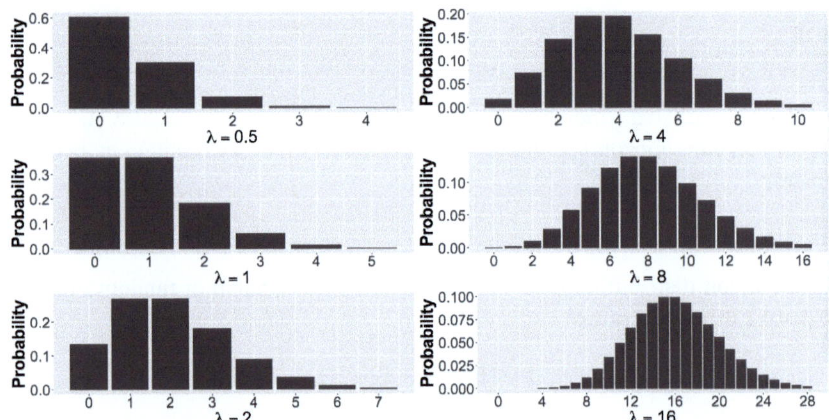

Figure 3.6 As λ increases, so the Poisson distribution becomes more symmetric.

When $\lambda < 1$ the distribution has mode at $x = 0$ and is very skewed. As λ increases, so the distribution takes on a more symmetrical appearance (see Figure 3.6).

Note that the diagrams are truncated—in each case it is *possible* for the Poisson random variable to take a larger value than those shown. However, although the range of possible values is infinite, in practice most values will lie between $\lambda - 3\sqrt{\lambda}$ and $\lambda + 3\sqrt{\lambda}$.[3]

[3] We will see later (Sections 4.2 and 4.4.3) that this corresponds to the mean $\pm$ 3 standard deviations.

Put theory into practice: Providing a road is not so busy that queues of traffic form, traffic flow may be modelled by a Poisson process. To investigate this, choose a reasonably busy road and count the numbers of cars (or lorries, or bicycles, or whatever) that pass in a particular direction in a period of one minute. Repeat for a complete half-hour. Represent your results using a bar chart. Calculate the sample mean and variance. If the stream of cars does form a Poisson process, then the mean and variance should be quite similar. If the variance is much larger than the mean, then this will suggest that there is appreciable clumping of the cars due to slower cars holding up faster ones or to the presence of a nearby roundabout or traffic light.

Put theory into practice: A chessboard and a packet of small sweets are required for this project. The idea is to toss the sweets one at a time onto the chessboard and, when all are on the board (you may need several attempts!), to count the numbers of sweets in each of the 64 squares. Providing the board is reasonably large, and you didn't cheat, the arrangement of the sweets should approximate a spatial Poisson process. Check by drawing a bar chart of the 64 observations and calculating their mean and variance, which should be reasonably similar.

3.12 Sums of Poisson random variables

If X and Y are independent Poisson random variables with parameters λ and μ, respectively, then the random variable Z, defined by $Z = X + Y$, is a Poisson random variable with parameter $\lambda + \mu$.

A direct proof of this result is tedious, but the result is obvious once we consider the Poisson process background to the distribution: mixing together two random patterns in space or time simply results in another random pattern.

Example 3.11

An observer is standing beside a road. Both cars and lorries pass the observer at random points in time. On average, there are 300 cars per hour, while the mean time between lorries is five minutes. Determine the probability that exactly 6 vehicles pass the observer in a one-minute period.

Since the question refers to 'random points in time' a Poisson distribution is appropriate both for cars and for lorries. The mean rate for lorries is 12 an hour, so the combined rate is 312 vehicles per hour, which corresponds to 5.2 vehicles per minute. The required probability is therefore:

$$\frac{(5.2)^6 e^{-5.2}}{6!} = 0.151 \text{ (to 3 d.p.)}.$$

Exercises 3e

1. Customers enter a shop at random points in time. During the first hour of the day, on average 0.7 customers arrive. The rate then increases to 1.3 an hour for the next three hours. Determine the mean and standard deviation of the total number arriving in these four hours.
2. The numbers of emissions per minute from two radioactive objects, A and B, are independent Poisson variables with means 0.65 and 0.45, respectively. Find the probability that, in a two-minute period, there is a total of exactly four emissions.

3.13 The Poisson approximation to the binomial

If X has a binomial distribution with parameters n and p, and if n is large and p is near 0, then the distribution of X is closely approximated by a Poisson distribution with mean np. If p is near 1 then the approximation applies to the distribution of failures.

Example 3.12

The discrete random variable X *has a binomial distribution with* $n = 60$ *and* $p = 0.02$. *Determine* $P(X = 1)$,

(a) *exactly,*

(b) *using a Poisson approximation.*

(a) The exact binomial probability is given by

$$P(X = 1) = \binom{60}{1}(0.02)^1(0.98)^{59} = 0.364 \text{ (to 3 d.p.).}$$

(b) Since n is quite large and p is small, we can expect the Poisson approximation to be quite accurate. Setting $\lambda = np = 60 \times 0.02 = 1.2$, we have

$$P(X = 1) \simeq \frac{1.2}{1!}e^{-1.2} = 0.361 \text{ (to 3 d.p.).}$$

The approximation is indeed an accurate one.

Exercises 3f

1. Of the photos taken by a photographer, just 5% are really good. One day the photographer takes 100 photos. Use the Poisson approximation to determine the probability that exactly 3 of these are really good. Compare your value with that obtained using the binomial distribution.
2. A machine produces resistors of which 99% are up to standard. They are packed in boxes containing 200 resistors. Using the Poisson approximation, determine the probability that a randomly chosen box contains at least 198 resistors that are up to standard. Compare your approximate value with the exact value.

3. Fred has cancer. As a consequence, deformed blood corpuscles occur in Fred's blood at random at the rate of 10 per 1000 corpuscles.
   (a) Using an appropriate approximation, determine the probability that a random sample of Fred's blood, consisting of 200 corpuscles, contains none that are deformed.
   (b) How large a sample should be taken in order to be 99% certain of there being at least one deformed corpuscle in the sample?

Group project: Ordinary packs of playing cards are required for this practical. The event of interest is that a single card drawn from a pack is the ace of spades. Each class member should have 26 attempts at striking lucky, with the card chosen being returned to the pack, and the pack being shuffled between attempts. Since $n = 26$ and $p = 1/52$, the approximating Poisson distribution has $\lambda = 1/2$. About 60% of the class should not see the ace of spades at all, but about 9% (where do these percentages come from?) should see the ace more than once.

3.13.1 *Derivation of the Poisson distribution from the binomial

We will demonstrate the connection between the distributions using the case of x events occurring at random in the time interval $(0, t)$. Assume that the events are occurring at a constant rate of λ per unit time, so that, on average, there are λt events in the time interval.

We begin by dividing the time interval into N equi-sized subintervals that are so short that they can contain at most one event. For each subinterval we have a Bernoulli situation (see Section 3.4): here success means the subinterval contains an event, while failure means that it does not,. For each subinterval, the probability of a success is $p = \lambda t/N$. The probability that the time interval $(0, t)$ contains exactly x events is therefore

$$\binom{N}{x} p^x (1-p)^{N-x} = \binom{N}{x} \left(\frac{\lambda t}{N}\right)^x \left(1 - \frac{\lambda t}{N}\right)^{N-x}.$$

We now explore the consequences of increasing N, starting with the first two terms:

$$\begin{aligned} \binom{N}{x}\left(\frac{\lambda t}{N}\right)^x &= \frac{N(N-1)(N-2)\cdots(N-x+1)}{x!}\left(\frac{1}{N}\right)^x \times (\lambda t)^x \\ &= 1\left(1-\frac{1}{N}\right)\left(1-\frac{2}{N}\right)\cdots\left(1-\frac{x-1}{N}\right) \times \frac{(\lambda t)^x}{x!}. \end{aligned}$$

As N increases, each of the first terms decreases to 1, so that this product simplifies to $(\lambda t)^x/x!$.

Now

$$\left(1-\frac{\lambda t}{N}\right)^{N-x} = \left(1-\frac{\lambda t}{N}\right)^N \times \left(1-\frac{\lambda t}{N}\right)^{-x}.$$

As N increases, the second term on the right-hand side again reduces to 1. However, the first term, as we now show, does not:

$$\begin{aligned} \left(1-\frac{\lambda t}{N}\right)^N &= 1 - N\frac{\lambda t}{N} + N(N-1)\frac{(\lambda t)^2}{2!N^2} - N(N-1)(N-2)\frac{(\lambda t)^3}{3!N^3} + \cdots \\ &= 1 - \lambda t + \left(1-\frac{1}{N}\right)\frac{(\lambda t)^2}{2!} - \left(1-\frac{1}{N}\right)\left(1-\frac{2}{N}\right)\frac{(\lambda t)^3}{3!} + \cdots, \end{aligned}$$

and so, as N increases, in this case we arrive at

$$1 - \lambda t + \frac{(\lambda t)^2}{2!} - \frac{(\lambda t)^3}{3!} + \cdots = e^{-\lambda t}.$$

Putting this together with the previous outcome, we see that, as N increases,

$$\binom{N}{x} p^x (1-p)^{N-x} \longrightarrow \frac{(\lambda t)^x}{x!} e^{-\lambda t}.$$

3.14 The negative binomial distribution

Whereas, for the geometric distribution, the random variable of interest was the number of tosses up to and including the first head, for the negative binomial distribution, the random variable X is the number of tosses required until the rth success.

The outcome $X = n$ implies that the first $n-1$ tosses included precisely $r-1$ successes, with the nth toss providing the final success. Thus

$$P(X = n) = \binom{n-1}{r-1} p^r (1-p)^{n-r}, \qquad n = r, r+1, \ldots. \tag{3.6}$$

The case $r = 1$ evidently corresponds to the geometric distribution. To find the mean of this distribution, which is r/p, requires some algebraic sleight of hand:

$$\begin{aligned}
E(X) &= \sum_{n=r}^{\infty} n \times \binom{n-1}{r-1} p^r (1-p)^{n-r} \\
&= \sum_{n=r}^{\infty} n \times \frac{(n-1)!}{(r-1)!(n-r)!} p^r (1-p)^{n-r} \\
&= \sum_{n=r}^{\infty} r \times \frac{n!}{r!(n-r)!} \frac{1}{p} p^{r+1} (1-p)^{n-r} \\
&= \frac{r}{p} \sum_{n=r}^{\infty} \binom{n}{r} p^{r+1} (1-p)^{n-r} \\
&= \frac{r}{p} \sum_{n=r}^{\infty} \binom{(n+1)-1}{(r+1)-1} p^{r+1} (1-p)^{\{(n+1)-(r+1)\}} \\
&= \frac{r}{p} \times 1 = \frac{r}{p}.
\end{aligned}$$

In a similar fashion we find that $\mathrm{Var}(X) = r(1-p)/p^2$.

The distribution is often expressed in terms of Y, the number of extra tosses required beyond the minimum (r). Writing $Y = X - r$, the distribution is then written

$$P(Y = m) = \binom{r+m-1}{r-1} p^r (1-p)^m \qquad m = 0, 1, \ldots. \tag{3.7}$$

Written in this form, the distribution has mean $r(1-p)/p$, with the variance again being $r(1-p)/p^2$. Thus, by contrast with the Poisson distribution (which also has range from 0 to ∞), the variance of the negative binomial distribution is greater than the mean. For this reason the negative binomial may be used in preference to the Poisson when modelling situations where events come in groups.

The distribution is also called the **Pascal distribution** in honour of the pioneering work of Blaise Pascal.[4]

3.15 The hypergeometric distribution

We have a population of size N, with each member being classed as either a success or a failure. The hypergeometric distribution is the distribution of the number of successes when taking a sample of size n *without replacement* from this population. If there are initially s members classed as a success, then the probability that the sample contains r such individuals is given by

$$P(X = r) = \frac{\binom{s}{r}\binom{N-s}{n-r}}{\binom{N}{n}}, \qquad 0 \le r \le s; \quad 0 \le (n-r) \le (N-s). \tag{3.8}$$

Exercises 3g

1. A traffic census is monitoring the cars passing a certain point. Suppose that the probability of a randomly chosen car being blue is 0.1. Assuming that the colour of any particular car is independent of the colour of any other car, determine the probability that the second blue car is the tenth car that passes.
2. A car park contains 50 cars, with four of these cars being coloured blue. Two cars are damaged by large hailstones. Determine the probability that exactly one of these is a blue car.

[4] Blaise Pascal (1623–62), a French child prodigy, was a mathematician, physicist, and philosopher. Possibly best known now for Pascal's triangle which displays binomial coefficients, he was also an inventor of an early mechanical calculator (the so-called *Pascaline).*

Key facts

- A probability distribution is a set of possible values together with their probabilities. It may be summarized by a **probability function** (pf) or **cumulative distribution function** (cdf).

- The **discrete uniform distribution** with values $x_1, \ldots, x_k$ has pf:

  $$P(X = x_i) = 1/k.$$

- The **Bernoulli distribution** has pf:

  $$P(X = 0) = p, \; P(X = 1) = 1 - p.$$

- The **binomial distribution**, with parameters n and p, denoted by $B(n, p)$ has pf:

  $$P(X = r) = \binom{n}{r} p^r (1 - p)^{n-r} \text{ for } r = 1, 2, \ldots, n.$$

- Requirements for a binomial situation:

  A fixed number, n, of independent 'trials'.

  Each trial results in one of two outcomes (success or failure).

  The probability of success, p, is the same for each trial.

- The **geometric distribution** has pf:

  $$prob(X = n) = (1 - p)^{n-1} p, \text{ for } n = 1, 2, \ldots,$$

  where p is the probability of a success, and n is the number of trials required until a success is obtained.

- The requirement for a **Poisson** situation is that events occur at random in space or time. The distribution has pf:

  $$P(X = r) = \frac{\lambda^r e^{-\lambda}}{r!} \text{ for } r = 0, 1, \ldots.$$

- The distribution of a sum of independent Poisson variables also has a Poisson distribution.

- A binomial distribution, with n large and with p close to either 0 or 1, can be closely approximated by a Poisson distribution with $\lambda = np$.

- The aim is to observe r successes. The **negative binomial** distribution is often expressed in terms of Y, the number of extra attempts required to achieve this aim:

  $$P(Y = m) = \binom{r + m - 1}{r - 1} p^r (1 - p)^m \qquad m = 0, 1, \ldots.$$

- The distribution has a variance greater than its mean.

- The **hypergeometric distribution**

$$P(X = r) = \frac{\binom{s}{r}\binom{N-s}{n-r}}{\binom{N}{n}}, \qquad 0 \leq r \leq s; 0 \leq (n-r) \leq (N-s),$$

is the probability of obtaining r successes in a sample of n individuals taken, without replacement, from a population containing N individuals of which s are successes.

R

For a binomial distribution with parameters n and p:

- the probability of exactly x successes is given by dbinom(x, n, p).
- the probability of x successes, or fewer, is given by pbinom(x, n, p).

For a geometric distribution with success probability p:

- the probability of y failures before the first success is given by dgeom(y, p).
- the probability of *at most* y failures before the first success is given by pgeom(y, p).

For a Poisson distribution with parameter λ:

- the probability of the outcome x is given by dpois(x, λ),
- the probability of a value less than or equal to x is given by ppois(x, λ).

For a negative binomial distribution with success probability p, where the aim is to obtain r successes:

- the probability of requiring exactly n extra attempts is given by dnbinom(n, r, p),
- the probability of requiring at most n extra attempts is given by pnbinom(n, r, p).

For a hypergeometric distribution with a population containing s individuals classed as successes and $(N - s)$ individuals classed as failures:

- the probability of obtaining x successes is given by dhyper($x, s, N - s, n$),
- the probability of obtaining at most x successes is given by phyper($x, s, N - s, n$).

4

Expectations

The expectation of the random variable X, denoted by $E(X)$, can be interpreted as the long-term average value of X. The term expectation can be used with any function of one or more random variables. Thus we could consider $E(X^2)$, $E(X + Y)$, $E(2X + 3Y)$, etc. In every case it represents the notional long-term average value of the combination of random variables being considered.

For an individual random variable, X, it is customary to denote its expectation by μ. Thus μ will denote the mean of the distribution of X,

$$\mu = E(X) = \sum xP_x, \tag{4.1}$$

where the summation is over all possible values of X.

Example 4.1

Determine the expectation of the random variable X, which has probability distribution given below:

Value of X	*0*	*1*	*2*	*3*
Probability	*0.3*	*0.4*	*0.2*	*0.1*

$$\begin{aligned} E(X) &= (0 \times P_0) + (1 \times P_1) + (2 \times P_2) + (3 \times P_3) \\ &= 0 + 0.4 + 0.4 + 0.3 \\ &= 1.1 \end{aligned}$$

The expectation of X is 1.1.

Group project: Roll a die four times, recording your results using a tally chart. Calculate the sample mean. Compare your results with other members of your group. You should find that almost everyone has a sample mean between 2 and 5. Now roll the die a further 36 times and calculate the sample mean for the combined set of 40 values. How variable are people's results now? You should find that most people have obtained values in the range 3 to 4. As the sample size increases so the sample mean becomes less likely to deviate far from 3.5. Calculate a sample mean for the entire group.

Sometimes either **expected value** or **expected number** is used in place of 'expectation'—these are all synonyms for one another. Whichever term is used, the numerical value being sought can be thought of as being *the long-term average value.*

This is just one of several places where the subject of statistics has 'borrowed' a word from the ordinary English vocabulary but subtly altered its meaning—the 'expected value of X', using the statistical meaning of the phrase, does not have to be a value of X that is actually expected, using the everyday interpretation of the word 'expected'.

Example 4.2

In a multiple-choice paper, each question has four possible answers. The candidate is asked to choose one of these answers. If the chosen answer is correct, the candidate gains three marks, but otherwise the candidate loses one mark. Determine the expected value of the mark gained if:

(a) *the candidate chooses an answer at random,*

(b) *the candidate knows that one of the incorrect answers is incorrect and chooses at random from the remaining three possibilities.*

Comment on the results in each case.

Let X denote the number of marks gained.
(a) The probability distribution for X is

$$P_3 = 1/4, \qquad P_{-1} = 3/4,$$

so that

$$\mathrm{E}(X) = \{3 \times (1/4)\} + \{(-1) \times (3/4)\} = 3/4 - 3/4 = 0.$$

The examination marking scheme has been designed so that the expected mark obtained by someone who knows nothing and guesses every question will be zero.

(b) The revised probability distribution, after the elimination of one of the incorrect answers, is

$$P_3 = 1/3, \qquad P_{-1} = 2/3,$$

so that

$$\mathrm{E}(X) = \{3 \times (1/3)\} + \{(-1) \times (2/3)\} = 1 - 2/3 = 1/3.$$

Since $\mathrm{E}(X)$ is greater than 0, if one or more of the possibilities can be eliminated as being certainly incorrect, then there will be an advantage in guessing the answer.

4.1 Expectations of functions

We have seen that, essentially, $\mathrm{E}(X)$ is the long-term average value of the random variable X. In the same way $\mathrm{E}(X^2)$ is the long-term average value of X^2, $\mathrm{E}(X^3 + 2X)$ is the long-term average value of $X^3 + 2X$, and so on.

For a general function, $\mathrm{g}(X)$, the value of $\mathrm{E}[\mathrm{g}(X)]$ is calculated using

$$\mathrm{E}[\mathrm{g}(X)] = \sum \mathrm{g}(x)P_x, \tag{4.2}$$

where the summation is over all possible values of X and P_x is the probability of the random variable X taking the value x.[1]

Example 4.3

Calculate the expected value of 1/X, where X is the value obtained from rolling a fair six-sided die.

$$\begin{aligned} \mathrm{E}(1/X) &= \{(1/1) \times P_1\} + \{(1/2) \times P_2\} + \cdots + \{(1/6) \times P_6\} \\ &= 1/6 + 1/12 + \cdots + 1/36 \\ &= 49/120. \end{aligned}$$

The expected value of $1/X$ is about 0.4 and is *not* simply the reciprocal of $\mathrm{E}(X)$ (which would be $1/3.5 \approx 0.3$.).

Two simple functions of particular interest are $X+a$, and aX, where a is a constant. Using Equations (4.1) and (4.2), we find that

$$\mathrm{E}(X + a) = \sum(x + a)P_x = \sum xP_x + \sum aP_x = \mathrm{E}(X) + a,$$

since $\sum P_x = 1$.

[1] This is sometimes referred to as the **Law of the Unconscious Statistician (LOTUS)**, because, for many people it is intuitively obvious.

Similarly

$$E(aX) = \sum axP_x = a\sum xP_x = aE(X).$$

These two key results,

$$E(X + a) = E(X) + a, \tag{4.3}$$

$$E(aX) = aE(X), \tag{4.4}$$

are intuitively obvious. They simply state

- If the same constant is added to every value, then the average of those values increases by that constant.
- If every value is multiplied by the same constant, then their average is increased to the same extent.

4.2 The population variance

The population variance is a measure of the variability in the values of a random variable. It is the expected value of the squared difference between a value taken by the random variable and the mean (μ) of that random variable:

$$\text{Var}(X) = E[(X - \mu)^2] = E\left(X^2 - 2\mu X + \mu^2\right) = E(X^2) - 2\mu E(X) + \mu^2 = E(X^2) - \mu^2. \tag{4.5}$$

The population variance is usually denoted by σ^2.

We often simply refer to 'the variance of X', omitting the qualifier 'population'.

The variance can never be negative: $\sigma^2 \geq 0$.

If we add the same constant to every value in a collection, then the resulting values are just as variable as the original values. The corresponding result for the product aX is not so obvious. To determine Var(aX), we first note that

$$E(aX) = aE(X) = a\mu.$$

Now, substituting in Equation (4.5),

$$\text{Var}(aX) = E[(aX)^2] - (a\mu)^2 = a^2E(X^2) - a^2\mu^2 = a^2\{E(X^2) - \mu^2\} = a^2\text{Var}(X).$$

So, in summary, we have the important results that

$$\text{Var}(X + a) = \text{Var}(X), \tag{4.6}$$

$$\text{Var}(aX) = a^2\text{Var}(X). \tag{4.7}$$

Rearranging Equation (4.5), another useful result is

$$\text{E}(X^2) = \mu^2 + \sigma^2. \tag{4.8}$$

The **population standard deviation** is the square root of the population variance.

Example 4.4

The random variable X has probability distribution given by

Value of X	*2*	*5*
Probability	*0.4*	*0.6*

Determine the variance of X.

We first calculate the expectation of X:

$$\text{E}(X) = (2 \times 0.4) + (5 \times 0.6) = 3.8.$$

We next calculate $\text{E}(X^2)$:

$$\text{E}(X^2) = (2^2 \times 0.4) + (5^2 \times 0.6) = 16.6.$$

Finally, using Equation (4.5), we get

$$\text{Var}(X) = \text{E}(X^2) - \{\text{E}(X)\}^2 = 16.6 - 3.8^2 = 2.16.$$

Example 4.5

The random variable X has the Bernoulli distribution:

$$P_0 = 1 - p, \qquad P_1 = p.$$

Find the expectation and variance of X.

Finding $\mathrm{E}(X)$ is straightforward:

$$\mathrm{E}(X) = \{0 \times (1-p)\} + \{1 \times p\} = p.$$

To determine the variance of X we use Equation (4.5) and first find $\mathrm{E}(X^2)$:

$$\mathrm{E}(X^2) = \{0^2 \times (1-p)\} + \{1^2 \times p\} = p.$$

Hence,

$$\begin{aligned} \mathrm{Var}(X) &= \mathrm{E}(X^2) - \{\mathrm{E}(X)\}^2 \\ &= p - (p)^2 \\ &= p(1-p) \end{aligned}$$

A random variable having a Bernoulli distribution with parameter p has mean p and variance $p(1-p)$.

4.3 Sums of random variables

4.3.1 The expectation of a sum of random variables

A rather obvious result that is surprisingly difficult to prove is that for two random variables X and Y,

$$\mathrm{E}(X+Y) = \mathrm{E}(X) + \mathrm{E}(Y). \qquad (4.9)$$

The expectation of a sum is the sum of the expectations.

4.3.2 The variance of a sum of random variables

The general result is

$$\text{Var}(X+Y) = \text{Var}(X) + \text{Var}(Y) + \text{Cov}(X,Y),$$

where $\text{Cov}(X,Y)$ is shorthand for the **covariance** of X and Y.

The covariance is the two-variable equivalent of variance. Previously, we wrote μ as shorthand for $\text{E}(X)$, and defined $\text{Var}(X)$ by

$$\text{Var}(X) = \text{E}[(X-\mu)\times(X-\mu)],$$

which simplified to $\text{E}(X^2)-\mu^2$. In the same way, write μ_x for $\text{E}(X)$ and μ_y for $\text{E}(Y)$. Then

$$\text{Cov}(X,Y) = \text{E}[(X-\mu_x)\times(Y-\mu_y)] = \text{E}(XY) - \text{E}(\mu_x \times Y) - \text{E}(\mu_y \times X) + \text{E}(\mu_x \times \mu_y).$$

The last three terms are all equal to $\text{E}(X)\times\text{E}(Y)$, so that the expression simplifies to give

$$\text{Cov}(X,Y) = \text{E}(XY) - \text{E}(X)\times\text{E}(Y). \tag{4.10}$$

From the definition of variance we can write

$$\text{Var}(X+Y) = \text{E}[(X+Y)^2] - \{\text{E}(X+Y)\}^2.$$

We know that the second term on the right-hand side is equal to $\{\text{E}(X)+\text{E}(Y)\}^2$, so our problems centre on the first term:

$$\begin{aligned} \text{E}[(X+Y)^2] &= \text{E}(X^2+2XY+Y^2) \\ &= \text{E}(X^2) + 2\text{E}(XY) + \text{E}(Y^2), \end{aligned}$$

since $\text{E}(R+S+T) = \text{E}(R)+\text{E}(S)+\text{E}(T)$.

Hence

$$\begin{aligned} \text{Var}(X+Y) &= \{\text{E}(X^2)+2\text{E}(XY)+\text{E}(Y^2)\} - \{\text{E}(X)+\text{E}(Y)\}^2 \\ &= \{\text{E}(X^2)-\text{E}(X)^2\} + 2\{\text{E}(XY)-\text{E}(X)\text{E}(Y)\} \\ &\quad +\{\text{E}(Y^2)-\text{E}(Y)^2\} \\ &= \text{Var}(X) + 2\{\text{E}(XY)-\text{E}(X)\text{E}(Y)\} + \text{Var}(Y) \end{aligned}$$

Thus

$$\text{Var}(X+Y) = \text{Var}(X) + \text{Cov}(X,Y) + \text{Var}(Y).$$

An especially simple case occurs when $\text{Cov}(X,Y)=0$. The most common reason for this simplification is that X and Y are independent random variables (i.e. knowing the value of one tells us nothing about the value of the other). Hence, if X and Y are independent, then

$$\text{Var}(X+Y) = \text{Var}(X) + \text{Var}(Y). \tag{4.11}$$

Combining this result with the result for $\mathrm{Var}(aX + b)$ we get a more general result:

If X and Y are independent, then

$$\mathrm{Var}\,(aX + bY + c) = a^2\mathrm{Var}(X) + b^2\mathrm{Var}(Y). \tag{4.12}$$

Once again, these results can be extended to cases involving more than two random variables. For example, if R, S, T, and U are all mutually independent, then

$$\mathrm{Var}(R + S + T + U) = \mathrm{Var}(R) + \mathrm{Var}(S) + \mathrm{Var}(T) + \mathrm{Var}(U). \tag{4.13}$$

If X and Y are independent, then

$$\mathrm{E}(XY) = \mathrm{E}(X) \times \mathrm{E}(Y).$$

However, it is possible for $\mathrm{E}(XY)$ to be equal to $\mathrm{E}(X)\times\mathrm{E}(Y)$ without X and Y being independent.

A particular case of Equation (4.12) that should be noted is

$$\mathrm{Var}(X - Y) = \mathrm{Var}(X) + \mathrm{Var}(Y). \tag{4.14}$$

Put theory into practice: In order to verify that there really is a difference between $2X$ and $X_1 + X_2$, we can perform two simple experiments using dice:

1. Roll an ordinary die 25 times. On each roll the score should be doubled before recording on a tally chart. Calculate the values of the sample mean and variance.
2. Roll a pair of dice 25 times. On each roll the total of the two dice should be recorded on a second tally chart. Calculate the values of the sample mean and variance.

Verify that the two sample means are about equal, whereas the first sample variance is about twice the second.

Example 4.6

Two fair six-sided dice are rolled. One die has its sides numbered 0,0,0,1,1,2; the other die has its sides numbered 2,2,3,3,4,4. Determine the mean and variance of Z, the total of the numbers shown by the dice.

Let X and Y be the numbers shown by the two dice. We are interested in $Z = X + Y$. We require $\mathrm{E}(Z)$ and $\mathrm{Var}(Z)$. We can use the result $\mathrm{E}(Z) = \mathrm{E}(X) + \mathrm{E}(Y)$. Also, since the two dice are independent of one another, $\mathrm{Var}(Z) = \mathrm{Var}(X) + \mathrm{Var}(Y)$.

For X we have the probability distribution

$$\mathrm{P}(X = 0) = 3/6; \qquad \mathrm{P}(X = 1) = 2/6; \qquad \mathrm{P}(X = 2) = 1/6.$$

Hence,

$$\mathrm{E}(X) = \{0 \times (3/6)\} + \{1 \times (2/6)\} + \{2 \times (1/6)\} = 4/6 = 2/3.$$

Also

$$\mathrm{E}(X^2) = \{0^2 \times (3/6)\} + \{1^2 \times (2/6)\} + \{2^2 \times (1/6)\} = 1,$$

so that

$$\mathrm{Var}(X) = \mathrm{E}(X^2) - \mathrm{E}(X)^2 = 1 - (2/3)^2 = 5/9.$$

For Y we have the probability distribution

$$\mathrm{P}(Y = 2) = \mathrm{P}(Y = 3) = \mathrm{P}(Y = 4) = 1/3.$$

By symmetry $\mathrm{E}(Y) = 3$. Also

$$\mathrm{E}(Y^2) = \{2^2 \times (1/3)\} + \{3^2 \times (1/3)\} + \{4^2 \times (1/3)\} = 29/3.$$

so that

$$\mathrm{Var}(Y) = \mathrm{E}(Y^2) - \mathrm{E}(Y)^2 = 29/3 - (3)^2 = 2/3.$$

Thus,

$$\mathrm{E}(Z) = \mathrm{E}(X) + \mathrm{E}(Y) = 2/3 + 3 = 11/3,$$

and

$$\mathrm{Var}(Z) = \mathrm{Var}(X) + \mathrm{Var}(Y) = 5/9 + 2/3 = 11/9.$$

Exercises 4a

1. The independent random variables X and Y are such that $\mathrm{E}(X) = 5$, $\mathrm{E}(Y) = 7$, $\mathrm{Var}(X) = 3$, and $\mathrm{Var}(Y) = 4$. Determine the mean and variance for each of the variables U, V, and W which are defined by $U = 2X$, $V = X + Y$, $W = X - 2Y$.
2. The random variable X has mean 10 and standard deviation 5. The random variable Y is defined by $Y = \frac{1}{2}(X + 5)$. Find the mean and standard deviation of Y.
3. It costs \$30 to hire a car for a day, and there is a mileage charge of 10 cents per mile. The distance travelled in a day has expectation 200 miles and standard deviation 20 miles. Find the expectation and standard deviation of the cost per day.
4. The discrete random variable X has probability distribution given by

$$\mathrm{P}(X = x) = \begin{cases} kx^2 & x = 1, 2, 3, 4 \\ 0 & \text{otherwise.} \end{cases}$$

   (a) Determine the value of the constant k.

   (b) Show that X has expectation 10/3.

   (c) Determine the variance of Y, where $Y = 3 - 2X$.

5. A shop has two branches. On a Monday the number of customers at one branch has mean 100 and standard deviation 15. At the other branch, on a Monday, the number of customers has mean 50 and the standard deviation 20. Stating any necessary assumption, determine the mean and standard deviation of the total number of Monday customers.

4.3.3 Distinguishing between 2X *and* $(X_1 + X_2)$

Suppose that X, X_1, and X_2 are independent random variables, each with mean μ and variance σ^2. We now show that $2X$ and $(X_1 + X_2)$ have the same mean but different variances.

From previous results,

$$\mathrm{E}(2X) = 2\mathrm{E}(X) = 2\mu, \qquad \mathrm{E}(X_1) + \mathrm{E}(X_2) = \mu + \mu = 2\mu,$$

which is what we would expect. However,

$$\mathrm{Var}(2X) = 2^2\mathrm{Var}(X) = 4\sigma^2, \qquad \text{whereas} \qquad \mathrm{Var}(X_1 + X_2) = \sigma^2 + \sigma^2 = 2\sigma^2.$$

Example 4.7

The independent random variables X_1 *and* X_2 *each have the probability distribution* $P_2 = 0.4$, $P_3 = 0.6$. *Determine the value of* $\mathrm{Var}(2X_1 - 3X_2)$.

For each X variable we have

$$\mathrm{E}(X) = (2 \times 0.4) + (3 \times 0.6) = 2.6, \qquad \mathrm{E}(X^2) = (2^2 \times 0.4) + (3^2 \times 0.6) = 7.0.$$

Hence, for each of X_1 and X_2,

$$\mathrm{Var}(X) = \mathrm{E}(X^2) - \mathrm{E}(X)^2 = 7.0 - 2.6^2 = 0.24.$$

Thus $\mathrm{Var}(2X_1 - 3X_2) = 2^2 \times \mathrm{Var}(X_1) + (-3)^2 \times \mathrm{Var}(X_2) = 13 \times 0.24 = 3.12$.

4.4 Mean and variance of common distributions

We now use the previous results to find the mean and variance for each of the binomial, geometric and Poisson distributions.

4.4.1 Binomial distribution

The random variable Y has a binomial distribution with parameters n and p. The easiest way to find the expectation and variance of Y is to use the fact that

$$Y = X_1 + X_2 + \cdots + X_n,$$

where X_i is the Bernoulli random variable describing the number of successes (0 or 1) on the ith trial. Since the expectation of a sum is the sum of expectations, $\mathrm{E}(Y) = np$. Also, since the n trials are independent of one another, we can apply Equation (4.13) to obtain the answer $\mathrm{Var}(Y) = np(1-p)$.

A binomial distribution with parameters n and p has mean np and variance $np(1-p)$.

4.4.2 Geometric distribution

The random variable X has the geometric distribution given by

$$P_x = (1-p)^{x-1}p = q^{x-1}p \qquad (x = 1, 2, \ldots),$$

where $q = 1 - p$. To obtain the mean and variance of X. we use the result that

$$1 + q + q^2 + q^3 + \cdots = \frac{1}{1-q} = \frac{1}{p}.$$

The mean, $\mathrm{E}(X)$, is given by

$$\begin{aligned} \mathrm{E}(X) &= (1 \times p) + (2 \times pq) + (3 \times pq^2) + (4 \times pq^3) + \cdots \\ &= (1-q) + 2(1-q)q + 3(1-q)q^2 + 4(1-q)q^3 + \cdots \\ &= 1 - q + 2q - 2q^2 + 3q^2 - 3q^3 + 4q^3 - 4q^4 + \cdots \\ &= 1 + q + q^2 + q^3 + \cdots \\ &= \frac{1}{1-q} = \frac{1}{p}. \end{aligned}$$

We now calculate $E(X^2)$:

$$\begin{aligned} E(X^2) &= (1^2 \times p) + (2^2 \times pq) + (3^2 \times pq^2) + \cdots \\ &= (1-q) + 4(1-q)q + 9(1-q)q^2 + \cdots \\ &= 1 - q + 4q - 4q^2 + 9q^2 - 9q^3 + \cdots \\ &= 1 + 3q + 5q^2 + 7q^3 + \cdots . \end{aligned}$$

At this point we use a clever trick: we add on, and also take away, the quantity $(1+q+q^2+q^3+\cdots)$ which we know to be equal to $1/p$:

$$\begin{aligned} E(X^2) &= 1 + 3q + 5q^2 + 7q^3 + \cdots \\ &= 2 + 4q + 6q^2 + 8q^3 + \cdots - (1 + q + q^2 + q^3 + \cdots) \\ &= 2(1 + 2q + 3q^2 + 4q^3 + \cdots) - 1/p. \end{aligned}$$

Where have we seen something like $(1 + 2q + 3q^2 + 4q^3 + \cdots)$ before? In the expression for $E(X)$: it is $E(X)$ divided by p. So

$$E(X^2) = \left\{2 \times \frac{E(X)}{p}\right\} - \frac{1}{p} = \frac{2}{p^2} - \frac{1}{p}.$$

Finally,

$$\begin{aligned} \mathrm{Var}(X) &= E(X^2) - \{E(X)\}^2 \\ &= \left(\frac{2}{p^2} - \frac{1}{p}\right) - \left(\frac{1}{p}\right)^2 \\ &= \frac{1}{p^2} - \frac{1}{p} \\ &= \frac{1-p}{p^2}. \end{aligned}$$

A geometric distribution with success probability p has mean $1/p$ and variance $(1-p)/p^2$.

4.4.3 Poisson distribution

The random variable X has a Poisson distribution with parameter λ. We wish to find the expectation and variance of X. We start with $\mathrm{E}(X)$:

$$\begin{aligned}
\mathrm{E}(X) &= \sum_{r=0}^{\infty} rP_r \\
&= \sum_{r=0}^{\infty} r\frac{\lambda^r \mathrm{e}^{-\lambda}}{r!} \\
&= \left(0\times\frac{\lambda^0\mathrm{e}^{-\lambda}}{0!}\right)+\left(1\times\frac{\lambda^1\mathrm{e}^{-\lambda}}{1!}\right)+\left(2\times\frac{\lambda^2\mathrm{e}^{-\lambda}}{2!}\right)+\left(3\times\frac{\lambda^3\mathrm{e}^{-\lambda}}{3!}\right)+\cdots \\
&= \left(0+\lambda+\frac{\lambda^2}{1!}+\frac{\lambda^3}{2!}+\frac{\lambda^4}{3!}+\cdots\right)\mathrm{e}^{-\lambda} \\
&= \lambda\left(1+\frac{\lambda^1}{1!}+\frac{\lambda^2}{2!}+\frac{\lambda^3}{3!}+\cdots\right)\mathrm{e}^{-\lambda} \\
&= \lambda\mathrm{e}^{\lambda}\mathrm{e}^{-\lambda} \\
&= \lambda.
\end{aligned}$$

The usual method of calculating the population variance is by using the identity that

$$\mathrm{Var}(X) = \mathrm{E}(X^2) - \{\mathrm{E}(X)\}^2.$$

However, because of the form of the Poisson distribution, with its denominator of $r! = r(r-1)(r-2)\cdots$, it is easier to work with $\mathrm{E}[X(X-1)]$ than $\mathrm{E}(X^2)$:

$$\begin{aligned}
\mathrm{E}[X(X-1)] &= \sum_{r=0}^{\infty} r(r-1)P_r \\
&= \sum_{r=0}^{\infty} r(r-1)\frac{\lambda^r\mathrm{e}^{-\lambda}}{r!} \\
&= \left(0+0+\lambda^2+\frac{\lambda^3}{1!}+\frac{\lambda^4}{2!}+\frac{\lambda^5}{3!}+\cdots\right)\mathrm{e}^{-\lambda} \\
&= \lambda^2\left(1+\frac{\lambda^1}{1!}+\frac{\lambda^2}{2!}+\frac{\lambda^3}{3!}+\cdots\right)\mathrm{e}^{-\lambda} \\
&= \lambda^2\mathrm{e}^{\lambda}\mathrm{e}^{-\lambda} \\
&= \lambda^2.
\end{aligned}$$

Finally,

$$\begin{aligned}\text{Var}(X) &= \text{E}[X^2] - \text{E}[X]^2 \\ &= \text{E}[X(X-1)] + \text{E}(X) - \text{E}[X]^2 \\ &= \lambda^2 + \lambda - (\lambda)^2 \\ &= \lambda.\end{aligned}$$

A random variable having a Poisson distribution with parameter λ has mean and variance both equal to λ.

Exercises 4b

1. The random variable X has a discrete uniform distribution with the four possible values being 0, 1, 2, and 4. Determine the mean and variance of X.
2. The random variable Y is equally likely to take the values -1 and 1. Determine the mean and variance of Y.
3. A roll of cloth contains randomly placed flaws at the rate of 1 per 10 metres of cloth. A customer purchases a randomly chosen 40 metres of the cloth. State the mean and standard deviation of the number of flaws in the cloth that the customer purchases.
4. Ten per cent of the inhabitants of an island in the Outer Hebrides are called McTavish. A random sample of 90 of the islanders are selected to receive a cut-price ticket to England. State the mean and standard deviation of the number in the sample that are called McTavish.

4.5 The expectation and variance of the sample mean

A sample of n observations is taken from a population. Denote the ith of the n observations in the sample by x_i with the corresponding random variable being X_i. Denote the sample mean by $\bar{x}$ and the corresponding random variable by $\bar{X}$.

We assume that the random variables $X_1, X_2, ..., X_n$ are independent, each with mean μ and variance σ^2.

Thus,

$$\begin{aligned}\bar{X} &= \frac{1}{n}(X_1 + X_2 + \cdots + X_n) \\ &= \frac{1}{n}X_1 + \frac{1}{n}X_2 + \cdots + \frac{1}{n}X_n.\end{aligned}$$

Using the result concerning expectations of sums of random variables, we have

$$\begin{aligned} \mathrm{E}(\bar{X}) &= \mathrm{E}\left(\frac{1}{n}X_1 + \frac{1}{n}X_2 + \cdots + \frac{1}{n}X_n\right) \\ &= \frac{1}{n}\mathrm{E}(X_1) + \frac{1}{n}\mathrm{E}(X_2) + \cdots + \frac{1}{n}\mathrm{E}(X_n) \\ &= \frac{1}{n}\mu + \frac{1}{n}\mu + \cdots + \frac{1}{n}\mu \\ &= n\left(\frac{1}{n}\mu\right) \\ &= \mu. \end{aligned}$$

Since the random variables $X_1, X_2, ..., X_n$ are mutually independent,

$$\begin{aligned} \mathrm{Var}(\bar{X}) &= \mathrm{Var}\left(\frac{1}{n}X_1 + \frac{1}{n}X_2 + \cdots + \frac{1}{n}X_n\right) \\ &= \mathrm{Var}\left(\frac{1}{n}X_1\right) + \mathrm{Var}\left(\frac{1}{n}X_2\right) + \cdots + \mathrm{Var}\left(\frac{1}{n}X_n\right) \\ &= \left(\frac{1}{n}\right)^2 \mathrm{Var}(X_1) + \left(\frac{1}{n}\right)^2 \mathrm{Var}(X_2) + \cdots + \left(\frac{1}{n}\right)^2 \mathrm{Var}(X_n) \\ &= \left(\frac{1}{n}\right)^2 \sigma^2 + \left(\frac{1}{n}\right)^2 \sigma^2 + \cdots + \left(\frac{1}{n}\right)^2 \sigma^2 \\ &= n\left(\frac{1}{n}\right)^2 \sigma^2 \\ &= \frac{\sigma^2}{n}. \end{aligned}$$

The sample mean, $\bar{x}$, is an observation from a distribution having mean μ and variance σ^2/n.

This is an important result because, for $n > 1$, this tells us that the sample mean is much less variable than are the individual observations. We can also see that the variance decreases as n increases, so that the sample mean is increasingly likely to be close to the mean of the population being sampled.

Example 4.8

The discrete random variable X has probability distribution $\mathrm{P_x} = (4 - \mathrm{x})/10$ *for* $\mathrm{x} = 0, 1, 2, 3$. *Determine the variance of the sample mean:*

(a) *when the sample size is 2,*

(b) *when the sample size is 16.*

In tabular form the probability distribution of X is as follows:

x	0	1	2	3
P_x	0.4	0.3	0.2	0.1

The probabilities sum to 1, so it seems that we have interpreted the formula correctly! To answer the question we must first obtain the variance of a single observation on X. Now,

$$\begin{aligned} \mathrm{E}(X) &= (0 \times 0.4) + (1 \times 0.3) + (2 \times 0.2) + (3 \times 0.1) = 1.0, lyge \\ \mathrm{E}(X^2) &= (0^2 \times 0.4) + (1^2 \times 0.3) + (2^2 \times 0.2) + (3^2 \times 0.1) = 2.0, \end{aligned}$$

so that

$$\mathrm{Var}(X) = \mathrm{E}(X^2) - \{\mathrm{E}(X)\}^2 = 2.0 - (1.0)^2 = 1.0.$$

From the general formula for a sample of size n, we therefore have the answers
(a) $\mathrm{Var}(\bar{X}) = 1/2$, and
(b) $\mathrm{Var}(\bar{X}) = 1/16$.

The square root of the variance of the sample mean, $\sigma / \sqrt{n}$, is often called the **standard error of the mean**, or simply the **standard error**. The same terms may be used for the estimate $s / \sqrt{n}$.

Exercises 4c

1. A random variable has expectation 12 and standard deviation 3. A random sample of 81 observations are taken. Find the expected value and the variance of the sample mean.
2. An unbiased six-sided die, with sides numbered 1, 2, ..., 6, is thrown 100 times and the scores are noted. Find the expectation and standard error of the mean score.

Key facts

- The **expectation** (also called the **expected value**) of X is denoted by $\mathrm{E}(X)$.
- For a discrete random variable X the expectation of the function $g(X)$ is the average value of $g(X)$ given by

  $$\mathrm{E}(g(X)) = \sum g(x)\mathrm{P}(X = x).$$

  This may be referred to as the **law of the unconscious statistician, (LOTUS).**
- Useful results:

  $$\mathrm{E}(aX + b) = a\mathrm{E}(X) + b.$$
  $$\mathrm{E}(g(X) + h(X)) = \mathrm{E}(g(X)) + \mathrm{E}(h(X)).$$
- The **population variance** is

  $$\sigma^2 = \mathrm{E}\left((X - \mu)^2\right) = \mathrm{E}\left(X^2\right) - (\mathrm{E}(X))^2.$$
- For X_1 and X_2, the **covariance** is

  $$\mathrm{Cov}(X_1, X_2) = \mathrm{E}(X_1 X_2) - \mathrm{E}(X_1)\mathrm{E}(X_2).$$
- If X_1 and X_2 are independent, then $\mathrm{Cov}(X_1, X_2) = 0$.
- Further useful results:

  $$\mathrm{Var}(aX + b) = a^2\mathrm{Var}(X).$$
  $$\mathrm{Var}(X_1 + X_2) = \mathrm{Var}(X_1) + 2\mathrm{Cov}(X_1, X_2) + \mathrm{Var}(X_2).$$

  If X_1 and X_2 are independent
  $$\mathrm{Var}(X_1 + X_2) = \mathrm{Var}(X_1) + \mathrm{Var}(X_2).$$

5

Continuous random variables

Chapters 3 and 4 have focused on discrete random variables, quantities whose values are unpredictable, but for which a list of the possible values can be made. Continuous random variables differ in that no such list is feasible, though the range of values can be described. Here are some examples:

Continuous random variable	*Possible range of values*
The height of a randomly chosen 18-year old male student	1.3–2.3m
The true mass of a '1 kg' bag of sugar	990–1010g
The time interval between successive earthquakes of magnitude > 7 on the Richter scale	Any (positive) time

The measurements all refer to **physical** quantities. The number of distinct values is limited only by the inefficiency of our measuring instruments. Since there are an uncountable number of possible values that a continuous random variable might take, the probability of any particular value is zero—instead, we assign probabilities to (arbitrarily small) ranges of values.

If a continuous random variable is measured rather inaccurately, then we will treat it as though it is a discrete random variable:

Age of randomly chosen male measured to nearest 10 years $\longrightarrow$ Treat as a discrete random variable with 13 categories

Conversely, if a discrete random variable has a great many possible outcomes, then we may treat it as though it was a continuous random variable:

Mark on exam paper (an integer in the range 0 to 100) $\longrightarrow$ Treat as a continuous variable on the interval [0,100]

Because of this easy transition between the two types of variable, the ideas of expectation, and the formulae interrelating expectations, carry over straightforwardly to continuous variables.

5.1 The probability density function (pdf)

For a continuous random variable, X, the probability density function f is such that

$$P(a < X < b) = \int_a^b f(x)dx, \tag{5.1}$$

for all $a < b$.

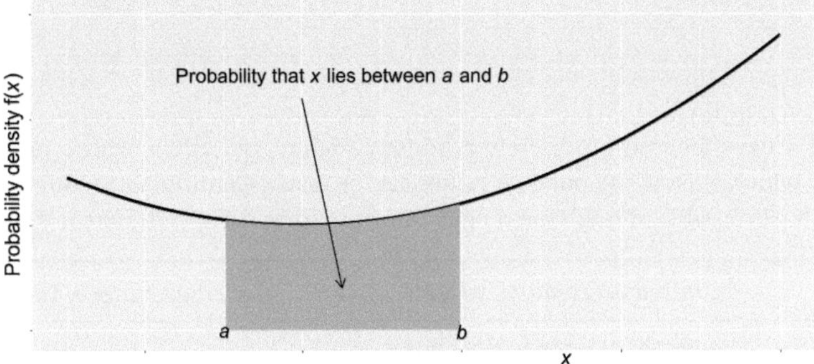

5.1.1 Properties of the pdf

1. Since we cannot have negative probabilities,

$$f(x) \geq 0, \text{for all } x. \tag{5.2}$$

2. As b approaches a, the value of the integral in Equation (5.1) approaches zero, so, in the limit, $P(X = a) = 0$. This result implies that we need not be fussy about whether we write $P(X < a)$ or $P(X \leq a)$, since the two have the same value.

3. If c and d are the lower and upper limits on the possible values of X, then

$$\int_c^d f(x)dx = 1. \tag{5.3}$$

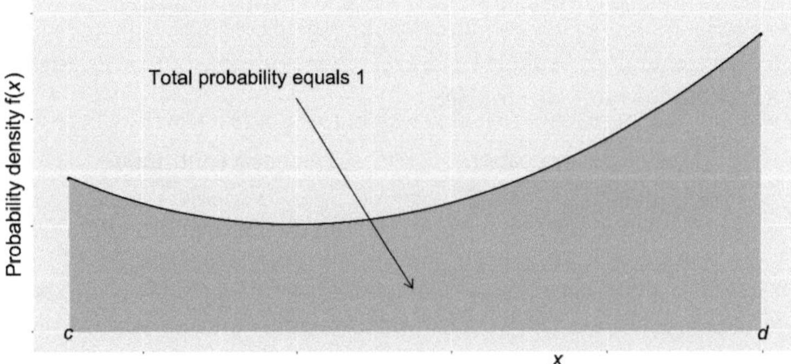

The function f measures probability density, not probability. Although f(x) often has values less than 1, this need not be the case. For example,

$$f(x) = \begin{cases} 2 & 0 < x < 1/2, \\ 0 & \text{otherwise,} \end{cases}$$

defines a proper probability density function that integrates to 1.

Example 5.1

The continuous random variable X *has pdf given by*

$$f(x) = \begin{cases} kx^2 & 1 < x < 3, \\ 0 & \textit{otherwise.} \end{cases}$$

Determine:

(a) *the value of the constant k,*

(b) P(X < 2)*.*

We start by drawing a sketch of the pdf to get a feel for what we have:

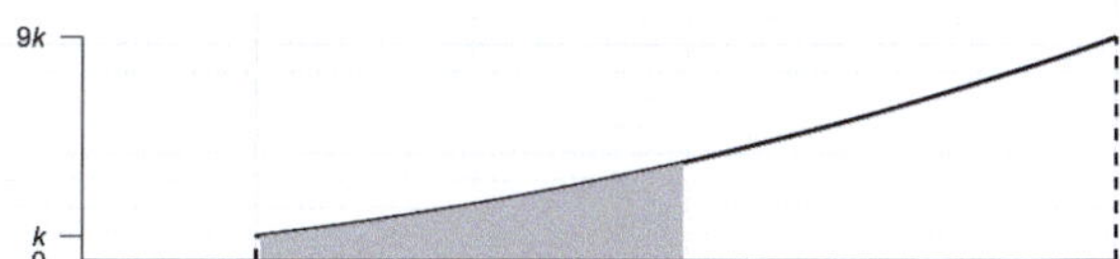

(a) To find k, we use the fact that f integrates to 1:

$$\int_1^3 kx^2 dx = \left[\frac{kx^3}{3}\right]_1^3 = \frac{k}{3}(27 - 1) = \frac{26k}{3}.$$

Since we know that the integral is equal to 1, it follows that $k = 3/26$.

(b)

$$P(X < 2) = \int_1^2 kx^2 dx = \left[\frac{kx^3}{3}\right]_1^2 = \frac{k}{3}(8 - 1) = \frac{7}{26}.$$

The probability that X takes a value less than 2 is 7/26, or 0.27 (to 2 d.p.).

Exercises 5a

1. The random variable X has probability density function given by

$$f(x) = \begin{cases} 1 - kx & 0 < x < 2, \\ 0 & \text{otherwise.} \end{cases}$$

(a) Sketch the graph of f.

(b) Find the value of the constant k.

(c) Determine $P(X \leq 1)$.

2. The random variable X has probability density function given by

$$f(x) = \begin{cases} 10k & -0.05 < x \leq 0.05, \\ 0 & \text{otherwise} \end{cases}$$

(a) Sketch the graph of f.

(b) Find the value of k.

(c) Determine $P(X > 0.1)$.

(d) Determine $P(X \leq 0.025)$.

3. A garage is supplied with petrol once a week. Its volume of weekly sales, X, in thousands of gallons, is distributed with probability density function $f(x)$ given by

$$f(x) = \begin{cases} kx(1-x)^2 & 0 < x < 1, \\ 0 & \text{otherwise.} \end{cases}$$

Determine the value of k and hence determine the probability that weekly sales are less than 700 gallons.

5.2 The cumulative distribution function, F

The cumulative distribution function is often referred to as the **distribution function**, or, more simply, as the **cdf**. The function is defined by

$$F(x) = P(X \leq x), \tag{5.4}$$

and is related to the function f by

$$F(b) = \int_{-\infty}^{b} f(x)dx. \tag{5.5}$$

The lower limit of the integral is given as $-\infty$, but is, in effect, the smallest attainable value of X. For every cdf it is true that

$$F(-\infty) = 0, \qquad F(\infty) = 1.$$

Strictly $F(-\infty)$ means 'the limiting value of $F(x)$ as x approaches $-\infty$, with $F(\infty)$ being similarly defined.

Useful relations are

$$P(c < X \leq d) = F(d) - F(c), \tag{5.6}$$

$$P(X > x) = 1 - F(x). \tag{5.7}$$

Example 5.2

We wish to find $F(x)$ in the case where the continuous random variable X has pdf $f(x)$ given by

$$f(x) = \begin{cases} 1 & 2 < x < 3, \\ 0 & \text{otherwise.} \end{cases}$$

For $b \leq 2$, $F(b) = 0$, since there is no chance that X will take such a low value. Similarly, for $b \geq 3$, $F(b) = 1$, since it is certain that X takes a value less than any b in this range.

For $2 \leq b \leq 3$, we must integrate the pdf:

$$F(b) = P(X \leq b) = \int_2^b 1 dx = b - 2.$$

Hence,

$$F(x) = \begin{cases} 0 & x \leq 2, \\ x - 2 & 2 \leq x \leq 3, \\ 1 & x \geq 3. \end{cases}$$

5.2.1 The median, m

The **median**, m, is the value that bisects the distribution, in the sense that X is equally likely to be smaller, or larger, than m. Hence,

$$\int_{-\infty}^{m} f(x)dx = \int_{m}^{\infty} f(x)dx = 0.5. \tag{5.8}$$

Percentiles and **quartiles** are defined similarly. For example, the 90th percentile is the solution of $F(x) = 0.90$, while the upper quartile is the solution of $F(x) = 0.75$.

Exercises 5b

1. It is given that X is a continuous random variable for which

$$f(x) = \begin{cases} 2(1-x) & 0 < x < k, \\ 0 & \text{otherwise.} \end{cases}$$

Find:

(a) the value of k,

(b) $F(x)$,

(c) the median of X.

2. It is given that X is a continuous random variable for which

$$f(x) = \begin{cases} 2x + k & 3 < x < 4, \\ 0 & \text{otherwise.} \end{cases}$$

Find:

(a) the value of k,

(b) $F(x)$,

(c) the lower and upper quartiles of X.

3. It is given that X is a continuous random variable for which

$$f(x) = \begin{cases} k(x+2) & -2 < x < 0, \\ \frac{1}{2}k(3-x) & 0 < x < 2, \\ 0 & \text{otherwise.} \end{cases}$$

Find:

(a) the value of k,

(b) $F(x)$,

(c) $P(-1 < X < 1)$,

(d) $P(1 < X < 3)$.

5.3 Expectations for continuous variables

For a discrete random variable, $\mu = E(X) = \sum xP_x$, where the summation is over all possible values of X. For a continuous random variable, we replace the summation by an integral, and the individual point probabilities by the pdf:

$$E(X) = \int_{-\infty}^{\infty} x f(x) dx. \tag{5.9}$$

The limits of the integral are given as $-\infty$ and ∞, but are, in effect, the largest and smallest attainable values of X. By a similar argument,

$$E[g(X)] = \int_{-\infty}^{\infty} g(x) f(x) dx. \tag{5.10}$$

In particular,

$$\mathrm{E}(X^2) = \int_{-\infty}^{\infty} x^2 \mathrm{f}(x)\mathrm{d}x, \tag{5.11}$$

and

$$\mathrm{Var}(X) = \mathrm{E}[(X-\mu)^2] = \int_{-\infty}^{\infty} (x-\mu)^2 \mathrm{f}(x)\mathrm{d}x, \tag{5.12}$$

though it is usually easiest to calculate $\mathrm{Var}(X)$ using

$$\mathrm{Var}(X) = \mathrm{E}(X^2) - \{\mathrm{E}(X)\}^2.$$

All the results of Chapter 4 continue to hold:

$$\begin{aligned} \mathrm{E}(X+a) &= \mathrm{E}(X) + a, \\ \mathrm{E}(aX) &= a\mathrm{E}(X), \\ \mathrm{E}[a\mathrm{g}(X)] &= a\mathrm{E}[\mathrm{g}(X)], \\ \mathrm{E}[\mathrm{g}(X) + \mathrm{h}(X)] &= \mathrm{E}[\mathrm{g}(X)] + \mathrm{E}[\mathrm{h}(X)], \end{aligned}$$

$$\begin{aligned} \mathrm{Var}(X+a) &= \mathrm{Var}(X), \\ \mathrm{Var}(aX) &= a^2\mathrm{Var}(X), \end{aligned}$$

$$\begin{aligned} \mathrm{E}(X+Y) &= \mathrm{E}(X) + \mathrm{E}(Y), \\ \mathrm{E}(aX+bY+c) &= a\mathrm{E}(X) + b\mathrm{E}(Y) + c, \\ \mathrm{E}[\mathrm{g}(X) + \mathrm{h}(Y)] &= \mathrm{E}[\mathrm{g}(X)] + \mathrm{E}[\mathrm{h}(Y)], \\ \mathrm{E}(R+S+T+U) &= \mathrm{E}(R) + \mathrm{E}(S) + \mathrm{E}(T) + \mathrm{E}(U). \end{aligned}$$

For independent random variables we also have

$$\begin{aligned} \mathrm{Var}(aX+bY+c) &= a^2\mathrm{Var}(X) + b^2\mathrm{Var}(Y), \\ \mathrm{Var}(R+S+T+U) &= \mathrm{Var}(R) + \mathrm{Var}(S) + \mathrm{Var}(T) + \mathrm{Var}(U). \end{aligned}$$

Most are easy to prove. For example, consider $\mathrm{E}[\mathrm{g}(X)+\mathrm{h}(X)]$, where $\mathrm{g}(X)$ and $\mathrm{h}(X)$ are two arbitrary functions of a continuous random variable X having pdf $\mathrm{f}(x)$:

$$\begin{aligned} \mathrm{E}[\mathrm{g}(X) + \mathrm{h}(X)] &= \int_{-\infty}^{\infty} \{\mathrm{g}(x) + \mathrm{h}(x)\}\mathrm{f}(x)\mathrm{d}x \\ &= \int_{-\infty}^{\infty} \mathrm{g}(x)\mathrm{f}(x)\mathrm{d}x + \int_{-\infty}^{\infty} \mathrm{h}(x)\mathrm{f}(x)\mathrm{d}x = \mathrm{E}[\mathrm{g}(X)] + \mathrm{E}[\mathrm{h}(X)]. \end{aligned}$$

Example 5.3

The continuous random variable X has pdf given by:

$$f(x) = \begin{cases} \frac{3}{4}(1 - x^2) & -1 < x < 1, \\ 0 & \text{otherwise.} \end{cases}$$

Determine E(X) and Var(X). Denoting these quantities by μ and σ^2, respectively, determine the probability that an observed value of X has a value in the interval $[\mu-\sigma, \mu+\sigma]$.

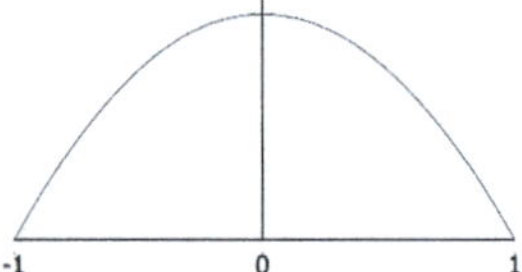

Since f is symmetric about the line $x = 0$, $\mu = \mathrm{E}(X) = 0$.

To calculate Var(X) we need E(X^2):

$$\begin{aligned} \mathrm{E}(X^2) &= \int_{-1}^{1} \frac{3}{4}x^2(1 - x^2)dx \\ &= \frac{3}{4}\left[\frac{x^3}{3} - \frac{x^5}{5}\right]_{-1}^{1} \\ &= \frac{1}{5}. \end{aligned}$$

Hence $\sigma^2 = \mathrm{Var}(X) = \mathrm{E}(X^2) - \{\mathrm{E}(X)\}^2 = 1/5$.

The probability of an observed value of X having a value in the interval $(\mu - \sigma, \mu + \sigma)$ is

$$\begin{aligned} \mathrm{P}[(\mu - \sigma) < X < (\mu + \sigma)] &= \mathrm{P}(-\sqrt{1/5} < X < \sqrt{1/5}) \\ &= \frac{3}{4}\int_{-\sqrt{1/5}}^{\sqrt{1/5}} (1 - x^2)dx \\ &= \frac{7}{5}\sqrt{\frac{1}{5}} = 0.626 \text{ (to 3 d.p.).} \end{aligned}$$

Example 5.4

The continuous random variable X has pdf given by

$$f(x) = \begin{cases} x & 0 < x < 1, \\ 2 - x & 1 < x < 2, \\ 0 & \text{otherwise.} \end{cases}$$

Determine E(X) and Var(X).

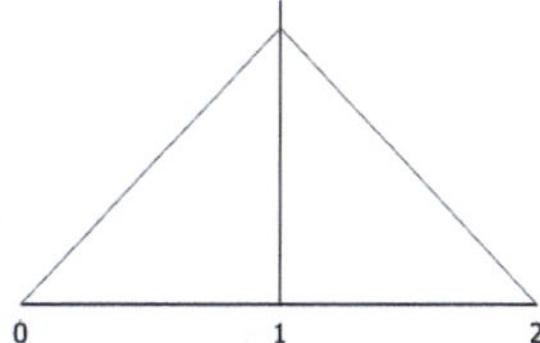

Since $f(x)$ is symmetric about the line $x = 1$, $\mu = E(X) = 1$.

To determine the value of the variance, we must first find the value of $E(X^2)$. Care is required, since the form of $f(x)$ depends upon the value of X:

$$\begin{aligned} E(X^2) &= \int_0^1 x^2 x dx + \int_1^2 x^2(2 - x)dx \\ &= \left[\frac{x^4}{4}\right]_0^1 + \left[2\frac{x^3}{3} - \frac{x^4}{4}\right]_1^2 \\ &= 7/6 \end{aligned}$$

Hence,

$$\sigma^2 = \text{Var}(X) = 7/6 - 1^2 = 1/6.$$

Exercises 5c

1. The continuous random variable X has pdf f given by

$$f(x) = \begin{cases} x/2 & 0 < x < 2, \\ 0 & \text{otherwise.} \end{cases}$$

Determine:

(a) $E(X)$,

(b) $\text{Var}(X)$,

(c) $P[X < E(X)]$.

2. The continuous random variable X has pdf f given by

$$f(x) = \begin{cases} 2x & 0 < x < 1, \\ 0 & \text{otherwise.} \end{cases}$$

The random variable Y is defined by $Y = 4X + 2$. Determine:

(a) $E(Y)$,

(b) $\text{Var}(Y)$.

3. It is given that

$$f(x) = \begin{cases} 1/2 & 1 < x < 2, \\ k & 2 < x < 4, \\ 0 & \text{otherwise.} \end{cases}$$

Determine:

(a) the value of k,

(b) $F(x)$,

(c) $E(X)$.

4. The continuous random variable X has probability density function given by

$$f(x) = \begin{cases} x & 0 \le x \le 1, \\ k - x & 1 \le x \le 2, \\ 0 & \text{otherwise.} \end{cases}$$

(a) Find k.

(b) Find also the mean, μ and show that the variance, σ^2, is equal to $\frac{1}{6}$.

(c) Determine the probability that a future observation lies in the interval $(\mu - \sigma, \mu)$.

5.4 Obtaining f from F

Since F can be obtained by integrating f, f can be obtained by differentiating F. The value of f(b) is therefore the slope of F at the point where $x = b$.

Example 5.5

The random variable X has cdf given by

$$F(x) = \begin{cases} 0 & x \leq 1 \\ \frac{1}{8}(x-1)^3 & 1 \leq x \leq 3 \\ 1 & x \geq 3 \end{cases}$$

Find the form of the pdf of X.

Evidently f(x) is equal to 0 for $x < 1$ and for $x > 3$, since F(x) is unchanging in these regions. Differentiating $\frac{1}{8}(x-1)^3$ we find that

$$f(x) = \begin{cases} \frac{3}{8}(x-1)^2 & 1 < x < 3, \\ 0 & \text{otherwise.} \end{cases}$$

5.5 The uniform (rectangular) distribution

We encountered a random variable having a uniform distribution in Example 5.2. Its characteristic is that, for the entire range of possible values of X (from a to b, say), f is constant:

$$f(x) = \begin{cases} 1/(b-a) & a < x < b, \\ 0 & \text{otherwise.} \end{cases} \tag{5.13}$$

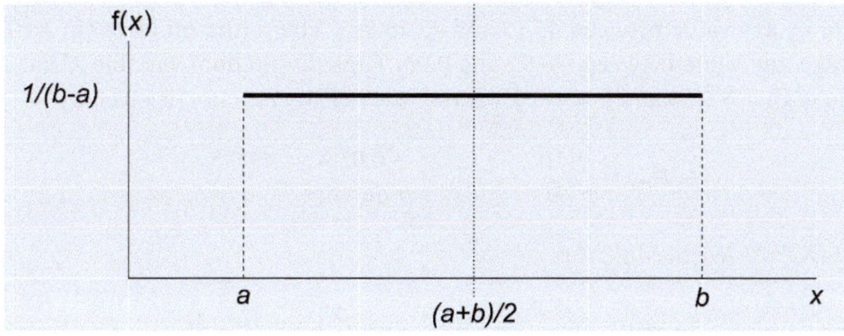

Between a and b the probability density is **uniform** and the resulting shape is **rectangular**. The rectangle has width $(b - a)$ and height $\frac{1}{b-a}$, so that its area is equal to 1, as required.

Since the probability density is symmetrical about the line $x = \frac{1}{2}(a + b)$, the mean, E(X), and the median, m, are both equal to $\frac{1}{2}(a + b)$. The cumulative distribution function, F, is given by

$$F(c) = P(X \leq c) = \int_a^c \frac{1}{b-a} dx = \left[\frac{x}{b-a}\right]_a^c = \frac{c-a}{b-a}.$$

Formally, therefore, we have

$$F(x) = \begin{cases} 0 & x \leq a, \\ \frac{x-a}{b-a} & a \leq x \leq b, \\ 1 & x \geq b. \end{cases} \tag{5.14}$$

After some integration and a little algebra we find that $\text{Var}(X) = \frac{1}{12}(b - a)^2$.

A distribution that is uniform between a and b has mean $\frac{1}{2}(a+b)$ and variance $\frac{1}{12}(b-a)^2$.

Example 5.6

(a) *The distance between two points, A and B, is to be measured correct to the nearest tenth of a kilometre. Working in km, determine the mean and standard deviation of the associated round-off error.*

(b) *There are four points A, B, C, and D. The lengths of the distances AB, BC, and CD are each to be measured correct to the nearest tenth of a kilometre. Determine the mean and standard deviation of the difference between the total of the three measured lengths and the true overall length.*

(a) To see what is required, suppose that the length of AB is given as 45.2 km. The true length of AB could be any value between 45.15 and 45.25 km. The round-off error (in km), X, could therefore take any value between −0.05 and 0.05. Thus the random variable X has a uniform distribution with $b = 0.05$ and $a = -0.05$, and so the density function is given by

$$f(x) = \begin{cases} 10 & -0.05 < x < 0.05, \\ 0 & \text{otherwise.} \end{cases}$$

Evidently $E(X) = 0$. We need $E(X^2)$:

$$E(X^2) = \int_{-0.05}^{0.05} 10x^2 dx = 10\left[\frac{x^3}{3}\right]_{-0.05}^{0.05} = \frac{1}{1200}.$$

Hence the standard deviation of the round-off error is $\sqrt{\frac{1}{1200}} = 0.029$ (to 3 d.p.).

(b) Denote the three round-off errors by X, Y, and Z. These variables have independent uniform distributions, each with mean 0 and variance 1/1200. Their sum therefore has mean 0 and variance $3 \times 1/1200 = 1/400$. Thus the required mean and standard deviation are 0 and $1/20 = 0.05$.

5.6 The exponential distribution

When events occur at random points in time according to a Poisson process (see Section 3.10), the number of events in an interval of time has a Poisson distribution. Suppose that these events are occurring at a rate λ per unit time (where λ is some positive constant). Obviously the *mean* time between events will be $1/\lambda$, but the individual time intervals will vary about this mean. The distribution of the lengths of these intervals is known as the **exponential distribution**. The pdf is given by

$$\mathrm{f}(x) = \begin{cases} \lambda \mathrm{e}^{-\lambda x} & x > 0, \\ 0 & \text{otherwise.} \end{cases} \tag{5.15}$$

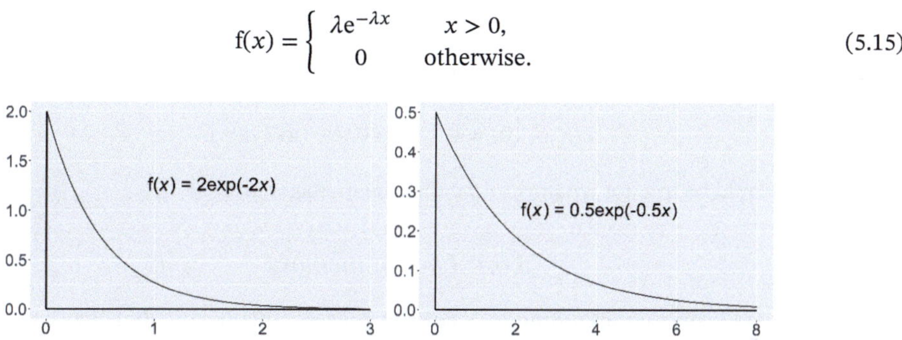

Figure 5.1 All exponential distributions have the same shape.

Like the geometric distribution, which is its discrete analogue, the exponential distribution has mode 0. All exponential distributions have the same basic shape (see Figure 5.1).

For $b \geq 0$, the distribution function is given by

$$\mathrm{F}(b) = \int_0^b \lambda \mathrm{e}^{-\lambda x} \mathrm{d}x = \left[-\mathrm{e}^{-\lambda x}\right]_0^b = 1 - \mathrm{e}^{-\lambda b}$$

so:

$$\mathrm{P}(X < x) = 1 - \mathrm{e}^{-\lambda x}, \tag{5.16}$$

$$\mathrm{P}(X > x) = \mathrm{e}^{-\lambda x}, \tag{5.17}$$

$$\mathrm{P}(a < X < b) = \mathrm{e}^{-\lambda a} - \mathrm{e}^{-\lambda b}. \tag{5.18}$$

Example 5.7

During the morning, a busy office receives on average 20 telephone calls an hour. The office manager arrives at 10 minutes past 10. Determine the probability that there are no calls during the next 10 minutes.

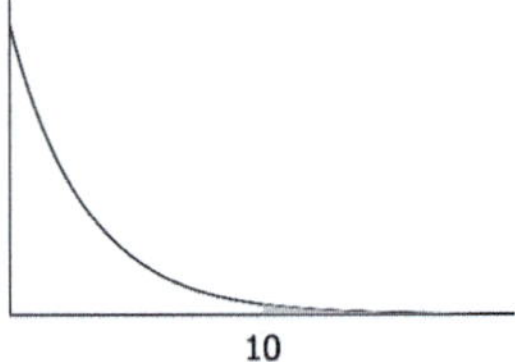

A rate of 20 calls an hour implies a rate of $20/60 = 1/3$ per minute. There are no calls during the next 10 minutes if the time to the next call is at least 10 minutes. Denoting the time (in mins) to the next call by X, we require $P(X > 10)$. Since the calls occur at random points in time, X has an exponential distribution with $\lambda = 1/3$. Now,

$$P(X > 10) = e^{-10/3} = 0.036 \text{ (to 3 d.p.)}.$$

The same would be true for whatever time that the manager arrived.

5.6.1 *Lack of memory*

When events are occurring according to a Poisson process, this means that they are occurring at random points in time. The implication is that the times at which events occurred in the past, will have no influence on when future events occur. The fact that a gambler has had a long run of disappointments does not influence the probability of success for the next gamble. Expressed in terms of probability statements, the implication is that, with X denoting the time to the next event,

$$P[X > (a + b)|X > a] = P(X > b).$$

5.6.2 *Mean and variance of the exponential distribution*

To obtain the mean and variance of the exponential distribution, we use integration by parts:

$$\begin{aligned} E(X) &= \int_0^\infty \lambda x e^{-\lambda x} dx \\ &= \lambda \left\{ \left[\frac{-x e^{-\lambda x}}{\lambda} \right]_0^\infty + \frac{1}{\lambda} \int_0^\infty e^{-\lambda x} dx \right\} \\ &= (-\infty \times e^{-\lambda \times \infty} - 0) - \frac{1}{\lambda}(e^{-\lambda \times \infty} - 1). \end{aligned}$$

The first bracket is 0 and the second bracket is -1, so we have the result

$$E(X) = \frac{1}{\lambda}. \quad (5.19)$$

Using integration by parts again, gives

$$E(X^2) = \frac{2}{\lambda^2}.$$

Using $\text{Var}(X) = E(X^2) - \{E(X)\}^2$, we therefore have that

$$\text{Var}(X) = \frac{1}{\lambda^2}. \quad (5.20)$$

5.6.3 *Connection with a Poisson process*

Consider a sequence of independent events occurring at random points in time at a rate λ; in other words, a Poisson process with parameter λ. We start examining this process at an arbitrary time $t = 0$ and denote the random variable 'the time to the first event' by X. Then

$$P(X > x) = P[\text{No events occur in the time interval } (0, x)].$$

To find this probability, we note that the mean number of events occurring in a time interval of length x is λx, and that the probability of obtaining the value 0 from a Poisson distribution with mean λx is

$$\frac{(\lambda x)^0 e^{-\lambda x}}{0!} = e^{-\lambda x}.$$

So

$$P(X > x) = e^{-\lambda x},$$

and

$$F(x) = 1 - P(X > x) = 1 - e^{-\lambda x}.$$

Finally, differentiating with respect to x, we obtain the pdf of X:

$$f(x) = \lambda e^{-\lambda x}.$$

In a Poisson process having events occurring at rate λ per unit time, the time to the first occurrence is an observation from an exponential distribution with mean $1/\lambda$.

Example 5.8

A bargain-hunter discovers a large roll of material that is being sold at a greatly reduced price because it contains flaws. These flaws occur at random locations down the length of the roll. The mean length of cloth between successive flaws is 0.5 metres. Determine the probability that the first 1.5 metres of the roll contain no flaws. Determine the probability that the first flaw occurs at between 0.6 and 0.8 metres from the start of the roll.

Random locations imply a Poisson process. The mean length between flaws of 0.5 metres implies a rate of occurrence (λ) of 2 per metre. The probability of no flaws in the first 1.5 metres may be obtained using either the Poisson or the exponential distributions. Using the latter with $\lambda = 2$ we have

$$P(X > 1.5) = e^{-3} = 0.050 \text{ (to 3 d.p.)}.$$

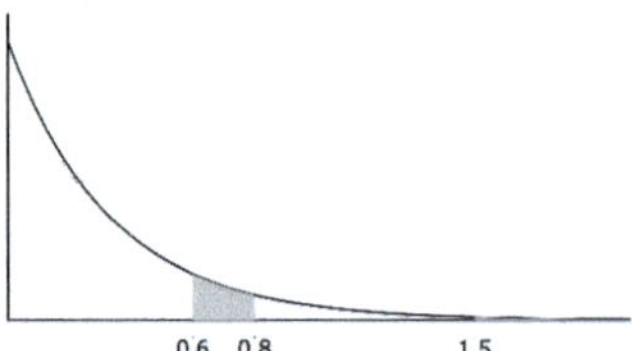

To determine the probability of the first flaw occurring between 0.6 and 0.8 metres from the start we need

$$F(0.8) - F(0.6) = \left(1 - e^{-1.6}\right) - \left(1 - e^{-1.2}\right) = 0.099 \text{ (to 3 d.p.)}.$$

Exercises 5d

1. The lifetime of a new type of lightbulb has an exponential distribution with mean 4000 hours. Determine the probability that:

   (a) a randomly chosen bulb will last more than 3000 hours,

   (b) of ten randomly chosen bulbs, none will last for less than 1000 hours.

2. Faults occur at random locations along a nylon line. The average rate of occurrence is 2 per 100 m. Determine the probability that:

   (a) 200 m of the line includes exactly three faults,

   (b) the length of line between the second and third faults exceeds 100 m.

3. The lifetime of a toaster element has an exponential distribution with mean nine years. A particular element is still working after four years. Find the probability that it will still be working after five more years.

4. A motorist joins a motorway. The distance, X miles, that she travels before seeing a police car has an exponential distribution with mean 50 miles. This is also the mean distance travelled between subsequent sightings.

   (a) Find $\mathrm{P}(X > 100)$.

   (b) The distance that she travels between seeing the first and second police cars is Y miles. Determine $\mathrm{P}(Y > 50)$.

5.7 *The beta distribution

The beta distribution is a two-parameter distribution in the range (0, 1). It will be used in Chapter 20. The pdf is given by

$$\mathrm{f}(x) = \frac{1}{\mathrm{B}(\alpha, \beta)} x^{\alpha-1}(1-x)^{\beta-1}, \tag{5.21}$$

where $0 < x < 1$, the two parameters α and β are positive, and B is the so-called **beta function** given by

$$\mathrm{B}(\alpha, \beta) = \frac{\Gamma(\alpha)\Gamma(\beta)}{\Gamma(\alpha+\beta)}, \tag{5.22}$$

and, in turn, Γ is the so-called **gamma function** given by

$$\Gamma(x) = \int_0^{\infty} s^{x-1}\mathrm{e}^{-s}\mathrm{d}s. \tag{5.23}$$

If x is an integer, then $\Gamma(x) = (x-1)!$.

Within its limited range the pdf of the beta distribution can take a wide variety of shapes, as evidenced by Figure 5.2. If both parameters are greater than 1, then the distribution has mode at $(\alpha-1)/(\alpha+\beta-2)$; if both parameters are less than 1, then the distribution is U-shaped; if only one parameter is less than 1, then the distribution is J-shaped.

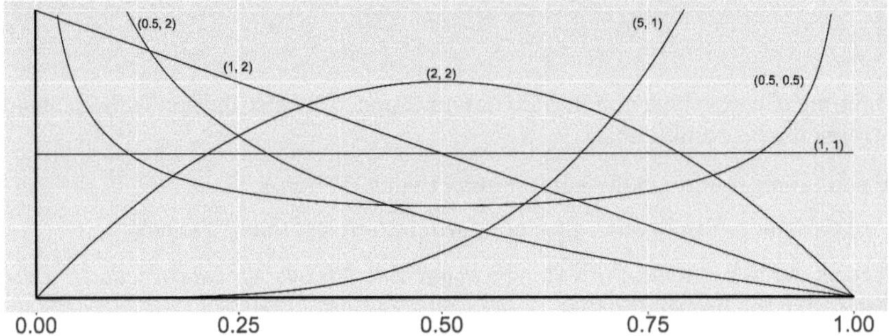

Figure 5.2 Examples of beta distributions. The values of α and β are shown in brackets for each curve.

The distribution has mean $\alpha/(\alpha+\beta)$ and variance $\alpha\beta/\{\alpha+\beta)^2(\alpha+\beta+1)\}$. A good approximation to the value of the median is $(3\alpha-1)/(3\alpha+3\beta-2)$.

5.8 *The gamma distribution

This is another two-parameter distribution. It has pdf given by

$$f(x) = \frac{1}{\Gamma(\alpha)} x^{\alpha-1} e^{-\beta x} \beta^{\alpha}, \tag{5.24}$$

where $x > 0$, the two parameters α and β are also positive, and $\Gamma(x)$ is the gamma function defined in Section 5.7.

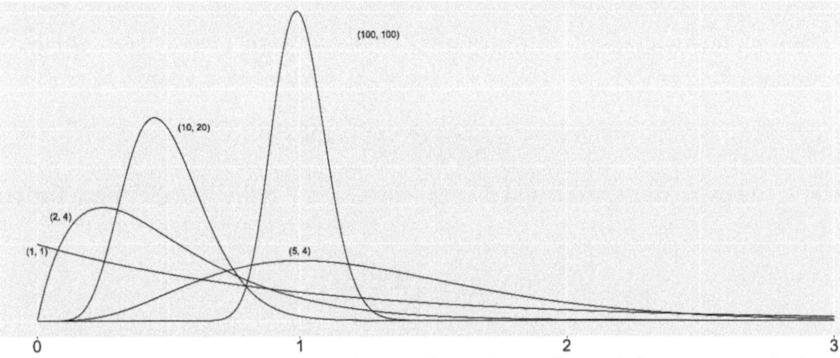

Figure 5.3 Examples of gamma distributions. The values of α and β are shown in brackets for each curve.

Some possible shapes for the gamma distribution are illustrated in Figure 5.3. The distribution has mean α/β and variance α/β^2. The mode is at 0 if $\alpha < 1$, and is otherwise at $(\alpha-1)/\beta$.

The gamma distribution is another distribution related to random events. Consider a sequence of such events, with the time between every pair of consecutive events being an observation from an exponential distribution with rate λ. The overall waiting time until the nth such event is an observation from a gamma distribution with $\alpha = n$ and $\beta = \lambda$.

5.9 *Transformation of a random variable

The random variable X has a known pdf f(x). Suppose that we wish to find the pdf of the random variable Y, where $Y = g(x)$. One approach is to find the distribution function of X, F(X), use this to deduce the distribution function of Y, G(y), and then differentiate G(y) to obtain the pdf of Y, g(y).

Example 5.9

Suppose that the random variable X *takes values between* 0 *and* c *with pdf given by* $f(x) = ax^2$, *where* a *is a constant. We wish to find the pdf of* Y, *where* $Y = X^4$.

The distribution function, F(x) is given by

$$F(x) = \begin{cases} 0 & x \leq 0, \\ \frac{ax^3}{3} & 0 \leq x \leq c, \\ 1 & x \geq c. \end{cases}$$

The case $X = c$ implies that $\frac{ac^3}{3} = 1$. Hence $a = \frac{3}{c^3}$ and the distribution function becomes

$$F(x) = \begin{cases} 0 & x \leq 0, \\ \left(\frac{x}{c}\right)^3 & 0 \leq x \leq c, \\ 1 & x \geq c. \end{cases}$$

For any value b in the interval $(0, c)$, $P(X \leq b) = P(Y \leq b^4)$. Thus the distribution function for Y is given by

$$G(y) = \begin{cases} 0 & x \leq 0, \\ c^{-3}y^{3/4} & 0 \leq y \leq c^4, \\ 1 & y \geq c^4. \end{cases}$$

Differentiating we find that $g(y) = \frac{3}{4}c^{-3}y^{-1/4}$ for $0 \leq y \leq c^4$ and 0 otherwise.

Example 5.10

The random variable X is uniformly distributed between -1 and 2. We wish to find the pdf of $Y = X^2$.

In this case $f(x) = 1/3$ for x in the interval $(-1, 2)$. For the calculation of the distribution function for Y, $P(Y \leq y)$, we have two cases to consider:

$$\begin{array}{lll} y \leq 1 & \text{corresponds to} & -\sqrt{y} \leq x \leq \sqrt{y}, \\ y \geq 1 & \text{corresponds to} & -1 \leq x \leq \sqrt{y}. \end{array}$$

Thus

$$G(y) = \begin{cases} 0 & y \leq 0, \\ \frac{2}{3}\sqrt{y} & 0 \leq y \leq 1, \\ \frac{1}{3}(1 + \sqrt{y}) & 1 \leq y \leq 4, \\ 1 & y \geq 4. \end{cases}$$

Differentiating we have

$$g(y) = \begin{cases} 1/(3\sqrt{y}) & 0 \leq y \leq 1, \\ 1/(6\sqrt{y}) & 1 \leq y \leq 4. \\ 0 & \text{otherwise.} \end{cases}$$

Key facts

- The **probability density function**, f, is such that an area contained between f and the x-axis corresponds to a probability and can be evaluated by integration:

  $$P(a \le X \le b) = \int_a^b f(x)dx.$$

- When $a = b$ the area is zero. Hence,

  $$P(X \ge x) = P(X > x),$$
  $$P(X \le x) = P(X < x).$$

- The **cumulative distribution function**, F, is defined by

  $$F(x) = P(X \le x) = \int_{-\infty}^{x} f(t)dt.$$

  Conversely, f(x) is found by differentiating F:

  $$f(x) = \frac{d}{dx}F(x).$$

- The **median**, m, is the value of x at which $F(x) = 0.5$.

- **Expectation**

  $$E(g(X)) = \int_{-\infty}^{\infty} g(x)f(x)dx.$$

- **Population mean**, μ, is given by

  $$\mu = \int_{-\infty}^{\infty} xf(x)dx$$

- The **continuous uniform (rectangular) distribution** has $f(x) = 1/(b - a)$ for the range $a \le x \le b$.

- The **exponential distribution** has $f(x) = \lambda e^{-\lambda x}$. The distribution has mean $1/\lambda$ and variance $1/\lambda^2$.

- The distribution of the intervals between events occurring at random points in time (a **Poisson process**) is an exponential distribution.

R

- The command `dexp(x, λ)` evaluates $f(x)$ where X has an exponential distribution with parameter λ.

- For an exponential distribution, the probability that $X \le x$ is given by `pexp(x, n, p)`.

- The command `dgamma(x, α, β)` evaluates $f(x)$ where X has a gamma distribution with parameters α and β.

- The command `dbeta(x, α, β)` evaluates $f(x)$ where X has a beta distribution with parameters α and β.

6

The normal distribution

The normal distribution describes the situation in which very large values are rather rare, very small values are rather rare, but middling values are rather common. This is a good description of lots of things: hence the description 'normal'.[1] Here are some examples:

- Heights and weights,
- The precise volumes of lager in 'pints' of lager at the local pub.

The distribution can also be applied as an approximation in the case of some discrete variables:

- Marks obtained by students on an exam paper,
- The IQ scores of the population.

We will see later that the distribution can also be used as an approximation to the binomial and Poisson distributions.

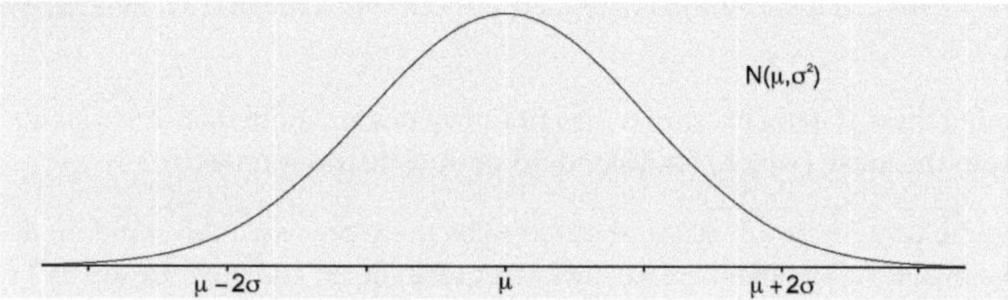

A normal distribution is a unimodal symmetric continuous distribution having two parameters: μ (the mean) and σ^2 (the variance). Because of the symmetry, the mean is equal to both the mode and the median. As a shorthand we refer to a $\mathrm{N}(\mu, \sigma^2)$ distribution.

Since, for any distribution, changes in μ and σ can be regarded as changes of location and scale, all normal distributions can be related to a single reference distribution, the so-called **standard normal** distribution, which has mean 0 and variance 1. The random variable having this distribution is usually denoted by Z. Hence,

$$Z \sim \mathrm{N}(0, 1).$$

[1] At one time, if a set of data did not appear to be well approximated by a normal distribution it was thought that the data must be in error.

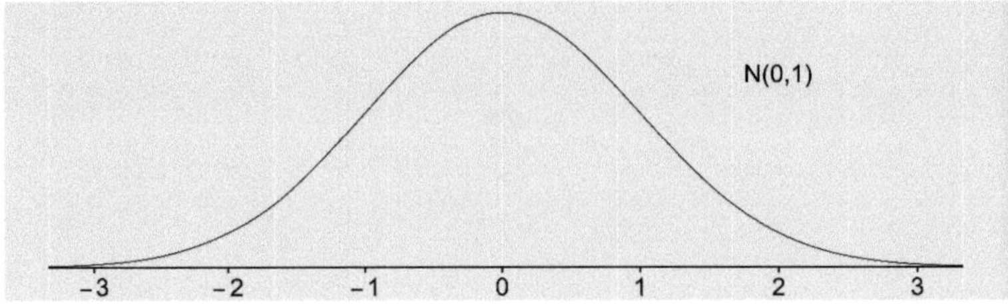

The pdf for Z is usually designated by ϕ (a lower-case Greek letter, pronounced 'fie'):

$$\phi(z) \propto e^{-z^2/2} \qquad -\infty < z < \infty.$$

The corresponding distribution function is denoted by Φ, (the capital letter version of ϕ and also pronounced 'fie'):

$$\Phi(a) = P(Z \leq a) = \int_{-\infty}^{a} \phi(z)dz.$$

For the standard normal distribution, the following table gives the values of $\Phi(z)$ corresponding to selected values of z:

z	0	1.645	1.960	2.576	3.090
$\Phi(z)$	0.5	0.95	0.975	0.995	0.999

Using these figures we can make more general statements concerning the values that we are likely to observe:

- Since 95% values are less than 1.645, it follows that 5% are greater than 1.645.
- Because the normal distribution is symmetric about 0, this means that another 5% are less than −1.645.
- Combining these statements shows that 10% of values are more than 1.645 standard deviations away from the mean (which, for a standard normal distribution, is 0).
- In the same way, 5% of values are about 2 (to be more precise, 1.96) standard deviations away from the mean, 1% are about 2.5 s.d. away from the mean, and 0.2% are about 3 s.d. away from the mean.

These results generalize to any normal distribution. Thus, for the random variable, X, having a normal distribution with mean μ and variance σ^2,

$$P[(\mu - 1.96\sigma) \leq X \leq (\mu + 1.96\sigma)] = 0.95.$$

This is sometimes referred to as the '**2-sigma rule**'.

Although the nominal range for X is infinite, most values (about 99.7%) of X fall in the interval

$$(\mu - 3\sigma, \quad \mu + 3\sigma).$$

6.1 The general normal distribution

The pdf of a random variable, X, having a general normal distribution is given by

$$\mathrm{f}(x) = \frac{1}{\sigma\sqrt{2\pi}} \mathrm{e}^{-\frac{1}{2\sigma^2}(x-\mu)^2} \qquad -\infty < x < \infty. \tag{6.1}$$

Here π and e have their usual values (3.14... and 2.78...) and the two parameters μ and σ^2 are the mean and the variance of the distribution. The derivation of the pdf first appears in a 1733 publication by de Moivre.[2]

An alternative name for the normal distribution is the **Gaussian distribution**, which honours the German mathematician and astronomer, Gauss, who popularized the use of the distribution in the context of studying errors of measurements.[3] Gauss believed that:

- A positive error of given magnitude should be as probable as a negative error of the same magnitude.
- Large errors should be less likely than small errors.
- The mean of the observations should be the most likely value of the quantity being measured.

The normal distribution has these properties and is also a natural choice because of the central limit theorem, which is discussed in Section 6.4.

[2] Abraham De Moivre (1667–1754) was a Huguenot (a French protestant) mathematician who emigrated to London because of persecution. He was a friend of the astronomer Edmond Halley, and also of Sir Isaac Newton, to whom he dedicated his book, *The Doctrine of Chances*, in which the formula appears.

[3] Carl Friedrich Gauss (1777–1855) published, in 1809, *The Theory of the Motion of Heavenly Bodies Moving about the Sun in Conic Sections*. In this context he introduced the method of least squares that will be discussed in Section 19.3.

6.2 The use of tables

If the computer is not available then it is necessary to use a table such as the following:

z	0.0	0.1	0.2	0.3	0.4	0.5	0.6	0.7	0.8	0.9
$\Phi(z)$	.500	.540	.579	.618	.655	.692	.726	.758	.788	.816
z	1.0	1.1	1.2	1.3	1.4	1.5	1.6	1.7	1.8	1.9
$\Phi(z)$	.841	.864	.885	.903	.919	.933	.945	.955	.964	.971
z	2.0	2.1	2.2	2.3	2.4	2.5	2.6	2.7	2.8	2.9
$\Phi(z)$	.977	.982	.986	.989	.992	.994	.995	.996	.997	.998

This table refers to the standard normal distribution, but it can be used for any normal distribution. For example, suppose Y has a normal distribution with mean μ and variance σ^2, then $Y - \mu$ has mean 0, and $(Y - \mu)/\sigma$ has mean 0 and variance 1, which is the situation to which the table applies.

Example 6.1

The random variable X has a normal distribution with mean 3.4 and variance 0.09. Determine the probability that X takes a value less than 3.1.

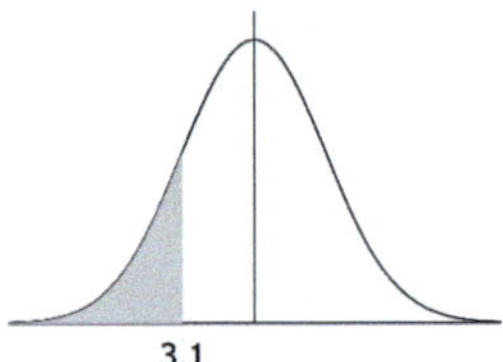

Now $X < 3.1$ corresponds to $Z < (3.1 - 3.4)\big/\sqrt{0.09}$, where Z is a standard normal variable. Also $(3.1 - 3.4)\big/\sqrt{0.09} = -1$. Therefore, because the normal distribution is symmetric:

$$\begin{aligned} P(Z < -1) &= P(Z > 1) \\ &= 1 - P(Z < 1) \\ &= 1 - 0.841 \\ &= 0.159. \end{aligned}$$

Example 6.2

The mass of a bag of flour can be considered to be an observation from a normal distribution with mean 454 g and standard deviation 2 g. Determine the probability that a randomly chosen bag has a mass of less than 450 g.

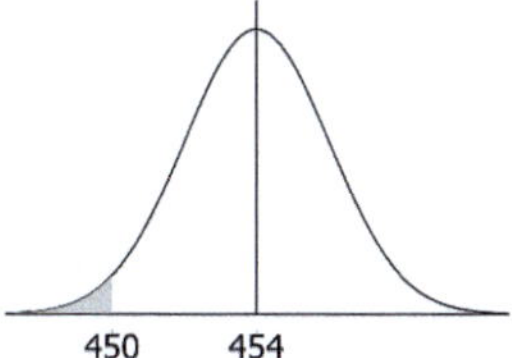

Denoting the mass by M we are asked to determine $P(M < 450)$ with $M \sim N(454, 2^2)$. Now,

$$P(M < 450) = P\left(\frac{M - 454}{2} < \frac{450 - 454}{2}\right) = P(Z < -2).$$

By symmetry, $P(Z < -2) = P(Z > 2) = 1 - P(Z < 2)$. The table reports that $P(Z < 2) = 0.977$, so the probability of a bag of flour having a mass less than 450 g is $1 - 0.977 = 2.3\%$.

Example 6.3

The random variable $X \sim N(50, 36)$. *Denoting* $E(X)$ *by* μ, *determine* $P(|X - \mu| > 4.8)$.

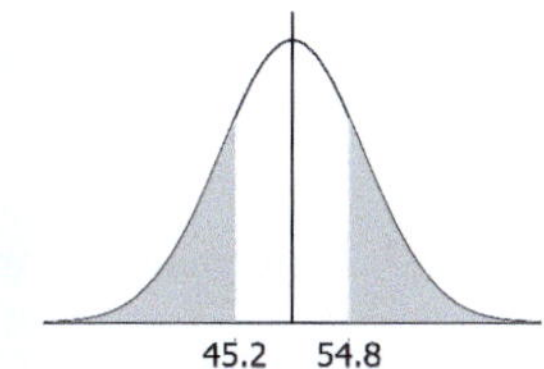

$$\begin{aligned} P(|X - \mu| > 4.8) &= P\left(\frac{|X - \mu|}{\sqrt{36}} > \frac{4.8}{\sqrt{36}}\right) \\ &= P(|Z| > 0.8) \end{aligned}$$

Because of symmetry, $P(Z > 0.8) = P(Z < -0.8)$. The required probability is therefore $2P(Z > 0.8) = 2\{1 - P(X < 0.8)\} = 2(1 - 0.788) = 0.424$.

Example 6.4

> *The random variable Y has a normal distribution with mean μ and variance σ^2. Given that 5% of the values of Y exceed 18.39 and that 0.1% of the values of Y are less than 12.23, find the values of μ and σ.*

The values of μ and σ are the solutions of

$$\frac{18.39 - \mu}{\sigma} = 1.645,$$
$$\frac{12.23 - \mu}{\sigma} = -3.090.$$

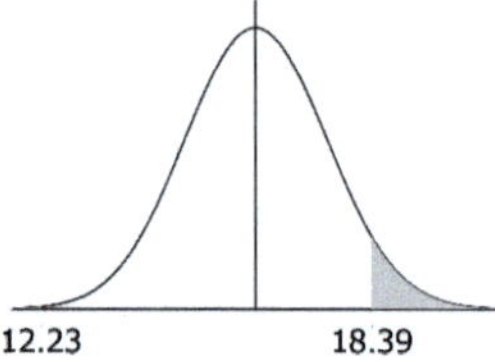

Multiplying through by σ, and subtracting, we get

$$\begin{aligned} 18.39 - \mu &= 1.645\sigma, \\ 12.23 - \mu &= -3.090\sigma, \\ \hline 6.16 &= 4.735\sigma. \end{aligned}$$

Thus $\sigma = 6.16/4.735 = 1.3$ and hence $\mu = 12.23 + 3.090 \times 1.3 = 16.25$.

Exercises 6a

1. Determine the probability that a standard normal random variable, Z, takes a value between -0.9 and 1.3.
2. The normal random variable X has mean 8 and variance 9. Determine the probability that $X > 11$.
3. The normal random variable X has mean 4 and variance 25.

    (a) Determine $P(X < 7)$.

    (b) Determine $P(X < 2)$.

4. Determine the probability that the absolute value of a normal random variable differs by more than 1.5 standard deviations from the mean of that variable.

Put theory into practice: This requires some accurate kitchen scales (capable of reporting weights in grams) and lots of 'identical' kitchen ingredients. Perhaps tins of baked beans, tins of soup, or packets of crisps. Weigh each tin or packet separately and do not be surprised when you find that they have different weights. If allowed, you could accumulate a lot of information from weighing tins in the local supermarket. Alternatively, if you have access to a high precision balance, then you could investigate how individual biscuits vary in weight. With enough data, you will find that the sample distribution resembles a normal distribution.

6.3 Linear combinations of independent normal random variables

If X and Y are two independent normally distributed random variables, and if a and b are constants, then $aX + bY$ also has a normal distribution.

$$\begin{aligned} \mathrm{E}(aX + bY) &= a\mathrm{E}(X) + b\mathrm{E}(Y), \\ \mathrm{Var}(aX + bY) &= a^2\mathrm{Var}(X) + b^2\mathrm{Var}(Y). \end{aligned}$$

Example 6.5

The random variables X *and* Y *are independent with* $\mathrm{X} \sim N(3, 1)$ *and* $\mathrm{Y} \sim N(7, 5)$. *The random variable* W *is defined by* $\mathrm{W} = \mathrm{Y} - 2\mathrm{X}$. *Determine the distribution of* W.

Since W is a linear combination of random variables having normal distributions, the distribution of W is normal. The mean and variance of W are given by

$$\begin{aligned} \mathrm{E}(W) &= \mathrm{E}(Y - 2X) = \mathrm{E}(Y) - 2\mathrm{E}(X) = 7 - (2 \times 3) = 1, \\ \mathrm{Var}(W) &= \mathrm{Var}(Y - 2X) = \mathrm{Var}(Y) + (-2)^2\mathrm{Var}(X) = 5 + (4 \times 1) = 9, \end{aligned}$$

so $W \sim \mathrm{N}(1, 9)$.

Example 6.6

> *The diameter, X mm, of the circular mouth of a bottle, has a normal distribution with mean 20 and standard deviation 0.1. The diameter Y mm, of the circular cross-section of a glass stopper, has a normal distribution with mean 19.7 and standard deviation 0.1. Determine the probability that a randomly chosen stopper will fit in the mouth of a randomly chosen bottle.*

We require $\text{P}(X > Y)$, which looks like a rather difficult quantity to calculate. However, $\text{P}(X > Y) = \text{P}(X - Y > 0)$ and $X - Y$ is a linear combination of independent normal random variables and therefore has a normal distribution. Let $G = X - Y$, where G mm is the gap between the diameters of the stopper and the mouth. Now $\text{E}(G) = \text{E}(X) - \text{E}(Y) = 20 - 19.7 = 0.3$ and $\text{Var}(G) = \text{Var}(X) + \text{Var}(Y) = 0.1^2 + 0.1^2 = 0.02$, so

$$G \sim \text{N}(0.3, 0.02).$$

The computer reports that $\text{P}(G > 0)$ is 0.983 (to 3 d.p.).

6.3.1 Extension to more than two variables

This follows immediately. Suppose that W, X, and Y are independent normal random variables, and that a, b, and c are constants. Consider the random variable U defined by

$$U = aW + bX + cY,$$

and let $V = aW + bX$. From the previous result we know that V also has a normal distribution. Thus we can write

$$U = V + cY,$$

and, since the right-hand side is once again a linear combination of independent normal random variables, it follows that U has a normal distribution.

This argument can be extended indefinitely.

> A linear combination of any number of independent normal random variables has a normal distribution.

Example 6.7

The total mass (in g) of a packet of biscuits is made up of the mass (in g) of the packaging, Y, and the masses (in g) of the fifteen biscuits $X_1, \ldots, X_{15}$. The mass of a biscuit has a normal distribution, with mean 30 and standard deviation 1, while the mass of the packaging has a normal distribution with mean 5 and standard deviation 0.2. Determine the distribution of the total mass of a packet of biscuits.

Let W denote the total mass. So

$$W = Y + X_1 + \cdots + X_{15},$$

and hence

$$\mathrm{E}(W) = 5 + (15 \times 30) = 455.$$

Assuming that the masses of the 16 items are independent of one another, we also have

$$\mathrm{Var}(W) = 0.2^2 + (15 \times 1^2) = 15.04,$$

so that $W \sim \mathrm{N}(455, 15.04)$.

Exercises 6b

1. The independent normal random variables T and U both have mean and variance equal to 8. Determine $\mathrm{P}(T - U > 2)$.
2. Small bananas have masses that are normally distributed wth mean 80 g and standard deviation 5g. A box, with a mass that is an observation from a normal distribution with mean 2000g and standard deviation 25g, contains 100 small bananas. Determine the distribution of the filled box.
3. Three women and four men enter a lift. Assume that the women have masses that are normally distributed with mean 60 kg and standard deviation 10 kg and the men have masses that are normally distributed with mean 80 kg and standard deviation 15 kg. Assuming that the masses of these seven people are mutually independent, determine the probable range (use ± 3 standard deviations) of the total mass of the people in the lift.
4. Sweets have weights that are normally distributed with mean 4 g and standard deviation 0.1 g.
   (a) Determine the probability that a random sample of 36 sweets has a total weight that is less than 143 g.
   (b) A sample of 36 sweets has a weight of exactly 143 g. Eleven sweets are randomly chosen from those in the bag and are eaten. Determine the probability that the weight of the remaining sweets is less than 100 g.
5. A toy company manufactures plastic nuts and bolts. The nuts have internal diameters that are normally distributed with mean 50 mm and standard deviation 4 mm. The bolts have external diameters that are normally distributed with mean 48 mm and standard deviation 3 mm. State the distribution of the gap between the outside of a bolt and the inside of a nut.
6. The amount of jam in a standard jar has a normal distribution with mean 340 g and standard deviation 10 g. The mass of the jar has a normal distribution with mean 150 g and standard deviation 8 g. Determine the distribution of the combined mass of the jam and jar.

6.4 The central limit theorem

An informal statement of this extremely important theorem is as follows:

> Suppose $X_1, X_2, \ldots, X_n$ are n independent random variables, each having the *same* distribution. Then, as n increases, the distributions of their sum $X_1 + \cdots + X_n$ and their average $(X_1 + \cdots + X_n)/n$ come increasingly to resemble normal distributions.

The paper in which Laplace[4] derived the central limit theorem was read to the French Royal Academy of Sciences in 1810, and was a direct consequence of the work by Gauss in the previous year. In France the normal distribution is often referred to as the **Laplacean distribution**.

The importance of the **central limit theorem** lies in the facts that:

- The common distribution of the X-variables is not stated—it can be almost *any* distribution.
- In most cases the resemblance to a normal distribution holds for remarkably small values of n.
- Many naturally occurring variables have distributions closely resembling the normal.

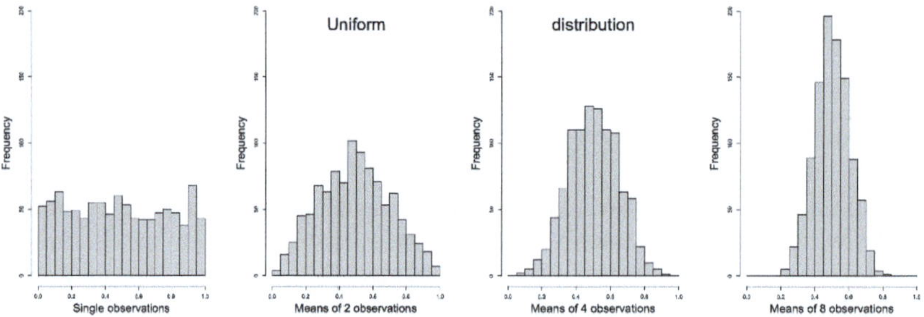

As an example, consider the results for random samples of observations on a random variable having a continuous uniform distribution in the interval (0,1). The successive diagrams show histograms of samples from the original distribution, and, successively, the histograms for the means of two, four, and eight observations from this distribution. As the group size increases, so the means become increasingly clustered in a symmetrical fashion about 0.5 (the mean of the original population).

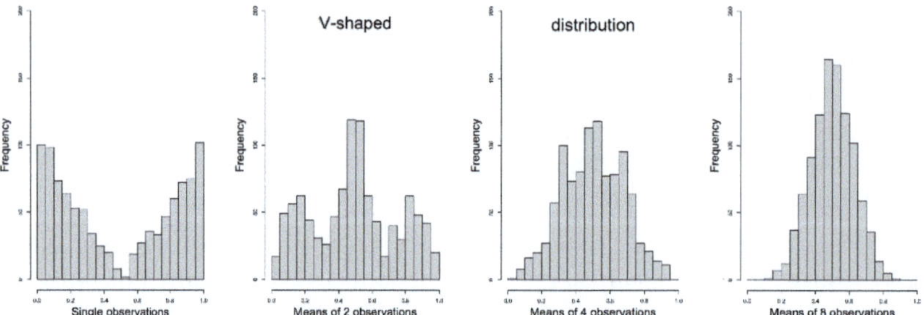

As a second example, consider successive averages of observations from a V-shaped distribution.

[4] Pierre Simon Laplace (1749–1827) was eulogized by Poisson as being 'the Newton of France'. Elected to membership of the French Royal Academy of Sciences by the age of 24, he was a professor at the Ecole Militaire when Napoleon was a student there. His greatest interest was in celestial mechanics.

Here the original distribution has a trough in the middle, whereas the normal distribution has a peak, but once we start looking at averages of even as few as $n = 2$ observations, a peak starts to appear.

As a final example we look at a triangular distribution, which is quite heavily skewed. In this case the histograms of the means of two observations and four observations still appear skewed, but this skewness has almost vanished by the time we are working with means of eight observations.

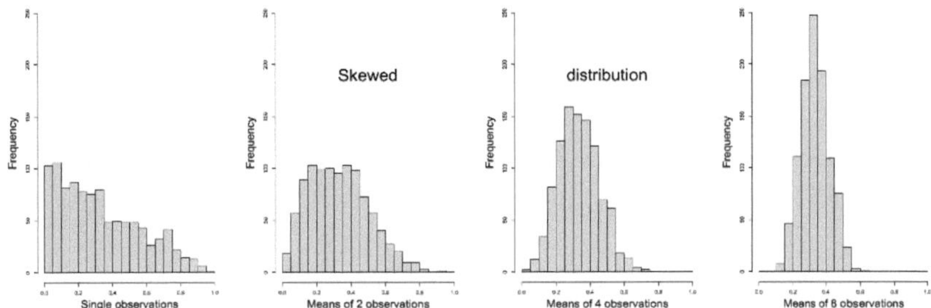

It is clear that the practical consequences of the central limit theorem were understood well before the time of Laplace. A sixteenth-century German treatise on surveying instructs the surveyor to establish the length of a rood (a standard unit of length at the time) in the following manner:

> 'Stand at the door of a church on a Sunday and bid 16 men to stop, tall ones and small ones, as they happen to pass out when the service is finished; then make them put their left feet one behind the other, and the length thus obtained shall be a right and lawful rood to measure and survey the land with, and the 16th part of it shall be a right and lawful foot.'

6.4.1 The distribution of the sample mean, $\bar{X}$

Denote the ith observation in a sample by x_i. Different samples will give different values for x_i. Thus x_i is an observation on a random variable that we will denote by X_i. Suppose that X_1 has mean μ and variance σ^2. The same will be true for each of $X_2, X_3, ..., X_n$, which are therefore identically distributed random variables, each with mean μ and variance σ^2. We define $\bar{X}$ by

$$\bar{X} = \frac{1}{n}(X_1 + X_2 + \cdots + X_n),$$

and we showed earlier (Section 4.5) that $\bar{X}$ has mean μ and variance σ^2/n. By the central limit theorem, $\bar{X}$ has an approximate normal distribution for large n, and so

$$\bar{X} \sim \mathrm{N}\left(\mu, \frac{\sigma^2}{n}\right).$$

The equivalent result for a sum is that

$$\Sigma X_i \sim \mathrm{N}(n\mu, n\sigma^2).$$

If the distribution of the individual X variables is normal, then, since $\bar{X}$ is then a linear combination of normally distributed random variables, the result is true even for small values of n.

Example 6.8

The continuous random variable X has mean 5 and variance 25. A random sample of 100 observations are taken on X. Determine the distribution of the sample mean.

The random variable $\bar{X}$, corresponding to the sample mean, has expectation 5 and variance $25/100 = 1/4$. By the central limit theorem it has a N(5, 1/4) distribution.

Example 6.9

The discrete random variable X has probability distribution given by $P(X = 0) = 1/4$, $P(X = 1) = 1/2$, $P(X = 2) = P(X = 3) = 1/8$. Let T denote the total of a random sample of 500 observations on X. Determine the approximate distribution of T.

x	0	1	2	3
P_x	$\frac{1}{4}$	$\frac{1}{2}$	$\frac{1}{8}$	$\frac{1}{8}$

We begin by determining the mean and variance of X:

$$E(X) = \left(0 \times \frac{1}{4}\right) + \left(1 \times \frac{1}{2}\right) + \left(2 \times \frac{1}{8}\right) + \left(3 \times \frac{1}{8}\right) = 9/8,$$

and:

$$E(X^2) = \left(0^2 \times \frac{1}{4}\right) + \left(1^2 \times \frac{1}{2}\right) + \left(2^2 \times \frac{1}{8}\right) + \left(3^2 \times \frac{1}{8}\right) = 17/8,$$

so that $\text{Var}(X) = 17/8 - 81/64 = 55/64$.

Hence $E(T) = (500 \times 9/8) = 562.5$ and $\text{Var}(T) = (500 \times 55/64) = 6875/16 \approx 429.7$. By the central limit theorem, the distribution of T is therefore approximately N(562.5, 429.7).

Exercises 6c

1. The random variable X has mean 15 and variance 25. The random variable $\bar{X}$ is the mean of a random sample of 70 observations on X. State the approximate distribution of $\bar{X}$.
2. The masses of size 2 eggs have a mean of 67.5 g and a standard deviation of 1.5 g. Determine the approximate probability that a random sample of 100 size 2 eggs has a mean mass between 67.2 g and 67.8 g.
3. The random variable Y has mean 50 and standard deviation 20. The random variable $\bar{Y}$ is the mean of a random sample of 100 observations. Determine the probable range (use ± 3 standard deviations) of the values of $\bar{Y}$.

6.5 The normal distribution used as an approximation

This section demonstrates that the normal distribution can be used to approximate the binomial and Poisson distributions. When we have a computer available, it is no longer necessary to follow through the calculations set out here, since a one-line instruction to the computer gives an immediate response. For example, if we ask the computer to calculate the probability of getting exactly 5,000 heads from 10,000 tosses of a fair coin, we immediately get the answer 0.007978646. To arrive at this answer the computer has not evaluated $\binom{10,000}{5,000}$ by laboriously multiplying 10,000 by 9,999 by 9,998 ..., it has used approximations. Not the normal approximations used here, but very sophisticated approximations that lie well outside the scope of this book. Nevertheless, since the mean number of heads is $10,000 \times 0.5 = 5,000$ and the standard deviation is $\sqrt{10,000 \times 0.5 \times 0.5} = 50$, we can use the simple approximation, $\mu \pm 2\sigma$, to assert that about 95% of the time we will get between 4,900 and 5,100 heads.[5]

6.5.1 Approximating a binomial distribution

Suppose $X \sim \mathrm{B}(n, p)$. In Section 4.4.1 we noticed that we could write $X = Y_1 + Y_2 + \cdots + Y_n$, where the Y-variables had independent Bernoulli distributions, each with parameter p. By the central limit theorem, the sum of independent identically distributed random variables has an approximate normal distribution. For large n, therefore, the binomial distribution must resemble a normal distribution. This was deduced by Pascal,[6] with the case $p = 1/2$ being studied at length by Abraham De Moivre.[7]

This is illustrated in the diagram, which shows three binomial distributions having the same value of p (0.3) but differing values of n.

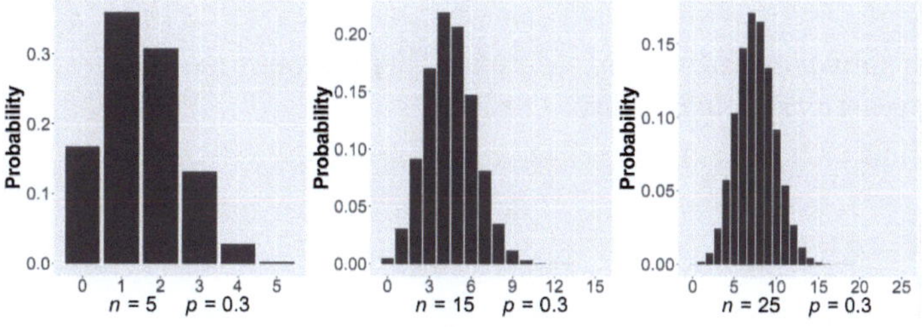

[5] The computer reports that the probability of a value in that range is 0.9556.
[6] Already encountered in Section 3.14.
[7] This appears in the second edition of his book, *The Doctrine of Chances*.

The limiting normal distribution must have the same mean and variance as its binomial counterpart, and hence, if we denote the normal counterpart of X by W, then

$$X \sim \mathrm{B}(n, p) \rightarrow W \sim \mathrm{N}(np, npq),$$

where $q = 1 - p$.

The normal distribution is continuous with probabilities associated with all small intervals between $-\infty$ and ∞. The binomial is discrete, with 'chunks' of probability, like slices of a slab of butter, associated with each integer between 0 and n, inclusive. If the bars of binomial probability really were made of butter, what would happen if we trod on them? They would spread out sideways—an equal amount on each side. This is precisely how we deal with the move from X to W—we imagine that the probability originally associated with the single point value x becomes identified with the interval $(x - 1/2, x + 1/2)$.

Hence the normal approximation is

$$\mathrm{P}(X = x) \simeq \mathrm{P}[(x - 1/2) < W < (x + 1/2)], \tag{6.2}$$

where $W \sim N(np, npq)$. The adjustment by 1/2 in each direction is referred to as using the **continuity correction**.

To calculate the approximate probability, we must transform to the standard normal distribution by writing

$$Z = \frac{W - np}{\sqrt{npq}}.$$

Hence

$$\mathrm{P}(X = x) \simeq \Phi\left(\frac{(x + 1/2) - np}{\sqrt{npq}}\right) - \Phi\left(\frac{(x - 1/2) - np}{\sqrt{npq}}\right). \tag{6.3}$$

This is sometimes referred to as the **de Moivre–Laplace theorem**.

Example 6.10

Using the normal approximation, determine the probability that exactly 30 heads are obtained when a fair coin is tossed 64 times.

Here $n = 64$, $p = q = 0.5$, so

$$\begin{aligned} \mathrm{P}(X = 30) &\simeq \Phi\left(\frac{30.5 - 32}{4}\right) - \Phi\left(\frac{29.5 - 32}{4}\right) \\ &= \Phi(-0.375) - \Phi(-0.625) \\ &= 0.3538302 - 0.26659855 = 0.0878447. \end{aligned}$$

The final calculations have been reported to an unreasonable number of decimal places, in order to properly assess the accuracy of the approximation to the true value, which (to the same level of accuracy) is 0.0878360.

6.5.2 Approximating a Poisson distribution

We saw in Section 3.10 that the shape of a Poisson distribution depends upon the value of its parameter λ. Although the distribution is always skewed, as λ increases this skewness becomes less visible and the distribution comes increasingly to resemble the normal in appearance.

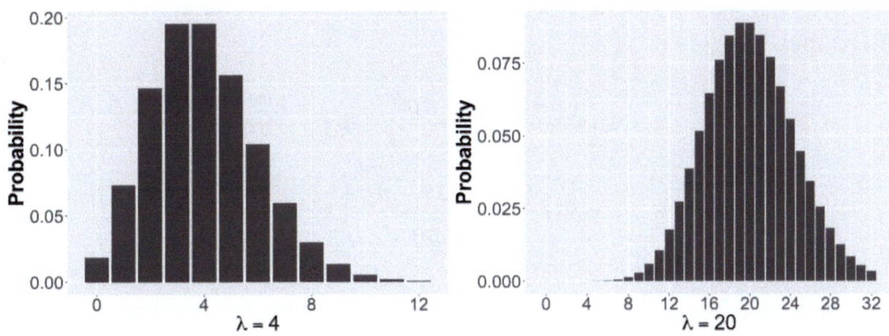

Since a Poisson random variable, X, with parameter λ has mean and variance both equal to λ, the approximating normal random variable, Y, therefore has a $\mathrm{N}(\lambda, \lambda)$ distribution. As in the case of the approximation to a binomial distribution, a **continuity correction** is required:

$$\mathrm{P}(X = x) \simeq \mathrm{P}[(x - 1/2) < Y < (x + 1/2)].$$

Example 6.11

> *The random variable* X *has a Poisson distribution with mean 9. Compare the exact values with those given by the normal approximation for the following probabilities:*
>
> ***(a)*** $\mathrm{P}(X = 9)$,
>
> ***(b)*** $\mathrm{P}(X \leq 2)$.

(a) Using the computer the normal approximation gives 0.1324 compared to the exact value of 0.1318 (with both results given to four decimal places for comparison).

(b) Here the exact value is 0.0062 whereas the approximation is very poor: 0.0151. This is because we are looking at the tail of a skewed distribution.

6.5.3 Approximating a range of integer values

To see how the approximation works for a range of values, consider $\mathrm{P}(X \leq 1)$, where X has a $\mathrm{B}(n, p)$ distribution, and the normal approximation is provided by W with a $\mathrm{N}(np, npq)$ distribution (where $q = 1 - p$). We will have

$$\begin{aligned} \mathrm{P}(X \leq 1) &= \mathrm{P}(X = 0) + \mathrm{P}(X = 1) \\ &\simeq \mathrm{P}(-1/2 < W < 1/2) + \mathrm{P}(1/2 < W < 3/2) \\ &= \mathrm{P}(-1/2 < W < 3/2). \end{aligned}$$

In general, for $\mathrm{P}(X \leq x)$, with x taking one of the values 0, 1, ..., n, the approximation is $\mathrm{P}[-1/2 < W < (x + 1/2)]$, so that

$$\begin{aligned} \mathrm{P}(X \leq x) &\simeq \Phi\left(\frac{(x+1/2) - np}{\sqrt{npq}}\right) - \Phi\left(\frac{-1/2 - np}{\sqrt{npq}}\right), \\ \mathrm{P}(X \geq x) &\simeq 1 - \Phi\left(\frac{(x-1/2) - np}{\sqrt{npq}}\right) + \Phi\left(\frac{-1/2 - np}{\sqrt{npq}}\right). \end{aligned}$$

It is usually the case that $\Phi\left(\frac{-1/2-np}{\sqrt{npq}}\right)$ is close to 0 and can be neglected, so that then

$$\mathrm{P}(X \leq x) \simeq \Phi\left(\frac{(x+1/2) - np}{\sqrt{npq}}\right), \tag{6.4}$$

$$\mathrm{P}(X \geq x) \simeq 1 - \Phi\left(\frac{(x-1/2) - np}{\sqrt{npq}}\right). \tag{6.5}$$

Example 6.12

> *It is given that* X $\sim$ *B*(25, 0.6). *Using a normal approximation, determine P*(X $\leq$ 16).

A binomial distribution with $n = 25$ and $p = 0.6$ has mean $np = 15$ and variance $npq = 6$. The normal approximation is

$$\begin{aligned} \mathrm{P}(X \leq 16) &\simeq \Phi\left(\frac{16.5 - 15}{\sqrt{6}}\right) - \Phi\left(\frac{-0.5 - 15}{\sqrt{6}}\right) \\ &= \Phi(0.612) - \Phi(-6.328) \\ &= 0.7298 - 0.0000 = 0.7298. \end{aligned}$$

The probability that $X \leq 16$ is approximately 0.730 (to 3 d.p.).[a]

[a] The exact value, using the binomial directly, is 0.726.

6.5.4 *Usefulness for quick calculations*

We noted previously that, for a normal distribution, approximately 95% of values lie within 2 standard deviations from the mean, while 99% lie within 2.5 standard deviations. Values more than 3 standard deviations away from the mean are extremely unlikely. Since the normal distribution can closely approximate the binomial and Poisson distributions, we can apply the general 'rules' to these distributions also.

Example 6.13

The random variable X has B(100, 0.8) distribution. What is the likely range of observed values for X?

The random variable X has mean $100 \times 0.8 = 80$ and variance $100 \times 0.8 \times 0.2 = 16$, so a standard deviation of 4. If we take the likely range to imply the central 95% of values, then this will be $80 \pm 2 \times 4$ which is (72, 88).[a] We can also deduce that virtually every observed value will lie in the interval (68, 92).

[a] The actual probability for values in this range is 95.3%.

Example 6.14

Travelling along a motorway an observer notices cars with a particular characteristic occur roughly once every 8 miles. Give a range (based on 2 standard deviations) for the number of cars, on a separate 120-mile motorway journey, that display this characteristic.

Assuming that these cars occur at random, we are dealing with a Poisson distribution with a mean, for a 120-mile journey, of $120/8 = 15$ cars. For the Poisson distribution the mean equals the variance, so the standard deviation is $\sqrt{15} = 3.87$. The likely number of cars seen is therefore $15 \pm 2 \times 3.87$ or between 7 and 23 cars.[a] It is almost certain that the number seen will be between 3 and 27.

[a] The exact probability for this range is 96.3%.

Exercises 6d

1. The random variable H is the number of heads obtained during 100 tosses of a biased coin, with P(Head)= 0.6. Using a normal approximation to the binomial (with a continuity correction) determine the approximate probability that fewer than 50 heads are obtained.
2. The number of faulty light bulbs returned to a shop in a week has a Poisson distribution with mean 0.7. Determine a range for the probable number of returns in a period of 50 weeks. (Use 3 standard deviations.)

3. A lake contains a large number of fish. Assuming that 60% of the fish are roach, determine the range for the likely number of roach contained in a random sample of 100 fish. (Use 3 standard deviations.)
4. At a certain university there are many rabbit holes in the grounds of its extensive campus. The rate of occurrence of these holes is 1 per 100 square metres.
   (a) Determine the probability that a randomly chosen 400 m^2 section of the grounds contains exactly two rabbit holes.
   (b) Determine the probability that a randomly chosen 200 m^2 section contains at least one rabbit hole.
   (c) Using a normal approximation, determine the probability that a randomly selected 10,000 m^2 section contains at least 80 rabbit holes.

6.6 *Proof that the area under the normal curve is 1

This proof makes use of a **bivariate normal distribution** (bivariate distributions are discussed later in Chapter 10). The general bivariate normal distribution allows for the variables X and Y to be dependent, but here we take them to be independent standard normal random variables. Thus their joint distribution is given by

$$\mathrm{f}_{XY}(x,y) = \mathrm{f}(x)\mathrm{f}(y) == \frac{1}{\sqrt{2\pi}}\mathrm{e}^{-x^2/2} \times \frac{1}{\sqrt{2\pi}}\mathrm{e}^{-y^2/2} = \frac{1}{2\pi}\mathrm{e}^{-(x^2+y^2)/2}.$$

If we can show that this joint function integrates to 1, then it must be the case that the two marginal functions also integrate to 1. So we wish to show that

$$\frac{1}{2\pi}\int_{-\infty}^{\infty}\int_{-\infty}^{\infty} \mathrm{e}^{-(x^2+y^2)/2}\mathrm{d}x\mathrm{d}y$$

integrates to 1. This is a double integral over the whole plane; the trick is to change from Cartesian coordinates (x, y) to polar coordinates (r, θ):

$$\frac{1}{2\pi}\int_{0}^{2\pi}\int_{0}^{\infty} \mathrm{e}^{-r^2/2}r\mathrm{d}r\mathrm{d}\theta.$$

Since this is a function of θ times a function of r we can separate the integrals:

$$\frac{1}{2\pi}\int_{0}^{2\pi} \mathrm{d}\theta \times \int_{0}^{\infty} r\mathrm{e}^{-r^2/2}\mathrm{d}r.$$

Since each component integrates to 1, their product is 1, and we have thereby demonstrated that the normal function does indeed define a probability distribution.

Key facts

- A **normal distribution** with mean μ and variance σ^2 is denoted by $N(\mu, \sigma^2)$.
- The **standard normal** random variable, denoted by Z, has mean 0, variance 1, and distribution function Φ , where

$$\Phi(z) = P(Z \leq z) = P(Z < z), \text{ with } \Phi(-\infty) = 0 \text{ and } \Phi(\infty) = 1.$$

- Some useful relations involving Z

$$P(Z > z) = 1 - P(Z \leq z) = 1 - \Phi(z),$$
$$P(a \leq Z \leq b) = \Phi(b) - \Phi(a),$$
$$P(Z > -z) = P(Z < z).$$

- Some more general relations:

If $X \sim N(\mu, \sigma^2)$, then $(X - \mu)/\sigma \sim N(0, 1)$.

If X_1 and X_2 are independent normal random variables then, for any constants a and b, $aX_1 + bX_2$ has a normal distribution.

$\overline{X} \sim N\left(\mu, \frac{\sigma^2}{n}\right)$.

- The **central limit theorem**:

The sum (or mean) of n independent identically distributed (i.i.d.) random variables has an approximate Normal distribution *whatever* the distribution of the random variables.

- The **continuity correction**: the approximation of a discrete random variable X, taking values 0, 1, 2, etc, by a continuous random variable, Y, taking ranges of values, −0.5 to 0.5, 0.5 to 1.5, 1.5 to 2.5, etc'

$$P(X < x) \approx P(Y < x - 0.5),$$
$$P(X \leq x) \approx P(Y < x + 0.5).$$

- Binomial and Poisson distributions may be approximated by normal distributions having the same means and variances (i.e. np and npq or λ and λ).

R

- The probability of a value less than x for a normal distribution with mean μ and standard deviation σ is given by `pnorm(x, μ, σ)`.

7

Distributions related to the normal distribution

7.1 The t-distribution

The family of distributions known as **Student's t-distributions**[1] will be used in Chapters 14 and 15 when investigating the value of a population mean.

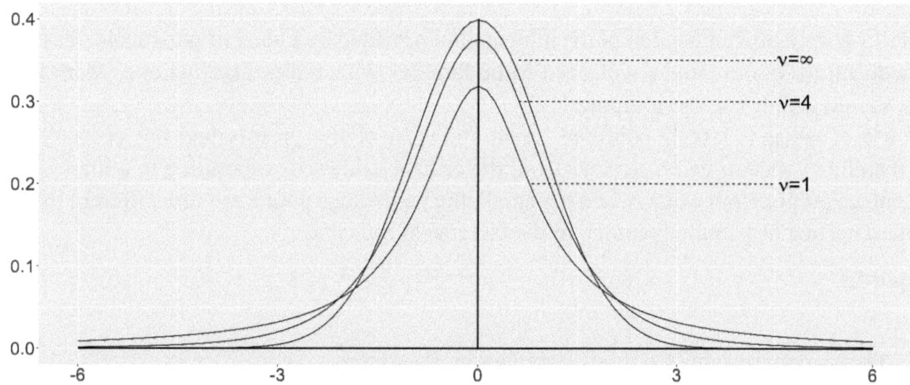

Figure 7.1 t-distributions having 1 and 4 degrees of freedom compared to the limiting normal distribution. The t-distributions have lower peaks and fatter tails.

All t-distributions are symmetric about 0 and have a single parameter, ν ('nu'), which is a positive integer. known as the number of **degrees of freedom** of the distribution. As a shorthand we replace the phrase 'a t-distribution with ν degrees of freedom' by 'a t_ν-distribution'.

As ν increases, so the corresponding t_ν-distribution increasingly resembles the limiting standard normal distribution (which corresponds to $\nu = \infty$).

[1] William Sealy Gossett (1876–1937) joined the staff of Arthur Guinness Son & Co. Ltd. as a 'brewer'. To assess the quality of the product, just a few samples of the drink were used. Gossett correctly mistrusted the application of the existing (large-sample) theory and, in a paper published in 1908, entitled 'The Probable Error of a Mean', he conjectured the form of the t-distribution relevant for small samples. Guinness company policy at the time meant that he was obliged to publish under a pseudonym: he chose the pen name 'Student'.

Table 7.1: Percentage points for t-distributions

	Values of t, such that $P(T < t) = P$								
P	.75	.90	.95	.975	.99	.995	.9975	.999	.9995
$\nu = 1$	1.000	3.078	6.314	12.71	31.82	63.66	127.3	318.3	636.6
2	0.816	1.886	2.920	4.303	6.965	9.925	14.09	22.33	31.60
3	0.765	1.638	2.353	3.182	4.541	5.841	7.453	10.21	12.92
10	0.700	1.372	1.812	2.228	2.764	3.169	3.581	4.144	4.587
12	0.695	1.356	1.782	2.179	2.681	3.055	3.428	3.930	4.318
30	0.683	1.310	1.697	2.042	2.457	2.750	3.030	3.385	3.646
99	0.677	1.290	1.660	1.984	2.365	2.626	2.871	3.175	3.392
∞	0.674	1.282	1.645	1.960	2.326	2.576	2.807	3.090	3.291

Table 7.1 gives a brief indication of the information provided by a book of statistical tables. Readers faced with future examinations will need to be familiar with tables like this one. More fortunate readers should simply use the computer!

The use of tables is largely confined to situations involving prespecified tail probabilities: the tables therefore concentrate on summarizing the critical values corresponding to a limited number of percentage points. Notice that, when ν is small, the percentage points are very different to those for a standard normal distribution (shown in the last row of the table).

Example 7.1

The random variable T *has a* t_2*-distribution. Determine:*

(a) $P(T > -2.920)$,

(b) $P(-4.303 < T < 14.09)$.

(a) By symmetry, $P(T > -2.920) = P(T < 2.920)$. The required probability is therefore 0.95.

(b) We begin by rewriting the required probability in terms of tail probabilities:

$$P(-4.303 < T < 14.09) = P(T < 14.09) - P(T < -4.303).$$

Now

$$\begin{aligned} P(T < -4.303) &= P(T > 4.303) \\ &= 1 - P(T < 4.303) \\ &= 1 - 0.9750 \\ &= 0.0250. \end{aligned}$$

Since $P(T < 14.09) = 0.9975$, the required probability is $0.9975 - 0.0250 = 0.9725$.

Exercises 7a

1. Find:
   - **(a)** $P(t_2 > 9.925)$,
   - **(b)** $P(t_{10} < 4.144)$.
2. Find c such that:
   - **(a)** $P(t_3 > c) = 2.5\%$,
   - **(b)** $P(t_{12} < c) = 1\%$.

7.2 The chi-squared distribution

The family of chi-squared distributions will be used in Chapters 14 and 15 when investigating the value of a population variance.

'Chi' is the Greek letter χ, pronounced 'kye'. The distribution is continuous and has a positive integer parameter d which determines its shape (see Figure 7.2).

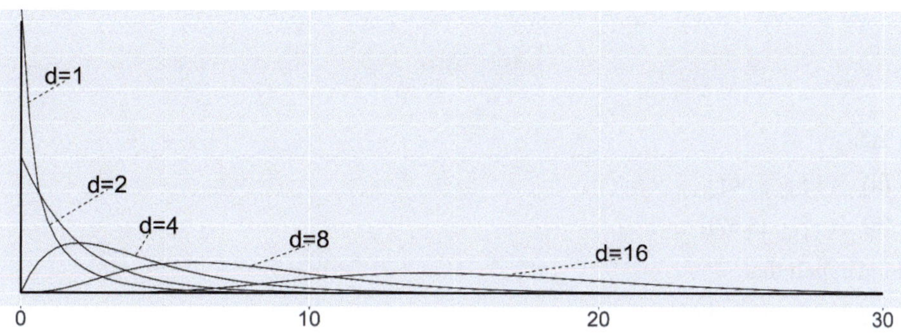

Figure 7.2 χ^2**-distributions having 1, 2, 4, 8, and 16 degrees of freedom.**

As its name implies, χ^2 cannot take a negative value. The parameter d is known as the **degrees of freedom** of the distribution and we refer to a 'chi-squared distribution with d degrees of freedom'. For simplicity, we write this as χ^2_d.

7.2.1 *Properties of the chi-squared distribution*

- A χ_d^2 distribution has mean d and variance $2d$.
- A χ_d^2 distribution has mode at $d - 2$ for $d \geq 2$. This is useful when doing a quick sketch.
- If Z has a N(0,1) distribution, then Z^2 has a χ_1^2 distribution.
- If U and V are independent random variables having χ_u^2 and χ_v^2 distributions, respectively, then their sum $U + V$ has a χ_{u+v}^2 distribution.
- The χ_2^2 distribution is an exponential distribution with mean 2.

The usual layout for a table for the chi-squared distribution uses individual rows referring to different values for d, with the entries in each row corresponding to selected percentage points. An example is provided by Table 7.2.

Table 7.2: Percentage points for χ^2-distributions

			p			
d	.900	.950	.975	.990	.995	.999
1	2.706	3.841	5.024	6.635	7.879	10.83
2	4.605	5.991	7.378	9.210	10.60	13.82
3	6.251	7.815	9.348	11.34	12.84	16.27
4	7.779	9.488	11.14	13.28	14.86	18.47
5	9.236	11.07	12.83	15.09	16.75	20.52

Thus $P(\chi_1^2 < 2.706) = 0.900$, $P(\chi_5^2 > 20.52) = 0.001$, and the upper 1% point of a χ_3^2 distribution is 11.34.

Exercises 7b

1. Find:
   - **(a)** $P(\chi_2^2 > 5.991)$,
   - **(b)** $P(\chi_4^2 < 18.47)$.
2. Find c such that:
   - **(a)** $P(\chi_4^2 > c) = 2.5\%$,
   - **(b)** $P(\chi_3^2 > c) = 0.5\%$.

7.3 The F-distribution

Formally, if the random variables U and V are independent, with

$$U \sim \chi^2_u \qquad \text{and} \qquad V \sim \chi^2_v,$$

then

$$\frac{U/u}{V/v} \sim F_{u,v}. \tag{7.1}$$

We describe the ratio as having 'an F-distribution with u and v degrees of freedom'.

All F-distributions have range $(0, \infty)$. They have mean equal to $v/(v-2)$, for $v > 2$. The precise shape of an F-distribution depends upon the values of u and v (see Figure 7.3) but the mode always lies in the range 0 to 1.

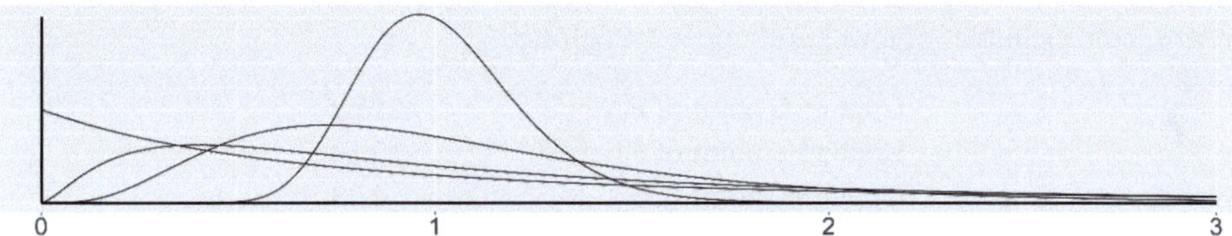

Figure 7.3 Examples of F-distributions. The mode lies in (0, 1).

> If $R = \frac{U/u}{V/v} \sim F_{u,v}$, then $1/R \sim F_{v,u}$.

The F-distribution is used in the comparison of variances (Chapters 16, 19). Tables were first provided by Snedecor,[2] who used the letter F as a tribute to Fisher,[3] who had developed the underlying theory. The distribution is sometimes referred to as **Snedecor's F- distribution**.

With two parameters, the tables are both more extensive and more limited than any previous tables that we have met. A separate table is provided for each significance level (typically, the upper 5, 2.5, and 1% levels), with the table's columns referring to the first of the degrees of freedom and the table's rows referring to the second. Table 7.3 shows a small extract relating solely to the upper 5% points.

Although the table refers only to the upper 5% points, it can also be used to obtain the lower 5% points, because of the relation

$$P(F_{u,v} < c) = P(F_{v,u} > 1/c). \tag{7.2}$$

[2] George Waddel Snedecor (1881–1974) was an American biometrician. His book (co-authored with W. G. Cochran) *Statistical Methods Applied to Experiments in Agriculture and Biology* sold more than 125,000 copies.

[3] Sir Ronald Aylmer Fisher (1860–1962) was an English biometrician who introduced many statistical methods that are now in common use. Examples include the method of maximum likelihood (Section 14.2.2), the methodology of significance tests (Chapter 15), the exact test (Section 17.7.2), and the methods of designed experiments.

Table 7.3: Percentage points for F-distributions

	Upper 5% points of $F_{u,v}$ distributions								
					u				
v	1	2	3	4	5	6	12	24	∞
1	161.4	199.5	215.7	224.6	230.2	234.0	243.9	249.1	254.3
2	18.51	19.00	19.16	19.25	19.30	19.33	19.41	19.45	19.50
3	10.13	9.55	9.28	9.12	9.01	8.94	8.74	8.64	8.53
4	7.71	6.94	6.59	6.39	6.26	6.16	5.91	5.77	5.63
20	4.35	3.49	3.10	2.87	2.71	2.60	2.28	2.08	1.84
40	4.08	3.23	2.84	2.61	2.45	2.34	2.00	1.79	1.51
∞	3.84	3.00	2.60	2.37	2.21	2.10	1.75	1.52	1.00

Here are some examples of lower and upper 5% points:

Upper 5% point of $F_{3,1}$	215.7
Upper 5% point of $F_{1,3}$	10.13
Upper 5% point of $F_{12,20}$	2.28
Upper 5% point of $F_{3,4}$	6.59
Lower 5% point of $F_{4,3}$	$1/6.59 = 0.152$
Lower 5% point of $F_{3,4}$	$1/9.12 = 0.110$

Note that, when taking a reciprocal to obtain a lower 5% point, the degrees of freedom are interchanged.

Exercises 7c

1. Find the upper 5% point of an $F_{a,b}$-distribution for each of the following cases:

   (a) $a = 2, b = 4$,

   (b) $a = 12, b = 40$.

2. Find the lower 5% point of an $F_{a,b}$-distribution for each of the following cases:

   (a) $a = 2, b = 4$,

   (b) $a = 4, b = 1$.

Key facts

- The parameters of the t-, χ^2-, and F-distributions are called the **degrees of freedom**.
- The distributions have complicated formulae not shown here!
- There are special tables for percentage points.
- The t-distribution is used when examining hypotheses concerning the values of means.
- The χ^2- and F-distributions are used when examining hypotheses concerning the values of variances.
- The F-distribution is used when examining the fit of models.

R

- For a t-distribution with d degrees of freedom:
  - The command pt(x, d) evaluates the probability of a value less than x.
  - The command qt(α, d) returns the value x for which $P(X < x) = \alpha$.
- For a χ^2-distribution with d degrees of freedom:
  - The command pchisq(x, d) evaluates the probability of a value less than x.
  - The command qchisq(α, d) returns the value x for which $P(X < x) = \alpha$.

  For an F-distribution with u and v degrees of freedom:
  - The command pf(x, u, v) evaluates the probability of a value less than x.
  - The command qf(α, u, v) returns the value x for which $P(X < x) = \alpha$.

8

*Generating functions

This is a chapter for the mathematical statistician and can be ignored by a reader simply wishing to apply statistics to the analysis of data.

In many branches of mathematics there are problems that can be solved by a variety of different methods. Using one method a problem can appear very difficult, whereas using another method it turns out to be quite easy. Sometimes the use of a generating function can make a hard problem much simpler.

8.1 The probability generating function, G

f

The **probability generating function** (**pgf** for short) is defined by the relation

$$\mathrm{G}(t) = \mathrm{E}(t^X) = \sum t^x \mathrm{P}(X = x), \tag{8.1}$$

where the summation is over all possible values of the discrete random variable X. Note that

$$\mathrm{G}(1) = 1.$$

One use of the pgf is as an alternative method of obtaining the mean and variance of a probability distribution. The method entails differentiating $\mathrm{G}(t)$ with respect to t. Differentiating a quantity involving a $\sum$ sign is not a problem, but on this first occasion we will write things out 'longhand'. Suppose the possible values for X are $x_1, x_2, \ldots$, so that

$$\mathrm{G}(t) = t^{x_1}\mathrm{P}(X = x_1) + t^{x_2}\mathrm{P}(X = x_2) + \cdots.$$

Differentiating with respect to t we get

$$\begin{aligned} \frac{\mathrm{dG}(t)}{\mathrm{d}t} &= x_1 t^{x_1 - 1}\mathrm{P}(X = x_1) + x_2 t^{x_2 - 1}\mathrm{P}(X = x_2) + \cdots \\ &= \sum x_i t^{x_i - 1}\mathrm{P}(X = x_i), \end{aligned}$$

where, as usual, the summation is over all possible values of X.

The notation $dG(t)/dt$ for the derivative is cumbersome, so we use the alternative notation $G'(t)$, with $G''(t)$ denoting the second derivative. Dispensing with the suffices on the x values, and using the usual notation, we get

$$G'(t) = \sum xt^{x-1}P_x,$$

with the second derivative becoming

$$G''(t) = \sum x(x-1)t^{x-2}P_x.$$

If we now set t equal to 1, we get

$$\begin{aligned} G'(1) &= \sum xP_x, \\ G''(1) &= \sum x(x-1)P_x. \end{aligned}$$

The right-hand sides of these two equations are simply expectations; hence,

$$G'(1) = E(X), \tag{8.2}$$

and

$$G''(1) = E[X(X-1)] = E(X^2) - E(X),$$

so that

$$\text{Var}(X) = G''(1) + G'(1) - \{G'(1)\}^2. \tag{8.3}$$

Example 8.1

The random variable X has probability distribution given by

$$P(X = 1) = P(X = 2) = P(X = 4) = 1/3.$$

Obtain the probability generating function for X.

From the definition of $G(t)$ we have

$$G(t) = \frac{1}{3}t^1 + \frac{1}{3}t^2 + \frac{1}{3}t^4 = t(1 + t + t^3)/3.$$

Example 8.2

The random variable Y *has pgf* G(t) *given by*

$$G(t) = k(1 + t^2 + t^5).$$

Determine:

(a) *the value of* k,

(b) *the probability that* Y *equals 2,*

(c) *the probability that* Y *equals 3,*

(d) *the expectation of* Y.

(a) To find k we use the fact that $G(1) = 1$:

$$k(1 + 1^2 + 1^5) = k(1 + 1 + 1) = 3k = 1,$$

giving $k = 1/3$.
(b) The probability that Y equals 2 is the coefficient of t^2. This is k, which we know to be 1/3.
(c) The probability that Y equals 3 is the coefficient of t^3. Since there is no t^3 in the expression for $G(t)$, the required probability is 0.
(d) $E(X)$ could be obtained directly from the probability distribution, but it is easier here to use Equation (8.2). The differential of $G(t)$ is:

$$G'(t) = k(0 + 2t + 5t^4).$$

Substituting for k and putting t equal to 1, we get

$$E(X) = G(1) = \frac{1}{3}(0 + 2 + 5) = \frac{7}{3}.$$

The expectation of Y is 7/3.

Example 8.3

The geometric random variable X *has distribution*

$$P_x = q^{x-1}p \qquad (x = 1, 2, ...),$$

where $q = 1 - p$. *Determine the pgf of this distribution and hence obtain its mean and variance.*

The pgf is given by

$$G(t) = \sum t^x q^{x-1} p = tp \sum (tq)^{x-1}.$$

The possible values for x are 1, 2, ..., so the right-hand side becomes $1 + tq + (tq)^2 + (tq)^3 + \cdots$, which is a geometric progression with sum equal to $1/(1 - tq)$ (taking $|t| < 1/q$, so that $|qt| < 1$). The pgf of the geometric distribution is therefore

$$G(t) = tp/(1 - tq).$$

Differentiating with respect to t we get

$$G'(t) = p/(1 - tq)^2.$$

Hence, putting t equal to 1, $G'(1) = p/(1 - q)^2$. But $q = 1 - p$ and $G'(1) = E(X)$, so we have that the expectation of X is equal to $1/p$.

The second derivative is easier to obtain, since t only appears in the denominator of $G'(t)$:

$$G''(t) = (-2)(-q)p/(1 - tq)^3.$$

Setting t equal to 1, this gives

$$G''(1) = 2qp/(1 - q)^3 = 2qp/p^3 = 2q/p^2.$$

Hence

$$\begin{aligned} \text{Var}(X) &= G''(1) + G'(1) - \{G'(1)\}^2 \\ &= 2(1 - p)/p^2 + 1/p - 1/p^2 \\ &= (2 - 2p + p - 1)/p^2 \\ &= (1 - p)/p^2. \end{aligned}$$

8.1.1 Pgf for the sum of random variables

Suppose U and V are two independent random variables, and suppose that the random variable W is defined by $W = U + V$. Let the probability generating functions associated with U, V, and W be denoted by $G_U(t)$, $G_V(t)$, and $G_W(t)$, respectively. Then

$$\begin{aligned} G_W(t) &= E(t^W) && \text{[by definition]} \\ &= E(t^{U+V}) \\ &= E(t^U \times t^V) \\ &= E(t^U) \times E(t^V) && \text{[since } U \text{ and } V \text{ are independent]} \\ &= G_U(t) \times G_V(t). \end{aligned}$$

This result generalizes straightforwardly. Let the random variable S be defined by

$$S = \sum_{i=1}^{n} X_i,$$

where $X_1, \dots, X_n$ are independent and identically distributed random variables, each with pgf $\mathrm{G}_X(t)$. Then $\mathrm{G}_S(t)$, the pgf for S, is given by

$$\mathrm{G}_S(t) = \{\mathrm{G}_X(t)\}^n. \tag{8.4}$$

From this expression, on differentiating both sides with respect to t, we obtain

$$\mathrm{G}'_S(t) = n\{\mathrm{G}_X(t)\}^{n-1}\mathrm{G}'_X(t).$$

Setting $t = 1$, we see that $\mathrm{G}'_S(1) = n\mathrm{G}'_X(1)$, since $\mathrm{G}_X(1) = 1$, and hence we have shown once again that

$$\mathrm{E}(S) = \mathrm{E}\left(\sum X_i\right) = n\mathrm{E}(X).$$

Being able to obtain this result using probability generating functions may be reassuring, but is not very exciting since we knew it already! However, we also have the entire probability distribution for S encapsulated in $\mathrm{G}_S(t)$, since $\mathrm{P}(S = s)$ is, by definition of a pgf, equal to the coefficient of t^s in $\mathrm{G}_S(t)$.

Example 8.4

Determine the probability that the sum of ten independent Bernoulli random variables is equal to exactly 7.

A Bernoulli random variable, X, takes values 1 and 0 with probabilities p and $(1-p)$, respectively. It therefore has pgf given by

$$\mathrm{G}_X(t) = (1-p)t^0 + pt^1 = (1-p) + pt.$$

Denoting the sum of 10 such variables by S, we have

$$\begin{aligned} \mathrm{G}_S(t) &= \{(1-p) + pt\}^{10} \\ &= (1-p)^{10} + \binom{10}{1}(1-p)^9 pt + \binom{10}{2}(1-p)^8 (pt)^2 + \cdots \\ &\quad + \binom{10}{3}(1-p)^3 (pt)^7 + \cdots + (pt)^{10}, \end{aligned}$$

using the binomial expansion. The probability that S equals 7 is the coefficient of t^7, which is $\binom{10}{3}(1-p)^3 p^7$.

We can use the probability generating function to show that, for example, the distribution of the sum of independent random variables having a Poisson distribution has itself got a Poisson distribution. For a Poisson random variable, X, with parameter λ, the pgf is

$$\mathrm{G}_X(t) = \sum t^x \times \frac{\lambda^x \mathrm{e}^{-\lambda}}{x!} = \mathrm{e}^{-\lambda} \times \sum \frac{(\lambda t)^x}{x!} = \mathrm{e}^{-\lambda+\lambda t}.$$

In the same way, the pgf for a Poisson random variable, Y, with parameter μ is $\mathrm{e}^{-\mu+\mu t}$. Using Equation (8.4), the pgf of the sum $X + Y$ is

$$\mathrm{e}^{-\lambda+\lambda t} \times \mathrm{e}^{-\mu+\mu t} = \mathrm{e}^{-(\lambda+\mu)+(\lambda+\mu)t}.$$

This has the same form as the pgf of X, but with λ replaced by $(\lambda + \mu)$: so it is the pgf of a random variable having a Poisson distribution with parameter $(\lambda + \mu)$.

Exercises 8a

1. The discrete random variable Z has probability generating function

$$\mathrm{G}(t) = \frac{1}{20}\left(8t + 5t^2 + 4t^3 + 3t^4\right).$$

   (a) Determine $\mathrm{P}(Z = 3)$.

   (b) Determine $\mathrm{E}(Z)$.

2. Using the fact that the pgf of a Bernoulli distribution is $(1 - p) + pt$, together with the fact that a binomial random variable can be considered as the sum of n independent Bernoulli random variables, obtain the pgf of a binomial random variable and hence deduce the form of the binomial distribution.

8.2 The moment generating function

The **moment generating function** (**mgf** for short) of a random variable, X, provides an alternative procedure for calculating **central moments** ($\mathrm{E}(X)$, $\mathrm{E}(X^2)$, ...) which is sometimes much easier to use than the direct method. It is also useful in establishing some standard results. The mgf of X is usually denoted by $\mathrm{M}_X(t)$, or, when there is no possibility of confusion, by $\mathrm{M}(t)$. It is defined by

$$\mathrm{M}(t) = \mathrm{E}\left(\mathrm{e}^{tX}\right), \tag{8.5}$$

where t is a variable that can be restricted to take values close to 0. The mgf can be used with either discrete or continuous random variables, though the pgf is often preferred for the discrete case. In either case, using the series expansion for e^c,

$$\mathrm{e}^c = 1 + \frac{c}{1!} + \frac{c^2}{2!} + \frac{c^3}{3!} + \cdots,$$

together with the result

$$\mathrm{E}[\mathrm{g}(X) + \mathrm{h}(X)] = \mathrm{E}[\mathrm{g}(X)] + \mathrm{E}[\mathrm{h}(X)],$$

we can write

$$\begin{aligned} M(t) &= 1 + E\left(\frac{tX}{1!}\right) + E\left(\frac{t^2X^2}{2!}\right) + E\left(\frac{t^3X^3}{3!}\right) + \cdots. \\ &= 1 + \frac{t}{1!}E(X) + \frac{t^2}{2!}E(X^2) + \frac{t^3}{3!}E(X^3) + \cdots. \end{aligned}$$

We see that $E(X^k)$ **is the coefficient of** $t^k/k!$ **in the series expansion of** $M(t)$.

The Maclaurin expansion for $M(t)$ is

$$M(0) + \frac{t}{1!}M'(0) + \frac{t^2}{2!}M''(0) + \frac{t^3}{3!}M'''(0) + \cdots,$$

where $M'(t)$, $M''(t)$, etc, are successive derivatives of $M(t)$ and each is being evaluated at $t = 0$. Comparing coefficients of $t^k/k!$ in the two formulae we see at once that

$$M'(0) = E(X), \tag{8.6}$$

$$M''(0) = E(X^2), \tag{8.7}$$

so that

$$\mathrm{Var}(X) = M''(0) - \{M'(0)\}^2. \tag{8.8}$$

Evidently putting $t = 0$ in the kth derivative of $M(t)$ will give us the value of $E(X^k)$.

Moment generating functions can also be calculated for discrete distributions, although probability generating functions are usually easier to use in such cases.

Example 8.5

Use the moment generating function to find the mean and variance of the continuous random variable X which has an exponential distribution with parameter λ.

Since the pdf of X is given by $f(x) = \lambda e^{-\lambda x}$, for $x > 0$, the mgf is given by

$$M(t) = E(e^{tX}) = \int_0^\infty e^{tx}\lambda e^{-\lambda x}dx = \lambda\int_0^\infty e^{-(\lambda-t)x}dx = \frac{\lambda}{\lambda - t}.$$

Differentiating, we get

$$M'(t) = \frac{\lambda}{(\lambda - t)^2},$$

$$M''(t) = \frac{2\lambda}{(\lambda - t)^3}.$$

Finally, putting t equal to 0, we get $M'(0) = 1/\lambda$ and $M''(0) = 2/\lambda^2$, so that $E(X) = 1/\lambda$ and $\mathrm{Var}(X) = 2/\lambda^2 - (1/\lambda)^2 = 1/\lambda^2$.

Example 8.6

Determine the moment generating function of the random variable Z which has a standard normal distribution.

The probability density function of Z is $\frac{1}{\sqrt{2\pi}}e^{-z^2/2}$, so the moment generating function is given by

$$M_Z(t) = \frac{1}{\sqrt{2\pi}}\int_{-\infty}^{\infty} e^{tz}e^{-z^2/2}dz.$$

We now combine the two exponents of e:

$$\begin{aligned} tz - \frac{1}{2}z^2 &= -\frac{1}{2}(z^2 - 2tz) \\ &= -\frac{1}{2}(z^2 - 2tz + t^2 - t^2) \\ &= -\frac{1}{2}(z - t)^2 + \frac{1}{2}t^2. \end{aligned}$$

Thus

$$\begin{aligned} M_Z(t) &= \frac{1}{\sqrt{2\pi}}\int_{-\infty}^{\infty} e^{-\frac{1}{2}(z-t)^2} \times e^{\frac{1}{2}t^2}dz \\ &= e^{\frac{1}{2}t^2} \times \frac{1}{\sqrt{2\pi}}\int_{-\infty}^{\infty} e^{-\frac{1}{2}(z-t)^2}dz. \end{aligned}$$

The integral is just a normal distribution centred on t rather than 0. It therefore integrates to 1. Hence:

For a standard normal distribution,

$$M_Z(t) = e^{\frac{1}{2}t^2}. \tag{8.9}$$

8.2.1 Mgf for the sum of random variables

Suppose U and V are two independent random variables, and suppose that the random variable W is defined by $W = U + V$. Let the moment generating functions associated with U, V, and W be denoted by $\mathrm{M}_U(t)$, $\mathrm{M}_V(t)$, and $\mathrm{M}_W(t)$, respectively. Then:

$$\begin{aligned}
\mathrm{M}_W(t) &= \mathrm{E}\left(\mathrm{e}^{tW}\right) && \text{[by definition]} \\
&= \mathrm{E}\left(\mathrm{e}^{tU+tV}\right) \\
&= \mathrm{E}\left(\mathrm{e}^{tU}\right) \times \mathrm{E}\left(\mathrm{e}^{tV}\right) && \text{[since } U \text{ and } V \text{ are independent]} \\
&= \mathrm{M}_U(t) \times \mathrm{M}_V(t).
\end{aligned}$$

This simple result obviously extends to more than two independent random variables. It comes in particularly handy when dealing with independent identically distributed random variables. Let the random variable S be defined by

$$S = \sum_{i=1}^{n} X_i,$$

where $X_1, ..., X_n$ are independent and identically distributed random variables, each with mgf $\mathrm{M}_X(t)$. The mgf for S is simply

$$\begin{aligned}
\mathrm{M}_S(t) = \mathrm{E}\left(\mathrm{e}^{tS}\right) &= \mathrm{E}\left(\mathrm{e}^{t\sum X_i}\right) \\
&= \mathrm{E}(\mathrm{e}^{tX_1}\mathrm{e}^{tX_2} \dots \mathrm{e}^{tX_n}) \\
&= \mathrm{E}(\mathrm{e}^{tX_1}) \times \mathrm{E}(\mathrm{e}^{tX_2}) \times \dots \times \mathrm{E}(\mathrm{e}^{tX_n})
\end{aligned}$$

because of the independence of the random variables. Thus

$$\mathrm{M}_S(t) = \{\mathrm{M}_X(t)\}^n. \tag{8.10}$$

From this expression, on differentiating both sides with respect to t, we obtain:

$$\mathrm{M}_S'(t) = n\{\mathrm{M}_X(t)\}^{n-1}\mathrm{M}_X'(t).$$

Setting $t = 0$, we see that $\mathrm{M}_S'(0) = n\mathrm{M}_X'(0)$, since $\mathrm{M}_X(0) = 1$, and hence we have shown once again that:

$$\mathrm{E}(S) = \mathrm{E}\left(\sum X_i\right) = n\mathrm{E}(X).$$

One final useful result concerns the mgf of a linear function of a random variable. If $Y = aX + b$, where a and b are constants, then:

$$\mathrm{M}_Y(t) = \mathrm{E}(e^{tY}) = \mathrm{E}(e^{atX+bt}) = \mathrm{e}^{bt}\mathrm{E}(e^{atX}) = \mathrm{e}^{bt}\mathrm{M}_X(at). \tag{8.11}$$

Example 8.7

> *The random variable X has moment generating function given by*
>
> $$\mathrm{M}(t) = 1 + 2t + 3t^2 + 4t^3 + \cdots.$$
>
> *Determine the mean and variance of X.*

The coefficient of t is 2, so $\mathrm{E}(X) = 2$. Writing $3t^2$ as $6t^2/2!$ reveals that $\mathrm{E}(X^2) = 6$ and hence $\mathrm{Var}(X) = 6 - 2^2 = 2$.

Alternatively, differentiating, we get $\mathrm{M}'(t) = 2 + 6t + \cdots$, so that $\mathrm{M}'(0) = 2$, while $\mathrm{M}''(t) = 6 + 24t + \cdots$ and $\mathrm{M}''(0) = 6$.

8.2.2 Proof of the central limit theorem

Let $X_1, X_2, \ldots, X_n$ be independent identically distributed random variables, each with mean 0, variance σ^2, and mgf $\mathrm{M}(t)$. Let

$$Z = \sum_{i=1}^{n} \frac{X_i}{\sigma\sqrt{n}}$$

so that Z has mean 0 and variance 1. Using Equation (8.11), the mgf of $\frac{X_i}{\sigma\sqrt{n}}$ is $\mathrm{M}\left(\frac{t}{\sigma\sqrt{n}}\right)$, for all i. Thus, using Equation (8.10),

$$M_Z(t) = \left\{\mathrm{M}\left(\frac{t}{\sigma\sqrt{n}}\right)\right\}^n.$$

We now examine the Maclaurin expansion of $\mathrm{M}\left(\frac{t}{\sigma\sqrt{n}}\right)$:

$$\mathrm{M}(0) + \frac{t}{\sigma\sqrt{n}}\mathrm{M}'(0) + \frac{1}{2}\left(\frac{t}{\sigma\sqrt{n}}\right)^2 \mathrm{M}''(0) + \frac{1}{6}\left(\frac{t}{\sigma\sqrt{n}}\right)^3 \mathrm{M}'''(0) + \cdots.$$

Here, since Z has mean 0 and variance σ^2, it follows that $\mathrm{M}'(0) = 0$ and $\mathrm{M}''(0) = \sigma^2$ and so

$$\mathrm{M}\left(\frac{t}{\sigma\sqrt{n}}\right) = 1 + \frac{t^2}{2n} + \frac{1}{6}\left(\frac{t}{\sigma\sqrt{n}}\right)^3 \mathrm{M}'''(0) + \cdots.$$

The third term is small compared to its predecessor (and will get smaller as n increases), so we will designate it (together with the remaining terms in the expansion) by ϵ hereafter.

Returning to $\mathrm{M}_Z(t)$ we therefore have

$$\mathrm{M}_Z(t) = \left(1 + \frac{t^2}{2n} + \epsilon\right)^n.$$

We are about to let n get infinitely large, so the question is what happens to something of the form $\left(1+\frac{w}{n}\right)^n$ as n increases. To see the answer we expand $\left(1+\frac{w}{n}\right)^n$ as a power series:

$$\begin{aligned}\left(1+\frac{w}{n}\right)^n &= 1+n\frac{w}{n}+\frac{n(n-1)}{2!}\left(\frac{w}{n}\right)^2+\frac{n(n-1)(n-2)}{3!}\left(\frac{w}{n}\right)^3+\cdots\\ &= 1+w+\left(1-\frac{1}{n}\right)\frac{w^2}{2!}+\left(1-\frac{1}{n}\right)\left(1-\frac{2}{n}\right)\frac{w^3}{3!}+\cdots.\end{aligned}$$

As n increases so $1/n$ tends to 0, so, in the limit

$$\left(1+\frac{w}{n}\right)^n \longrightarrow 1+w+\frac{w^2}{2!}=\frac{w^3}{3!}+\cdots=\mathrm{e}^w.$$

Finally, applying this to $\mathrm{M}_Z(t)$ and ignoring the ϵ terms that tend to 0 as n increases, we arrive at:

$$\mathrm{M}_Z(t) \longrightarrow \mathrm{e}^{\frac{1}{2}t^2},$$

which is the moment generating function of the standard normal distribution.

Exercises 8b

1. The continuous random variable, X, has the probability density function

$$\mathrm{f}(x) = k\mathrm{e}^{-3x} \qquad\qquad 0 \le x \le \infty.$$

Find the moment generating function of X and hence find the distribution's mean and variance.

2. The discrete random variable, Y, is such that

$$\mathrm{P}(Y=1)=0.2, \mathrm{P}(Y=2)=0.3, \text{ and } \mathrm{P}(Y=3)=0.5.$$

Determine the moment generating function for Y and hence find its mean and variance.

Key facts

- The **probability generating function** is $G(t) = E(t^X)$.
  With $'$ used to denote a differential with respect to t:

  $E(X) = G'(1)$, where $G(1)$ is $G'(t)$ evaluated at $t = 1$.

  $\mathrm{Var}(X) = G''(1) + G'(1) - \{G'(1)\}^2$.

- $P(X = x)$ is the coefficient of t^x in $G(t)$.
- The pgf of a sum of independent random variables is the product of their separate pgfs.
- The **moment generating function** is $M(t) = E\left(e^{tX}\right)$.

  $E(X) = M'(0)$.

  $\mathrm{Var}(X) = M''(0) - \{M'(0)\}^2$.

- The mgf of a sum of independent random variables is the product of their separate mgfs.

9

*Inequalities and laws

In this short chapter we collect together a number of useful general results, starting with a pair of inequalities.

9.1 Markov's inequality

If X is a random variable that takes only nonnegative values, then, for any positive constant k,

$$P(X \geq k) \leq \frac{E(X)}{k}.$$

We prove the inequality for the continuous case:

$$\begin{aligned} E(X) &= \int_0^\infty x f(x) dx = \int_0^k x f(x) dx + \int_k^\infty x f(x) dx \\ &\geq \int_k^\infty x f(x) dx \\ &\geq \int_k^\infty k f(x) dx = k \int_k^\infty f(x) dx = kP(X \geq k). \end{aligned}$$

An exactly equivalent procedure applies to a discrete variable. Markov's[1] inequality gives us bounds on probabilities when nothing is known about a distribution other than its mean. The bounds cannot be expected to be very close and are only useful for values of k greater than $E(X)$.

[1] Andrei Andreevich Markov (1856–1922) was a Russian mathematician who developed the theory associated with sequences of random events now known as Markov processes. He studied at St. Petersburg where Chebyshev was one of his teachers.

9.2 Chebyshev's inequality

This inequality, derived by Chebyshev,[2] provides a useful special case of Markov's inequality. It uses both the mean and variance of the unknown distribution.

If Y is a random variable with mean μ and variance σ^2, then, for any positive value c,

$$P(|Y - \mu| \geq c) \leq \sigma^2/c^2. \tag{9.1}$$

Since $(Y - \mu)^2$ is a non-negative random variable we can use that as the X in Markov's inequality and substitute c^2 for k:

$$P[(Y - \mu)^2 \geq c^2] \leq E[(Y - \mu)^2]/c^2.$$

But $E[(Y - \mu)^2] = \text{Var}(Y) = \sigma^2$, thus proving the result.

Example 9.1

In boxes that each contain 1000 beads, the number of red beads, X, is known to have mean 15. A box is chosen at random. Using Markov's inequality, the probability of observing more than 30 red beads is therefore at most $15/30 = 1/2$, with the probability of more than 60 red beads being at most $1/4$.

Now suppose that we know that variance of X is 5 and we are interested in the probability of there being between 10 and 20 red beads in the box. Applying Chebychev's inequality we know that

$$P(|X - 5| \geq 5) \leq 5/5^2 = 0.2.$$

Hence

$$P(|X - 5| \leq 5) = 1 - 0.2 = 0.8.$$

9.3 The weak law of large numbers

This law states that the sample average, $\bar{x}$, *converges in probability* towards the expected value, μ, as the sample size increases. To explain this statement, let ϵ be any small positive amount, then the weak law states that

$$\lim_{n \to \infty} P(|\bar{x} - \mu| < \epsilon) = 1. \tag{9.2}$$

The law has a wide application. For the particular case where the observations are independent and come from distributions with common variance, σ^2, the law follows immediately from Chebyshev's inequality, Equation (9.1). In this case the sample mean is an observation from a distribution having variance σ^2/n, so

$$P(|\bar{X} - \mu| \geq \epsilon) \leq \frac{\sigma^2}{n\epsilon^2}.$$

[2] Pafnuty Lvovich Chebyshev (1821–94) worked at St Petersburg University where he developed a mathematical school with an international reputation. Both he and Markov have lunar craters named in their memory.

Reversing the inequality in the probability statement gives

$$P(|\bar{X} - \mu| < \epsilon) = 1 - P(|\bar{X} - \mu| \geq \epsilon).$$

Using Chebyshev's inequality,

$$1 - P(|\bar{X} - \mu| \geq \epsilon) \geq 1 - \frac{\sigma^2}{n\epsilon^2}.$$

Whatever the size of ϵ, as n increases, the right-hand side approaches 1. This is the required convergence in probability.

9.4 The strong law of large numbers

This law makes (as one would expect) a slightly stronger statement about the convergence of the sample and population means. It states,

$$P\left(\lim_{n\to\infty} \bar{X} = \mu\right) = 1. \tag{9.3}$$

For the case of independent identically distributed random variables, the law was proved by Kolmogorov[3] and is sometimes referred to as **Kolmogorov's law**.

The proof of this law is omitted, since it is surprisingly difficult. In cases where the strong law holds, the weak law must also hold. However, there are cases where the weak law is true, but the strong law is not.

[3] Andrei Nikolaevich Kolmogorov (1903–87) was a Russian mathematician who was appointed to a professorship at Moscow University at the age of 28. He developed the theory in many aspects of mathematical statistics, including the Kolmogorov–Smirnov test (Section 17.4). Internationally recognized, he was elected a Fellow of the UK's Royal Society and was awarded membership of the National Academy of Sciences in the USA.

Key facts

Markov's inequality	$P(X \geq k) \leq E(X)/k$.
Chebyshev's inequality	$P(\|Y - \mu\| \geq c) \leq \sigma^2/c^2$.
The weak law of large numbers	$\lim_{n\to\infty} P(\|\bar{x} - \mu\| < \epsilon) = 1$.
The strong law of large numbers	$P\left(\lim_{n\to\infty} \bar{X} = \mu\right) = 1$.

10

Joint distributions

Previous chapters have focused on the distributions of individual random variables. Often, however, when we gather data, we are equally interested in several variables. For example, to determine whether a person is overweight one must know not only the weight. but also the person's height, while the age of the individual may also be relevant.

In this chapter we consider situations involving two variables simultaneously. We will denote the variables by X and Y.

10.1 Joint probability mass function

We begin by considering the case where both variables are discrete and we are interested in the probability that X equals x while simultaneously Y equals y. We will write this probability as $\mathrm{P}_{XY}(x, y)$, which is the joint probability mass function (pmf).

Suppose that there are m possible values for x and n possible values for Y, then the sum of $\mathrm{P}_{XY}(x, y)$ over the mn possible combinations of values will be 1. Thus $\mathrm{P}_{XY}(x, y)$ describes a **bivariate probability distribution**.

Example 10.1

> *Two fair coins, one marked X and one marked Y, are tossed. For each coin, the score is 1 if a head is obtained, and 0 otherwise. Determine the probability mass function for this situation.*

There are four possible (x, y) outcomes:

	0	1
0	(0,0)	(0,1)
1	(1,0)	(1,1)

All four outcomes are equally likely, so

$$\mathrm{P}_{XY}(x, y) = 1/4 \text{ for all } (x, y).$$

Example 10.2

Two fair six-sided dice are rolled. Each die is numbered 1 to 6. The values of the random variables X and Y, are defined as being, respectively, the smaller and the larger, of the two dice values. Determine the joint distribution of X and Y.

The 36 equally likely outcomes, together with the (x, y) values are shown in the table:

	1	2	3	4	5	6
1	(1,1)	(1,2)	(1,3)	(1,4)	(1,5)	(1,6)
2	(1,2)	(2,2)	(2,3)	(2,4)	(2,5)	(2,6)
3	(1,3)	(2,3)	(3,3)	(3,4)	(3,5)	(3,6)
4	(1,4)	(2,4)	(3,4)	(4,4)	(4,5)	(4,6)
5	(1,5)	(2,5)	(3,5)	(4,5)	(5,5)	(5,6)
6	(1,6)	(2,6)	(3,6)	(4,6)	(5,6)	(6,6)

The joint probability distribution is therefore

$$P_{XY}(x, y) = \begin{cases} 1/36 & x = 1, \dots, 6 \text{ and } y = x, \\ 2/36 & x = 1, \dots, 6 \text{ and } y \neq x. \end{cases}$$

Example 10.3

Two fair coins are tossed. If both show heads, then they are tossed a second time. If just one shows heads, then it is tossed again. If there are no heads on the first toss, then there is no second toss. The random variable X is the total number of coins tossed (2, 3, or 4). The random variable Y is the total number of heads obtained (from 0 to 4). Determine the joint distribution of X and Y.

If no heads are obtained on the first toss, then $X = 2$ and $Y = 0$; this occurs with probability 1/4. If one head is obtained on the first toss (probability 1/2), then Y is equally likely to take the values 1 and 2. The full pmf follows:

		Y				
		0	1	2	3	4
	2	1/4	0	0	0	0
X	3	0	1/4	1/4	0	0
	4	0	0	1/16	1/8	1/16

Thus $P(X = 4 \text{ and } Y = 2) = 1/16$; this corresponds to a first toss of two heads followed by a second of two tails.

10.2 Marginal distributions

The marginal distribution of X, $P_X(x)$, is obtained from the joint distribution of X and Y by summation over Y:

$$\mathrm{P}_X(x) = \sum_y \mathrm{P}_{XY}(x, y). \tag{10.1}$$

The corresponding summation over X gives the marginal distribution of Y.

Example 10.3 (cont.)

Determine the marginal distributions of X *and* Y.

The row totals are 1/4, 1/2, and, 1/4: these are the the marginal probabilities of X taking the values 2, 3, and 4, respectively. Similarly, using the column totals, we have

y	0	1	2	3	4
$\mathrm{P}_Y(y)$	1/4	1/4	5/16	1/8	1/16

If two variables, X and Y, are independent, then their joint distribution is the product of their marginal distributions:

$$\mathrm{P}_{XY}(x, y) = \mathrm{P}_X(x) \times \mathrm{P}_Y(y). \tag{10.2}$$

Example 10.4

The random variables X *and* Y *have joint distribution:*

		Y 0	1	2
	0	1/24	1/12	1/24
X	1	1/12	1/6	1/12
	2	1/8	1/4	1/8

Determine whether X *and* Y *are independent.*

In this case, for the outcomes 0, 1, and 2, the marginal probabilities for X are 1/6, 1/3, and 1/2, respectively, with the corresponding marginal probabilities for Y being 1/4, 1/2, and 1/4.

Since, in this case, $\mathrm{P}_{XY}(x, y) = \mathrm{P}_X(x) \times \mathrm{P}_Y(y)$ for all values of X and Y, the variables are independent.

10.2.1 Expectations

Suppose g is some function of X and Y, then

$$\mathrm{E}[g(X, Y)] = \sum_x \sum_y g(x, y)\mathrm{P}_{XY}(x, y). \tag{10.3}$$

In particular, with $g(X, Y) = XY$, we have

$$\mathrm{E}(XY) = \sum_x \sum_y xy\mathrm{P}_{XY}(x, y). \tag{10.4}$$

When X and Y are independent of one another, $\mathrm{P}_{XY}(x, y) = \mathrm{P}_X(x)\mathrm{P}_Y(y)$. In this case, therefore

$$\begin{aligned} \mathrm{E}[XY] &= \sum_x \sum_y xy\mathrm{P}_X(x)\mathrm{P}_Y(y) \\ &= \sum_x x\mathrm{P}_X(x) \sum_y y\mathrm{P}_Y(y). \end{aligned}$$

And so:

if X and Y are independent, then

$$\mathrm{E}(XY) = \mathrm{E}(X)\mathrm{E}(Y). \tag{10.5}$$

For a function of a single variable we need only consider the relevant marginal distribution. For example,

$$\mathrm{E}(X) = \sum_x x\mathrm{P}_X(x).$$

10.2.2 Covariance and correlation

We defined covariance in Equation (4.10) and repeat its definition here:

$$\mathrm{Cov}(X, Y) = \mathrm{E}(XY) - \{\mathrm{E}(X) \times \mathrm{E}(Y)\}.$$

The related statistic, the **correlation coefficient** (often referred to more simply as the **correlation**), is given by

$$\mathrm{Cor}(X, Y) = \frac{\mathrm{Cov}(X, Y)}{\sqrt{\mathrm{Var}(X) \times \mathrm{Var}(Y)}}. \tag{10.6}$$

The division by the square root of the product of the variances implies that correlation is scale-free and can only take values in the range $(-1, 1)$. When X and Y are independent, their correlation is 0.

Example 10.4 (cont.)

Using the marginal distribution we find that

$$\mathrm{E}(X) = (0 \times 1/6) + (1 \times 1/3) + (2 \times 1/2) = 0 + 1/3 + 1 = 4/3.$$

Since the distribution of Y is symmetric about $Y = 1$, we can immediately write that $\mathrm{E}(Y) = 1$. The calculation of $\mathrm{E}(XY)$ is a little more laborious:

$$\begin{aligned} \mathrm{E}(XY) &= (1\times 1\times 1/6) + (1\times 2\times 1/12) + (2\times 1\times 1/4) + (2\times 2\times 1/8) \\ &= 1/6 + 1/6 + 1/2 + 1/2 = 4/3. \end{aligned}$$

So $\mathrm{Cov}(XY) = 4/3 - 4/3 = 0$. The value 0 is another indication that X and Y are independent. Of course the correlation coefficient is also 0.

Example 10.5

The random variables X and Y have joint pmf given by

		Y			
		0	1	2	3
	0	0.1	0.1	0	0
X	1	0	0.3	0.3	0
	2	0	0	0.1	0.1

Determine the covariance of X and Y.

In this case it is easy to see that the variables are not independent. For example, if $Y = 3$, then X must equal 2.

Using the symmetry of the marginal distribution of X, $\mathrm{E}(X) = 1$. For Y, we have

$$\mathrm{E}(Y) = 0 + (1\times 0.4) + (2\times 0.4) + (3\times 0.1) = 1.5.$$

And, for the product,

$$\mathrm{E}(XY) = (1\times 1\times 0.3) + (1\times 2\times 0.3) + (2\times 2\times 0.1) + (2\times 3\times 0.1) = 1.9.$$

Hence $\mathrm{Cov}(X, Y) = 1.9 - 1\times 1.5 = 0.4$.

10.3 Conditional distributions

If we know that X takes a particular value, then we need only consider that row of the table. Algebraically,

$$\mathrm{P}\left(Y = y_j | X = x_i\right) = \frac{\mathrm{P}\left(X = x_i \text{ and } Y = y_j\right)}{\mathrm{P}\left(X = x_i\right)}, \tag{10.7}$$

with a corresponding outcome if it is Y that has a known value.

If the random variables X and Y are independent, then the conditional distribution of X is the same whatever value is taken by Y (and vice versa).

Example 10.5 (cont.)

Find the conditional distribution of X given that Y = 0, and given that Y = 2.

Given that $Y = 0$, then $\mathrm{P}_X(0) = \frac{0.1}{0.1} = 1$; $\mathrm{P}_X(1) = 0$; $\mathrm{P}_X(2) = 0$.
Given that $Y = 2$, then $\mathrm{P}_X(0) = 0$; $\mathrm{P}_X(1) = \frac{0.3}{0.4} = 3/4$; $\mathrm{P}_X(2) = \frac{0.1}{0.4} = 1/4$.

10.3.1 Conditional expectations

These are calculated using the relevant conditional distribution.

Example 10.5 (cont.)

Given that Y = 2, determine the expected value of X.

$\mathrm{E}(X|Y = 2) = 0 + 1 \times \frac{3}{4} + 2 \times \frac{1}{4} = 5/4$.

Exercises 10a

1. The discrete random variables X and Y have the following bivariate distribution:

		Y		
		1	2	3
X	0	0.1	0.05	0.05
	1	0.2	0.15	0.15
	2	0.1	0.1	0.1

   (a) Determine the marginal distribution of X.
   (b) Determine the conditional distribution of Y, given that $X = 0$.
   (c) Determine $\mathrm{E}(X|Y = 2)$.
   (d) Are X and Y independent?
   (e) Determine the expected value of the product XY.

2. The values of the random variables A and B are, respectively, the smaller value and the larger value of the variables X and Y, which were introduced in the previous question.

   (a) Determine the joint probability distribution of A and B.
   (b) Determine $\mathrm{E}(A)$.
   (c) Show that $\mathrm{E}(B - A) = 1$.

3. The random variable T is equally likely to take the values 0 and 1. The random variable U is such that $P(U = 3) = 0.6$ and $P(U = 4) = 0.4$. The variables T and U are independent. Determine their joint distribution.
4. The discrete random variables C and D have the following joint distribution:

		D		
		-1	0	1
	0	0.1	0.05	0.05
C	1	0.2	0.1	0.1
	2	0.1	0.05	0.25

   (a) Determine the conditional distribution of C, given that $D = 0$.
   (b) Determine the conditional distribution of C, given that D is less than 1.
   (c) Determine the distribution of E, where $E = (C + D)$.
   (d) Determine the joint distribution of D and E.

10.4 *Continuous variables

When X and Y are continuous variables the joint probability mass function is replaced by the **joint probability density function**, $\text{f}_X Y(x, y)$. Summations are replaced by integrals, so

$$\int_{\text{Range of } x} \int_{\text{Range of } y} \text{f}_{XY}(x, y)\text{d}x\text{d}y = 1.$$

If R is some region of interest in the (x, y) plane, then

$$\text{P}[(X, Y) \in \text{R}] = \int\int_{\text{R}} \text{f}_{XY}(x, y)\text{d}x\text{d}y.$$

Example 10.6

Suppose that the continuous random variables X and Y have the joint probability density function given by

$$\text{f}_{XY}(x, y) = (2 - x)y \text{ with } 0 \leq x \leq 2; 0 \leq y \leq 1.$$

We wish to determine the probability that X falls in the range $(1, 2)$ and simultaneously Y falls in the range $(0, 0.5)$.

The required probability is

$$\begin{aligned}
\int_{y=0}^{0.5}\int_{x=1}^{2}(2-x)y\,dx\,dy &= \int_{y=0}^{0.5}\left[(2x-\frac{x^2}{2})\right]_1^2 y\,dy \\
&= \int_{y=0}^{0.5}\left\{\left(4-\frac{4}{2}\right)-\left(2-\frac{1}{2}\right)\right\}y\,dy \\
&= \left(2-\frac{3}{2}\right)\int_{y=0}^{0.5} y\,dy \\
&= \frac{1}{2}\left[\frac{y^2}{2}\right]_0^{0.5} \\
&= \frac{1}{2}\times\left(\frac{1}{8}-0\right)=\frac{1}{16}.
\end{aligned}$$

A **marginal distribution** for one variable, is obtained from the joint pdf by integration over the other variable. For example

$$\mathrm{f}_X(x)=\int_{\text{Range of }Y}\mathrm{f}_{XY}(x,y)dy.$$

The continuous variables X and Y are independent if

$$\mathrm{f}_{XY}(x,y)=\mathrm{f}_X(x)\times\mathrm{f}_Y(y),$$

or, equivalently, if

$$\mathrm{f}_X(x|Y=y_i)=\mathrm{f}_X(x|Y=y_j)\text{ for all } i \text{ and } j.$$

Example 10.6 (cont.)

Here we have

$$\mathrm{f}_X(x)=\int_0^1(2-x)y\,dy=(2-x)\left[\frac{y^2}{2}\right]_0^1=(2-x)\left(\frac{1}{2}-0\right)=\frac{1}{2}(2-x),$$

and

$$\mathrm{f}_Y(y)=\int_0^2(2-x)y\,dx\,dy=y\left[2x-\frac{x^2}{2}\right]_0^2=y\left(4-\frac{4}{2}\right)=2y.$$

Notice that in this case, since $\mathrm{f}_{XY}(x,y)$ can be written as a product of a function of x alone and a function of y alone, it follows that X and Y are independent:

$$\mathrm{f}_{XY}(x,y)=(2-x)y=\frac{1}{2}(2-x)\times 2y=\mathrm{f}_X(x)\times\mathrm{f}_Y(y).$$

Finally, a **conditional distribution** is provided by the cross-section of the joint distribution at the relevant value. For example,

$$\mathrm{f}_X(x|Y=y_0)=\mathrm{f}_{XY}(x,y_0)\Big/\int_{\text{Range of }x}\mathrm{f}_{XY}(x,y_0)\mathrm{d}x.$$

Example 10.7

Suppose that the random variables X and Y have the joint probability density function given by

$$\mathrm{f}_{XY}(x,y)=6(x-y)\text{ for }0\le x\le 1;0\le y\le x.$$

We require the conditional distribution of X given that $Y=0.5$.

In this case X and Y are evidently not independent, since the range of Y depends upon the value of X. Furthermore, we know that, when $Y=0.5$, the minimum possible value for X is 0.5, since $y\le x$. Thus

$$\begin{aligned}\mathrm{f}_X(x|Y=0.5) &= 6(x-0.5)\Big/\int_{0.5}^{1}6(x-0.5)\mathrm{d}x\\ &= (x-0.5)\Big/\left[\frac{x^2}{2}-\frac{x}{2}\right]_{0.5}^{1}\\ &= (x-0.5)\Big/\left\{\left(\frac{1}{2}-\frac{1}{2}\right)-\left(\frac{1}{8}-\frac{1}{4}\right)\right\}\\ &= 8(x-0.5).\end{aligned}$$

Expectations are calculated as for the discrete case, but with integrals replacing sums.

Example 10.7 (cont.)

The expected value of X is given by

$$\begin{aligned}
\mathrm{E}(X) &= \int_{x=0}^{1}\int_{y=0}^{x} x \times 6(x-y)\mathrm{d}y\mathrm{d}x \\
&= 6\int_{0}^{1}\left[x^2y-\frac{1}{2}xy^2\right]_0^x \mathrm{d}x \\
&= 6\int_{0}^{1}\left(x^3-\frac{x^3}{2}\right)\mathrm{d}x = 6\int_0^1 \frac{x^3}{2}\mathrm{d}x \\
&= 6\left[\frac{x^4}{8}\right]_0^1 \\
&= \frac{3}{4}.
\end{aligned}$$

If it is known that $Y = 0.5$, then instead we use the marginal distribution

$$\begin{aligned}
\mathrm{E}(X|Y=0.5) &= 8\int_{0.5}^{1} x(x-0.5)\mathrm{d}x \\
&= 8\left[\frac{x^3}{3}-\frac{x^2}{4}\right]_{0.5}^1 \\
&= 8\left\{\left(\frac{1}{3}-\frac{1}{4}\right)-\left(\frac{1}{24}-\frac{1}{16}\right)\right\} \\
&= 8\left(\frac{1}{12}+\frac{1}{48}\right) \\
&= \frac{5}{6}.
\end{aligned}$$

Exercises 10b

1. The random variables X and Y have joint probability density function $\mathrm{f}_{XY}(x,y) = 4xy$, with $0 \le x \le 1$ and $0 \le y \le 1$.
   - **(a)** Are X and Y independent variables?
   - **(b)** Determine the probability that both X and Y are greater than 0.5.
   - **(c)** Determine the marginal distribution of X.
   - **(d)** Determine the expectation of X, given that $Y = 0.5$.
2. The random variables X and Y have joint probability density function $\mathrm{f}_{XY}(x,y) = k(x+y)$, with $0 \le x \le 1$ and $0 \le y \le 1$.
   - **(a)** Are X and Y independent variables?
   - **(b)** Determine the value of the constant k.
   - **(c)** Determine the probability that both X and Y are greater than 0.5.
   - **(d)** Determine the marginal distribution of X.
   - **(e)** Determine the expectation of X, given that $Y = 0.5$.

Key facts

- $\mathrm{P}(X = x \text{ and } Y = y)$ is a **bivariate probability distribution**; shorthand $\mathrm{P}_{XY}(x, y)$.
- Summation of $\mathrm{P}_{XY}(x, y)$ over the values of Y gives the **marginal distribution**, $\mathrm{P}_X(x)$.
- $\mathrm{E}[\mathrm{g}(X, Y)] = \sum_x \sum_y \mathrm{g}(x, y)\mathrm{P}(x, y)$ gives the expected value of the function g.
- The **correlation coefficient** is given by

  $$r = \mathrm{Cov}(X, Y)/\sqrt{\mathrm{Var}(X) \times \mathrm{Var}(Y)}.$$

- The **conditional distribution** of Y, given that $X = x_i$ is

  $$\mathrm{P}(Y = y_j | X = x_i) = \mathrm{P}(X = x_i \text{ and } Y = y_j) \big/ \mathrm{P}(X = x_i) \, .$$

- X and Y are **independent** if $\mathrm{P}_{XY}(x, y) = \mathrm{P}_X(x) \times \mathrm{P}_Y(y)$.
- For continuous variables, the probability mass function is replaced by the probability density function, with summations replaced by integrals.

PART II
Statistics

The focus of this part of the book is data. Beginning with simple data summaries, probabilistic models are used to draw inferences concerning the nature of the observed variation. These inferences can inform us concerning the nature of future data.

11

Data sources

We begin this part of the book with a brief discussion of where data come from.

11.1 Data collection by observation

This has been the standard method of data collection for millennia. The famous theories, such as Newton's theory of gravitation and Einstein's theory of relativity, all have their roots in numerical data collected by careful observation. On a more mundane level, decisions concerning local traffic flow (e.g. 'Would it help to replace this crossroadswith a roundabout?') are based on observations of flow made by video cameras or teams of observers.

The collection of data of a scientific nature (e.g. physical, chemical, biological, data) relies almost exclusively on observation (often remotely using machines). For example, rain gauges, anemometers, disdrometers (instruments for measuring raindrop sizes), and weather radars all give information about present or future weather, though it may not be straightforward to combine and understand the various readings.

In the past two centuries there has been increasing interest in the social sciences (e.g. sociology, politics, economics) for which other methods of data collection are relevant. The next sections examine some of the problems that arise in collecting reliable data of this type.

11.2 National censuses

The world's first major census is said to have taken place in the Babylonian Empire around 3800 years before the Christian Era. The census recorded both the numbers of people (with an eye to their possible military service) and the amounts of goods and food (to ensure that the military could be fed and clothed). These Babylonian records were made on clay tablets which no longer exist. The oldest extant census records date from 2 CE and give the population of China at that time as about 58 million.

In the USA, censuses have been carried out at 10-year intervals, beginning in 1790 when the 4 million recorded inhabitants included nearly three-quarters of a million slaves. By 2020 the population had risen to over 330 million.

Comprehensive censuses began to be taken in most countries in the nineteenth century: in the UK, the first to include named individuals was taken in 1841 (about 18.5 million people), with the most recent, in 2021, including about 59.5 million people.

One of the best known national surveys is preserved in the National Archives at Kew in west London. This survey, now referred to the *Domesday Book*, was commissioned by William the Conqueror, and dates to 1086. It gives details for about 270 thousand households.

11.3 Sampling

Censuses are time-consuming, very expensive, and rarely necessary. Often we can obtain a sufficiently accurate estimate of the quantity of interest by taking a relatively small number of observations. We call this small number of observations the **sample**, with the complete set of possible observations being the **population**.

In the case of a national census, the population being sampled (completely) is the human population. However, the statistical population need not be people; indeed it may not refer to living things. For example, suppose that a vehicle taking boxes of eggs to the warehouse is involved in an accident. The insurance assessor needs to obtain an idea concerning how many of the N egg boxes in the vehicle subsequently contain broken eggs. If N is large, then the assessor will not examine every box. Instead a sample of n boxes will be examined and an assessment of the damage will be based on the results for these n boxes. For this to be an appropriate procedure, a reliable sampling method is required. We now briefly introduce the most common methods.

11.3.1 The simple random sample

Most sampling methods endeavour to give every member of the population the same probability of being included in the sample. If each member of the sample is selected by the equivalent of drawing lots, then the sample selected is described as being a **simple random sample**.

One procedure for drawing lots is the following:

1. Make a list of all N members of the population.
2. Assign each member of the population a different number.
3. For each member of the population place a correspondingly numbered ball in a bag.
4. Draw n balls from the bag, without replacement. The balls should be chosen at random.
5. The numbers on the balls identify the chosen members of the population.

An automated version would use the computer to simulate the drawing of the balls from the bag.

The principal difficulty with this procedure is the first step: the creation of a list of all N members of the population. This list is known as the **sampling frame**. In many cases there will be no such central list. For example, suppose it was desired to test the effect of a new cattle feed on a random sample of Irish cows. Each individual farm may have a list of its own cows (Daisy, Buttercup, ...), but the government keeps no central list.

For the country as a whole there is not even a 100% accurate list of people (because of births, deaths, immigration, and emigration).

Because of the straightforward nature of the simple random sample, most analyses assume that this kind of sample has been used to obtain the data. The necessary adjustments that may be required when dealing with other methods of sampling are well beyond the scope of this book. However, the nature of these other methods of sampling 'needs discussion'.

11.3.2 *Cluster sampling*

Even if there was a 100% accurate list of the population of a country, simple random sampling would almost certainly not be performed because of expense: the intrepid interviewer would be a much travelled individual!

To avoid this problem, populations that are geographically scattered are usually divided into conveniently sized regions. A possible procedure is then:

1. Choose a region at random.

2. Choose individuals at random from that region.

The consequences of this procedure are that, instead of a random scatter of selected individuals, there are randomly scattered **clusters** of individuals. The selection probabilities for the various regions are not equal, but are adjusted to be in proportion to the number of individuals that those regions contain. If the ith region contains N_i individuals, then the chance that it is selected is chosen to be N_i/N, where $N = \sum N_i$.

The size of the chosen region is usually sufficiently small that a single interviewer can perform all the interviews in that region without incurring huge travel costs. In practice, in the UK, because of the sparse population and the difficulties of travel in the Highlands and islands of Scotland, studies of the British population are usually confined to the region south of the Caledonian Canal (which stretches from Inverness on the East coast of Scotland to Fort William on the West coast).

11.3.3 *Stratified sampling*

Most populations contain identifiable **strata**, which are distinctive nonoverlapping subsets of the population. For example, for human populations, useful strata might be 'males' and 'females', or 'receiving education', 'working' and 'retired', or combinations such as 'retired female'. From census data we might know the proportions of the population falling into these different categories. With stratified sampling, we ensure that these proportions are reproduced by the sample. Suppose, for example, that the age distribution of the adult population in a particular district is as given as follows:

Aged under 40	Aged between 40 and 60	Aged over 60
38%	40%	22%

A simple random sample of 200 adults would be unlikely to *exactly* reproduce these figures. If we were very unfortunate, over half the individuals in the sample might be aged under 40. If the sample were concerned with people's taste in music, then, by chance, the simple random sample might provide a misleading view of the population.

A **stratified sample** is made up of separate simple random samples for each of the strata. In the present case, we would choose a simple random sample of 76 adults aged under 40, a separate simple random sample of 80 adults aged between 40 and 60, and a separate simple random sample of 44 adults aged over 60.

Stratified samples exactly reproduce the characteristics of the strata and this almost always increases the accuracy of subsequent estimates of population parameters. Their slight disadvantage is that they are a little more difficult to organize.

11.3.4 Systematic sampling

Both cluster sampling and stratified sampling subdivide the population into components. In both cases the final stage consists of selecting a random sample from a portion of the population. One possible method of doing the final selection is by simple random sampling. An alternative is to use **systematic sampling**, which is described below for the case of a sample of n individuals to be drawn from a population of N individuals.

1. Choose one individual at random.

2. Choose every kth individual thereafter, returning to the beginning of the list when the end is reached. The value of k is not crucial, but should be chosen beforehand. A popular choice is a convenient value close to N/n. The use of this wide spacing guards against the list consisting of clusters of similar individuals.

For example, suppose we wish to choose six individuals from a list of 250. A convenient value for k might be 40. Suppose that the first individual selected is number 138. The remainder would be numbers 178, 218, 8, 48, and 88.

If the list has been ordered by some relevant characteristic (e.g. age, or years of service), then, with $k \simeq N/n$, this procedure produces a spread of values for the characteristic—a type of informal stratification.

11.3.5 Quota sampling

This is the method often used for street interviews. The interviewer is given a series of targets. For example, he or she might be instructed to interview equal numbers of men and women, of whom one quarter should be aged over 60 and one third should be in low-paid jobs. The instructions would be more detailed than these, with the idea being that each interviewer will select a representative cross section of the population. It is easy to see that an interviewer might have some difficulty in completing his or her **quota** on any given day.

The results of quota sampling must always be viewed with a little suspicion, since the interviewees were not chosen at random.

11.3.6 Self-selection

However bad quota sampling may be, it is wonderful by comparison with self-selection! The latter is exemplified by radio or television 'phone-ins' where listeners or viewers record their 'vote'. The views of the apathetic majority are seriously under-represented (though maybe they don't have any to represent!). The same is likely to be the case on websites that publish user reviews.

11.3.7 A national survey

To illustrate the methods of sampling discussed in the previous sections, we now give a brief outline of the method of selection for the households included in the BHPS (British Household Panel Study), which was an annual study carried out between 1991 and 2009.

The *sampling frame* for the BHPS was the Postcode Address File, a (computer-based) master list of all the 1.5 million postcodes in Britain. Each individual postcode (e.g. CO5 8JU) is a member of a so-called postcode sector (e.g. CO5 8). There are around 9000 postcode sectors, each of which identifies a *cluster* of about 2500 households.

A simplified version of the stages involved in the selection of the households is as follows.

1. **Selection of sectors.**

   Sector selection was accomplished by using *systematic* sampling of a cleverly reordered list. The reordering process was as follows:

   (a) The 9000 postcode sectors were subdivided into 18 geographical regions.

   (b) Within each region the postcode sectors were arranged in an ordered list. The first sector in the list was the postcode sector having the highest proportion of professional heads of households, with the last in the list being that sector with the lowest proportion. These proportions were determined using data from the 1981 national census.

   (c) Each regional group was now split into a 'high' half and a 'low' half, to form two subgroups.

   (d) Each subgroup was reordered by using descending order of percentage of pensioners as a criterion and was again split into two.

   (e) The process of reordering was repeated on one further occasion, so as to give a total of 144 subgroups, each separately ordered.

   The systematic selection of sectors from the reordered list will result in the selected sectors being spread across the country and across the characteristics used for the reordering. The effect is similar to that of *stratification*.

2. **Selection of households**
   These were a *simple random sample* of about 35 households chosen from each selected sector. In all, about 8000 households were selected.

It can be seen that a large survey is likely to combine together several different types of sampling procedure.

11.3.8 Pseudo-random numbers

Suppose we require a sample of 1000 random digits. A thousand draws of balls from a box would be feasible but very tedious. Instead therefore, we make use of computer-generated **pseudo-random numbers**. These are numbers which are generated by a mathematical formulae. They have the following properties:

- someone who did not know where they came from would be unable to deduce that they had not been generated using a 10-sided die, but
- the computer could generate exactly the same sequence time after time if this was required.

In practice the description 'pseudo-' is usually dropped and these numbers are also described as **random numbers**.

11.4 Questionnaires

The most common method for collecting social science data is by means of a **questionnaire** which consists of a series of questions concerning the facts of someone's life or their opinions on some subjects. The recipient of a questionnaire is usually referred to as the **respondent**.

The principal methods of collecting the data using a questionnaire are:

- Face-to-face interview.
- By post or on line.

11.4.1 The face-to-face interview

In this case the interviewer and the respondent communicate directly, either in a street interview (in which the interviewer selects passers-by for interview) or in an interview in the respondent's home.
Advantages

- *Complex structure.* The structure of the questionnaire (e.g. 'If answer is "Yes" then go to question 23c') can be relatively complicated, since only the interviewer needs to understand it.
- *Consistency.* If the interviewer does the writing, then the questionnaire will be completed in a consistent fashion.
- *Help.* If the respondent has difficulty understanding a question, then the interviewer is available to give an explanation.
- *Response rate.* The **response rate** is defined as the number of interviews completed divided by the number attempted. Assuming that the interviewer is friendly, this is likely to be quite high (say 70%).

Disadvantages

- *Expense.* The procedure uses up a lot of time for each interviewer. There may also be costs associated with the travelling between respondents.
- *Bias.* Although the questionnaire is completed in a consistent fashion, this consistency may contain bias (e.g. the interviewer consistently misinterprets an answer, or gives misleading guidance).
- *Lack of anonymity.* A respondent may refuse to answer questions because of being embarrassed by the presence of the interviewer.

11.4.2 The 'postal' questionnaire

Here we refer to any questionnaire that is given out for self-completion and return by (anonymous) respondents. Such a questionnaire may appear either in the post or online.

The principal advantage of this method of gathering information is:

- *Economy.* Since no interviewer is required, it is a cheap method of collecting data.

However, set against this advantage is a major disadvantage:

- *Non-response*. The response rate (measured as the proportion of questionnaires that are returned) can be very low indeed (e.g. 10%) and is rarely greater than 50%. This low level of response is a problem because the replies received are unlikely to be representative of those of the population as a whole. People who take the trouble to fill in and return a questionnaire are not typical (it is well known that 'apathy reigns K.O.')! If the response rate is very low, then the replies may be seriously misleading.

11.5 Questionnaire design

To ask someone a series of questions might seem to be a ridiculously simple task, but this is certainly not the case. It is easy to accidentally create unanswerable questions, while small changes to the wording can make a difference. Even the order of questions needs careful thought.

11.5.1 Some poor questions

1. Do you think that boys or girls have the better dress sense or is it simply the influence of their parents?
   Unanswerable! This 'question' is at least two questions and is unlikely to be understood by anyone (including its author!).

2. Does your family watch a lot of television?
   Unanswerable! Some family members may be TV addicts, whereas others scarcely ever watch. Also 'a lot' is not a well-defined quantity.

3. Do you think that statistics is:

   (a) a very interesting subject,

   (b) an interesting subject,

   (c) quite an interesting subject?

   A biased set of choices.

4. Are you alive?
   Not worth asking! Avoid questions that will be answered the same way by everyone (or almost everyone).

5. I am going to ask you about the Monarchy. Bertrand Russell once said ... [something long and rambling taking several minutes to read]. Do you agree?
   Avoid long questions—the respondent will forget what the question is about.

6. You are against the death penalty, aren't you?
   This is a leading question—the respondent is being pressurized into saying 'Yes'. Your barrister would object!

7. What do you think of the verisimilitude of this simulacrum?
   Avoid unfamiliar words.

8. Are you aged:

   (a) over 30,

   (b) under 21,

   (c) under 18?

   When giving a range of alternatives make sure that they are not overlapping and include all possibilities.

9. When they are not playing at home, Arsenal are not a good side at scoring goals. Do you agree or disagree?
   Avoid double (or multiple) negatives—some respondents will misunderstand the question.

10. Please don't be embarrassed by this question: do you pick your nose?
    But for the preamble, many respondents would have answered the question without worry. Don't invite respondents not to respond!

11. Where were you on March 7th?
    Unless this question is asked soon after that date it is unlikely to get a response! Questions about the distant past are likely to require the respondent to guess.

12. Are you a communist?
    Since communists are rather out of fashion at present, some supporters of communism are unlikely to own up. Respondents tend to give 'socially acceptable' answers.

11.5.2 Some good questions

The best questions are probably those that have been used in surveys conducted by market research or other organizations that specialize in asking questions. From their experience they will know which questions work well. A large public library may be able to help with this:

- books on survey methods may contain example questionnaires,
- the 'quality' newspapers may report questions asked in national surveys by an organization such as Gallup,
- the survey organizations themselves may publish questionnaire details.

Most good questions are *short* and *simple.*
The same applies to questionnaires!

11.5.3 The order of questions

Two general rules are:

- Start with easy questions.
  This encourages the respondent to participate.
- Ask general questions (e.g. 'How satisfied are you with school lunches?') first, and specific questions (e.g. 'What do you dislike most about school lunches?') afterwards. *This is to avoid the 'satisfaction' question being influenced by the subsequent 'dislike' question.*

Some questions occur naturally before others. For example, if one were investigating a respondent's history, it would be natural to begin with questions about childhood before questions about middle-age.

11.5.4 Question order and bias

The order in which questions are asked can influence a respondent's reply. Contrast:

1. 'Do you intend to be an organ donor?'
2. 'Did you know that dozens of people die each year because there are not enough organ donors?'

with:

1. 'Did you know that dozens of people die each year because there are not enough organ donors?'
2. 'Do you intend to be an organ donor?'

11.5.5 Filtered questions

Many questionnaires have what might be described as 'miss-out sections' (flagged by statements such as 'If NO then go to Q24'). Thus a question such as:

'How much money did you earn last week?'

should not precede:

'Were you employed last week?'

since, if the answer to the second question is 'No', then the first question is not asked (it has been *filtered* out).

11.5.6 Open and closed questions

An **open question** is one in which there are no suggested answers:

'What is your opinion of the prime minister?'

The advantage of this type of question is that the respondent can choose precisely how to answer. The disadvantage is that every respondent may answer in a different way, making it difficult to summarize the data obtained.

A **closed question** is one in which there is a prescribed set of alternative answers:

'How do you think the present government compares with others that we have had?
Is it (i) above average, (ii) average, (iii) below average?'

With a closed question the respondent may find difficulty because none of the alternatives offered is found to be suitable. However, this problem will not arise if all possibilities are covered (as in the preceding example).

11.5.7 *The order of answers for closed questions*

We noted earlier that question order can affect the responses obtained. The same is true of the alternative answers provided for closed questions.

- There is a bias towards the left-hand answer in 'postal' questionnaires.
  Because the respondent reads from left to right and may get bored before reaching the right-hand answers.

- There is a bias towards the right-hand answer in face-to-face interviews.
  Because this is the last answer read out and is therefore the one that the respondent remembers most easily.

- If there is a sequence of similar questions the respondent is likely to develop a 'habit' and answer each the same way.
  So it is a good idea to vary the questions—this also makes the questionnaire more interesting.

11.5.8 *The pilot study*

Before conducting a questionnaire it is essential to make sure that it 'works'. Are there any ambiguous questions? Are there closed questions that cause trouble because a possibility has been overlooked? Are there any questions that you have forgotten to ask? The **pilot study** uses the entire questionnaire with a small number of people who need not be chosen in any scientific way. The aim is simply to find and overcome any difficulties *before* conducting the real questionnaire.

Put theory into practice: Choose a single issue of a 'quality' newspaper and search for reports that include statistics. Try and decide what type of organization collected the reported statistics. Counting just one for all the sports reports, one for all the financial reports and one for all the weather information, how many different reports can you find in a single issue of the paper? How many different organizations appeared to have collected the statistics?

12

Summarizing data

The word 'data' is the plural of 'datum', which means a piece of information—so **data** are pieces of information (usually numbers). In this book the example data sets are deliberately small, but the methods presented are valid for use with larger data sets.

12.1 A single variable

12.1.1 Tally charts

A tally chart is a very simple summary method that makes use of so-called **five-barred gates**. It provides a convenient counting frame and gives a useful visualization of the numbers observed.

Example 12.1

Given below are the numbers of strokes taken in the last round by the players having the top 30 final scores of a golf championship:

62, 65, 63, 65, 70, 68, 65, 67, 67, 69, 70, 70, 70, 67, 68,
68, 66, 69, 74, 67, 68, 69, 69, 69, 71, 71, 72, 69, 72, 68.

It is easy to see that most scores are around 70 or a little less. But it is not so easy to see which score was most common.

SCORE	TALLIES
62.	I
63.	I
64.	
65.	III
66.	I
67.	IIII
68.	卌
69.	卌 I
70.	IIII
71.	II
72.	II
73.	
74.	I

Figure 12.1 Tally-chart for the golf score data

The tally chart is constructed on a single 'pass' through the data. For each score a vertical stroke is entered on the appropriate row, with a diagonal stroke being used to complete each group of five strokes. This is much easier than going through the data counting the number of occurrences of a 62 and then repeating this for each individual score.

The tally count for each outcome is called the **frequency** of that outcome. For example, the frequency of the outcome 65 was 3. The individual frequencies could be summarized in a **frequency table**:

Final round score	62	63	64	65	66	67	68	69	70	71	72	73	74
Number of golfers	1	1	0	3	1	4	5	6	4	2	2	0	1

12.1.2 Stem-and-leaf diagrams

Tally charts become uncomfortably long if the range of possible values is very large as with these individual scores from a low-scoring Sunday league cricket match:

22, 58, 12, 17, 4, 7, 26, 10, 13, 1, 39, 0, 1, 10, 6, 0, 11, 14, 1, 0.

A convenient alternative is the **stem-and-leaf diagram** (also called a **stemplot**), in which the stem represents the most significant digit (i.e. the 'tens') and the leaves are the less significant digits (the 'units'). The following stem-and-leaf chart has been created following the order of the data:

0	4, 7, 1, 0, 1, 6, 0, 1, 0
1	2, 7, 0, 3, 0, 1, 4
2	2, 6
3	9
4	
5	8
tens	*units*

If the original stem-and-leaf diagram had been created on rough paper then a tidied version could have the leaves neatly ordered as shown below:

0	0, 0, 0, 1, 1, 1, 4, 6, 7
1	0, 0, 1, 2, 3, 4, 7
2	2, 6
3	9
4	
5	8
tens	*units*

Stem-and-leaf diagrams retain the original data information, but present it in a compact and more easily understandable way: this is the hallmark of an efficient data summary.

It is often useful to provide an explanation (a 'key') with the diagram.

Example 12.2

The ages of the patients in one wing of a hospital were as follows:

Males	*24, 56, 71, 88, 55, 73, 32, 59, 66, 60, 90, 42, 77.*
Females	*40, 59, 93, 77, 86, 82, 60, 35, 76, 82, 84, 37, 61.*

Summarize the data using a back-to-back stem-and-leaf diagram.

In this case we have a central stem with leaves on either side depending on the gender of the patient:

Males		*Females*
4	20	
2	30	5,7
2	40	0
9,5,6	50	9
0,6	60	0,1
7,3,1	70	7,6
8	80	6,2,2,4
0	90	3

Key: 2|80 = 82 = 80|2

The older patients are predominantly female.

Exercises 12a

1. The numbers of absentees in a class over a period of 24 days were:

0	3	1	2	1	0	4	0
1	1	2	3	1	0	0	1
2	4	6	4	2	1	0	1

By first drawing up a tally chart, obtain a frequency table.

2. The total scores in a series of basketball matches were:

215	224	182	200	229
209	217	195	162	210
204	208	197	192	187

Construct a stem-and-leaf diagram to represent these data.

12.1.3 Bar charts

The lengths of the rows of a tally chart, or of a stem-and-leaf diagram, provide an instant picture of the data. This picture is neatened by using bars whose lengths are proportional to the numbers of observations of each outcome (i.e. to the frequencies). In the resulting diagram the bars may be either **horizontal** (like the tally chart) or **vertical**.

Bar charts are easier to read if the width of the bars is different from the width of the gaps between the bars. A bar chart in which the bars are simply lines may be called a **line graph**.

Example 12.1 (cont.)

The golf scores are illustrated using a bar chart in Figure 12.2. For these players there is a peak in the low 60s, with two exceptionally low scores.

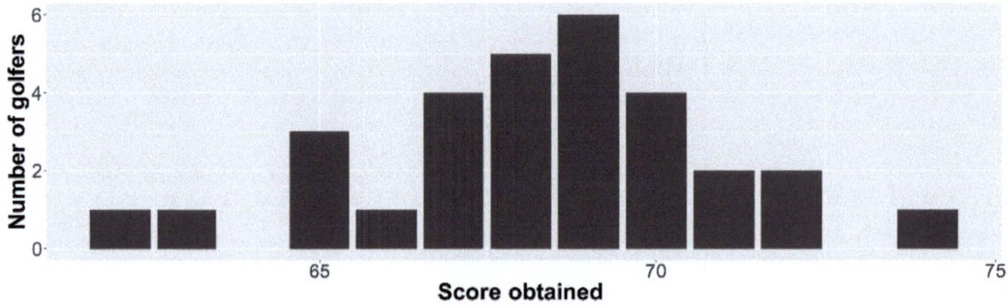

Figure 12.2 Bar chart illustrating the frequencies of the golfers' scores.

Put theory into practice: Roll an ordinary six-sided die 24 times, recording each outcome as it occurs (e.g. 3, 6, 2, 2, ...). Summarize the data using a tally chart and write down your frequency distribution.

Repeat the process, using a separate tally chart. Compare your two frequency distributions. There may be large differences due to random variation! Combine the two sets of results and illustrate them with a vertical bar chart. Does it look as though your die was fair?

12.1.4 Grouped frequency tables and histograms

The following data are the masses (in grams) of 30 brown pebbles chosen at random from those on one area of a shingle beach:

3.4, 12.3, 7.5, 8.2, 8.6, 15.4, 6.9, 7.0, 2.9, 5.0,
13.5, 8.4, 9.9, 11.8, 4.6, 7.7, 3.8, 7.7, 8.6, 14.6,
4.3, 7.9, 9.1, 11.9, 17.4, 6.3, 8.7, 10.1, 5.1, 10.2.

A bar chart of these data would look like a very old comb that had had an unfortunate accident! It is obviously sensible to work with ranges of values, which we call **classes**, rather than with the individual values. As a start we summarize the data (perhaps using a tally chart to help with the counting) in order to form a **grouped frequency table**:

Range of masses (g)	1.95–3.95	3.95–5.95	5.95–7.95	7.95–9.95	9.95–11.95	11.95–13.95	13.95–15.95	15.95–17.95
Frequency	3	4	7	7	4	2	2	1

Published tables frequently use the rounded figures in the grouped frequency table, and may give only the class mid-point or just one of the class boundaries (usually the lower). For example, the pebble data might be reported thus:

Range of masses (nearest 0.1g)	Frequency
2–	3
4–	4
6–	7
8–	7
10–	4
12–	2
14–	2
16–	1

Great care and some ingenuity is often needed to deduce the true class boundaries!

Many quantities that we measure are not really continuous, but are best treated as such.

Example 12.3

The following data set consists of the advertised prices (in £, in 2021) of 23 second-hand Nissan Leafs first registered in 2017 (in the UK).

14,291	15,400	15,495	11,498	12,495	14,000	14,500	14,795
14,799	15,000	15,595	16,500	123,40	13,895	13,998	14,990
14,299	12,577	12,999	13,995	12,500	7,995	10,450	

Although price in £ is not a continuous quantity it is sensible to treat it as such. The grouped frequency table is:

Price	7–	8–	9–	10–	11–	12–	13–	14–	15–	16–
Frequency	1	0	0	1	1	5	3	7	4	1

Price in tens of thousands of £

Bar charts are not appropriate for data with grouped frequencies for ranges of values. Instead the data can be illustrated using a **histogram**, which is a diagram using the areas of rectangles to represent frequency. It is the data-based equivalent of a probability density function.

Example 12.3 (cont.)

Figure 12.3 shows three possible histograms for the Leaf prices. Figure 12.3 (a) uses bins (groups) with widths of £100. The result is a very spiky outline which is not too easy to interpret. It is, however, preferable to Figure 12.3 (b) which uses bins that are so wide that most of the price information is effectively buried.

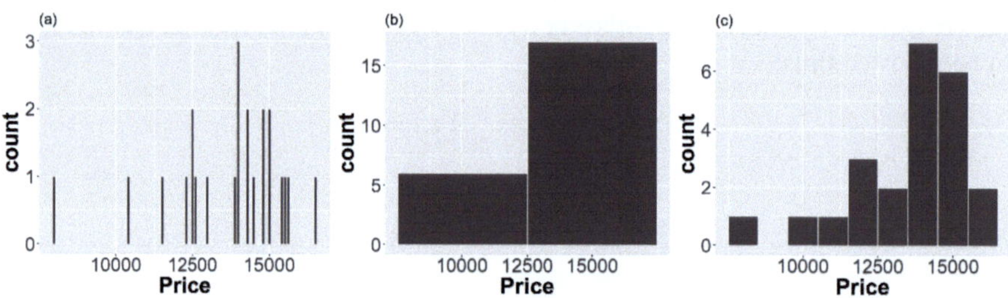

Figure 12.3 Three histograms: (a) using a bin width of £100 (too small); (b) using a bin width of £5000 (too large); (c) using a bin width of £1000 (reasonable).

Figure 12.3 (c) uses groups with the width (£1000) that was used for the grouped frequency table. In the diagram the counts range from, for example, £7,500 to £8,500, rather than from £7,000 to £8,000 as in the table. Any starting point could be used; all are equally valid.

Histograms often use wider rectangles for the tails of the data, with heights adjusted appropriately: it is their areas that must be proportional to frequencies.

Group project: How long can you get a 10p piece to spin on a flat surface? Using a timer from which seconds can be read accurately, note the lengths of times of four spins. Your personal value will be the length of the longest spin. Note the personal values for the whole group and represent the data using a histogram. Is the histogram roughly symmetrical, or is it skewed? Was your personal value typical, or was it unusually short or long?

Exercises 12b

1. A gardener plants 20 potatoes and weighs the crop obtained from each plant. The results, in g, are as follows:

853	759	891	923	755
885	896	821	911	789
854	861	915	835	784
853	891	942	758	867

Construct a frequency table with class boundaries at 750, 800, ..., 950. Illustrate the results using a histogram.

2. The masses (in g to the nearest g) of a random collection of offcuts taken from the floor of a carpenter's shop are summarized below:

0–19	20–39	40–59	60–99
4	17	12	6

Represent these data using a histogram.

12.1.5 Cumulative frequencies

These diagrams provide answers to questions such as 'What proportion of the data have values less than x?'. In such a diagram, cumulative frequency on the 'y-axis' is plotted against observed value on the 'x-axis'. The result is a graph in which, as the x-coordinate increases, the y-coordinate never decreases. It is the data-based equivalent of a distribution function.

With grouped data, the first step is to produce a table of cumulative frequencies. These are then plotted against the upper class boundaries. The successive points may be connected either by straight-line joins (in which case the diagram is called a **cumulative frequency diagram**) or by a curve (in which case the diagram is called an **ogive**).

A **cumulative proportion diagram** is identical to a cumulative frequency diagram, but the y-axis runs from 0 to 1.

Example 12.4

In order to study bird migration a standard technique is to put coloured rings around the legs of the young birds at their breeding colony. The source of a bird subsequently seen wearing coloured rings can therefore be deduced. The following data, which refer to recoveries of razorbills, consist of the distances (measured in hundreds of miles) between the recovery point and the breeding colony:

Distance (miles) (x)	Frequency	Cumulative frequency
$x < 100$	2	2
$100 \leq x < 200$	2	4
$200 \leq x < 300$	4	8
$300 \leq x < 400$	3	11
$400 \leq x < 500$	5	16
$500 \leq x < 600$	7	23
$600 \leq x < 700$	5	28
$700 \leq x < 800$	2	30
$800 \leq x < 900$	2	32
$900 \leq x < 1000$	0	32
$1000 \leq x < 1500$	2	34
$1500 \leq x < 2000$	0	34
$2000 \leq x < 2500$	2	36

Figure 12.4 plots distance travelled against the number of birds that travelled at least that far.

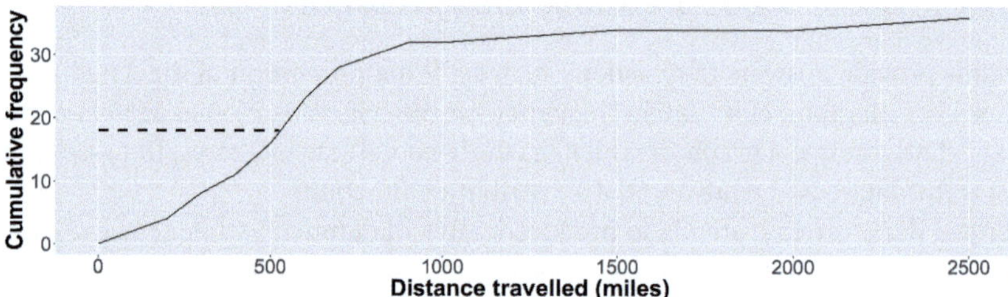

Figure 12.4 Cumulative frequency diagram of distances flown by 36 razorbills from their breeding colony. Half the birds flew further than 500 miles.

Rather than joining the cumulating frequencies directly, sometimes the data are presented as a series of steps creating a so-called **step diagram**. The steps help to emphasize that it is only at the end points of the distance intervals that the cumulative frequency is known for certain.

Example 12.4 (cont.)

Figure 12.5 shows the step diagram for the razorbill data.

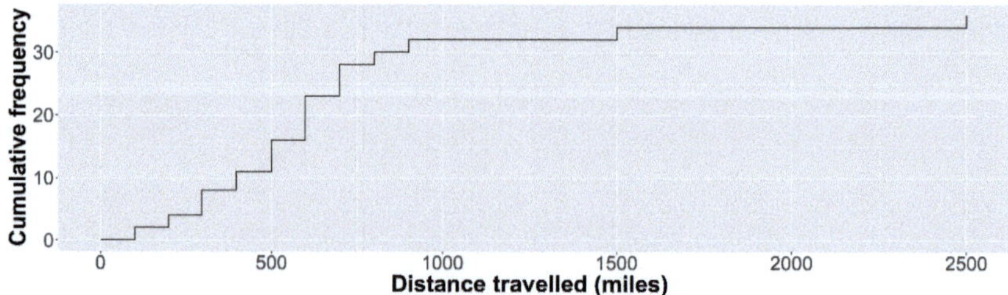

Figure 12.5 Step diagram of distances flown by 36 razorbills from their breeding colony.

12.2 Two variables

12.2.1 Multiple bar charts

When data occur naturally in groups and the aim is to contrast the variations within different groups a **multiple bar chart** may be used. This consists of groups of two or more adjacent bars separated from the next group by a gap having, ideally, a different width to the bars themselves. The diagram may be horizontal or vertical, with the values either specified on the diagram, or indicated using a standard axis.

Example 12.5

The following table shows estimates of the 2000 and 2020 populations (in millions) for five countries.

	France	Mexico	Nigeria	Pakistan	U. K.
2000	59	99	115	131	59
2020	65	128	217	207	67

Source: *Monthly Bulletin of Statistics Online*

The data show the differing rates of population growth of the two European nations and the three non-European countries and provide a graphic illustration of a world problem.

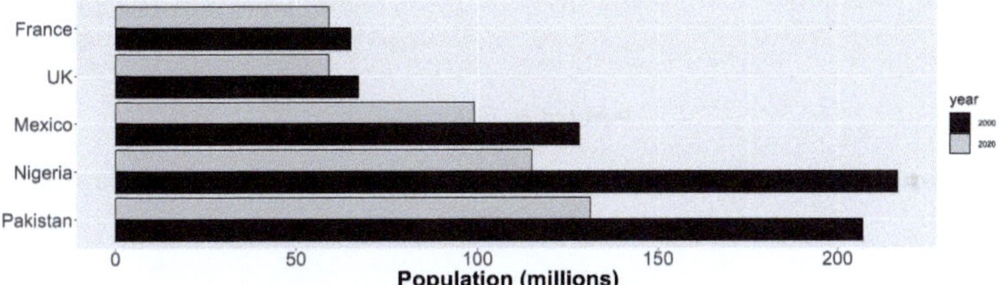

Figure 12.6 Multiple bar charts comparing populations in the years 2000 and 2020.

To increase visibility, in Figure 12.6, the countries are ordered in terms of their 2000 populations. The figure brings the numbers to life, showing the similarity between the two European countries. The population of Nigeria has nearly doubled in twenty years.

12.2.2 Compound bar charts for proportions

In a compound bar chart the length of a complete bar signifies 100% of the population. The bar is subdivided into sections that show the relative sizes of components of the populations. By comparing the sizes of the subdivisions of two parallel compound bars, differences can be seen between the compositions of the separate populations. The populations need not be populations of living creatures—they could be, for example, the populations of bricks in two builders' trucks!

Example 12.6

One consequence of the surge in population of the 'third world' countries is that there are many young people and relatively few old people. The World Bank[a] gives the following figures for the populations in 2020:

	France	Mexico	Nigeria	Pakistan	U. K.
% Under 15	18	26	43	35	18
% 15 to 64	62	67	54	61	64
% 65 and over	20	7	3	4	18

The data are conveniently presented in percentage form and, since comparisons are intended, a compound bar chart (Figure 12.7) is appropriate. The countries are reordered so as to best reflect the differences in the age structure.

[a] `https://data.worldbank.org/`

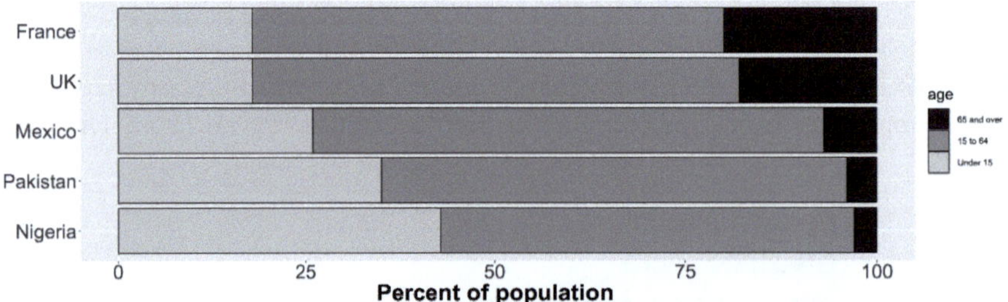

Figure 12.7 Compound bar charts contrasting the age structures of the populations of five countries.

Exercises 12c

1. The numbers of absentees in a class over a period of 24 days were:

0	3	1	2	1	0	4	0
1	1	2	3	1	0	0	1
2	4	6	4	2	1	0	1

Construct a bar chart for these data.

2. The table shows, for three neighbouring countries, the percentages of the population adhering to the main religions.

	Islam	Hindu	Buddhism	Christian	Other
Bangladesh	90.4	8.5	0.6	0.4	0.1
India	14.2	79.8	0.7	2.3	3.0
Pakistan	96.5	1.9	*	1.6	*

* indicates less than 0.1%

Display these results using a compound bar chart.

12.2.3 *Pie charts*

Pie charts provide an alternative to compound bar charts. The choice between them will be based on which form of presentation best brings out the feature of interest in the data.

Example 12.6 (cont.)

Figure 12.8 illustrates the age structure data of the previous example using pie charts. The similarity between the age structures of France and the United Kingdom is again very apparent, as is their difference from the age structure in the other countries considered. As a comparison of the age structures, compound bar charts and pie charts appear equally effective.

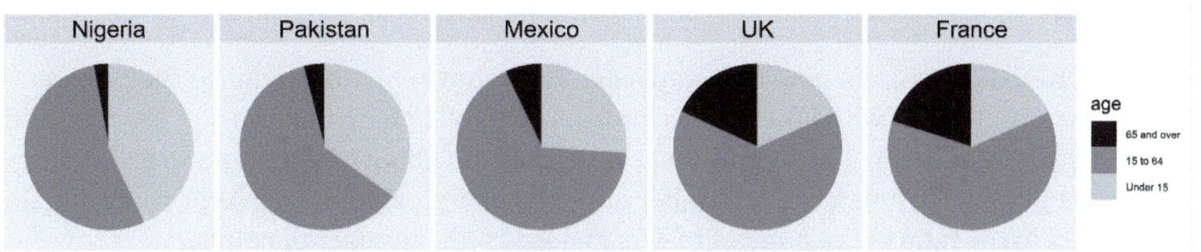

Figure 12.8 Pie charts comparing the populations of five countries.

12.2.4 *Population pyramids*

A **population pyramid** is used in examining age distributions. A typical pyramid consists of two multiple bar charts (one for males and one for females) placed back to back, with the bars referring to different age categories.

Example 12.7

The pyramids shown in Figure 12.9 contrast the age distribution of the United Kingdom and Nigeria. Each pyramid has the same area (representing 100% of the population): the contrast is between the age distributions rather than the population sizes.

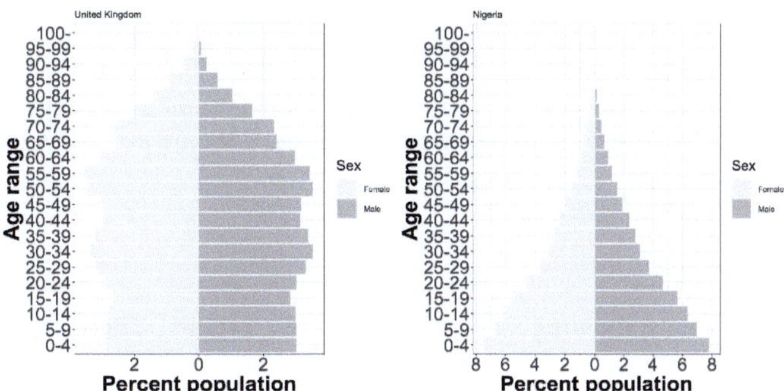

Figure 12.9 Population pyramids comparing the detailed age structures of the populations of the United Kingdom and Nigeria.

In the United Kingdom, the approximately constant birth rate and low mortality amongst the young, means that there are roughly equal numbers in each five-year age range up to the age of 60. In Nigeria with greater mortality rates, there are few people aged over 70. The data, which refer to 2021, were downloaded from the International Data Base of the United Census Bureau.

12.2.5 Time series

Time series graphs are probably the type of diagram most frequently encountered in newspapers. They are also possibly the most straightforward: time is plotted on the x-axis and the quantity of interest is plotted on the y-axis.

Example 12.8

Figure 12.10 uses data from `www.zap-map.com`. The top two sections refer to the numbers of electric cars (including hybrids) and the numbers of public charging points at the end of the calendar years 2016–2021. The numbers of cars grew from about 85 thousand to 650 thousand during this period. Fortunately the number of public charging points has kept pace. The final graph refers to the ratio of the number of cars divided by the number of points.

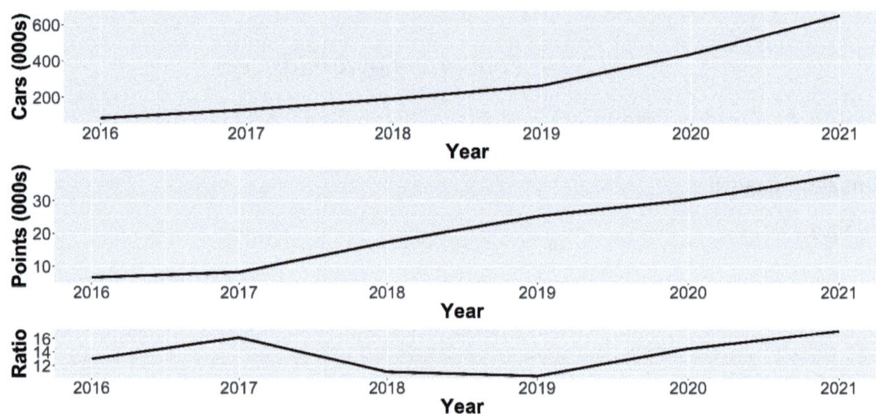

Figure 12.10 The numbers of electric cars (including hybrids) and public charging points at the ends of years 2016–2021. The final graph shows the numbers of cars per charging point.

> Beware advertisements showing time series that rise rapidly! There are two possible explanations:
>
> 1. The series was probably falling fast in the previous time period!
> 2. The vertical scale may be exaggerated—check where 0 would occur.

Exercises 12d

1. The numbers of persons per 1000 unmarried population who married in a specified year have decreased in recent years. The table gives figures for selected years.

	1992	1996	2000	2004	2008
England and Wales	36.2	30.9	27.8	27.6	22.2
Scotland	40.2	33.2	31.6	32.1	26.6

Display these two time series using a single graph.

12.2.6 Scatter diagrams

The time series graph was an example of using ordinary Cartesian coordinates to examine the relationship between two variables. Relationships between variables are particularly interesting since the variation in the values of one variable (x) may to some extent explain the variation in the other (y).

Time series data are ordered and their order is indicated on the plot by joining successive values. By contrast, in a scatter diagram, there is no order and the pairs of values are indicated by points (or crosses, or some other symbol).

Example 12.9

The following data relates soil erosion (in kg/day) to daily average wind velocity (in km/hr) in a region in the sandy plains of Rajasthan in India:

Wind velocity	13.5	13.5	14	15	17.5	19	20
Soil erosion	0	10	31	20	20	66	76
Wind velocity	21	22	23	25	25	26	
Soil erosion	137	71	122	188	300	239	

Figure 12.11 is a straightforward diagram with the x-coordinate indicating the daily average wind velocity and the y-coordinate shows the resulting estimated soil erosion (measured in kg/day).

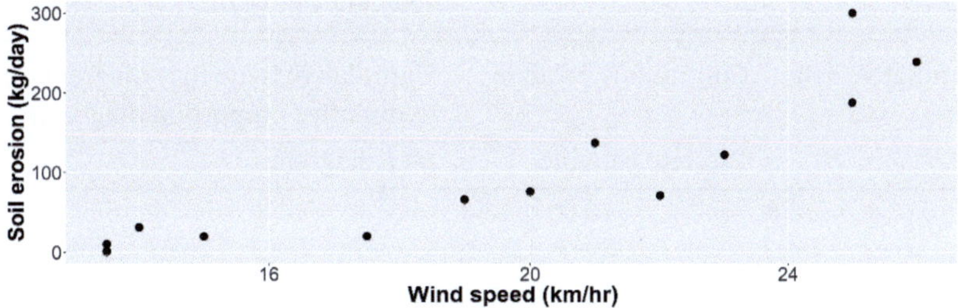

Figure 12.11 Soil erosion (in kg/day) plotted against wind velocity (in km/hr).

12.3 More than two variables

In practice there are often more than two variables of interest. Sometimes the most useful approach will be to look at the variables two at a time, using one of the displays in Section 12.2. With four variables (for example) that would mean $\binom{4}{2} = 6$ different diagrams. With more than four variables that would scarcely be a useful approach even if many of the displays proved to be uninformative and could be discarded.

With computers, however, it is relatively easy to produce displays that provide information on several variables simultaneously. The x- and y-coordinates, together with the sizes of plotting symbols, can illustrate three quantitative variables. These, taken together with varying colours and varying symbols for two categorical variables, provide a way of conveying information on five variables simultaneously on a single two-dimensional plot. For geographical data, superimposing information on a map can be very effective.[1]

12.4 Choosing which display to use

Sections 12.1 and 12.2 have presented a rather bewildering variety of types of diagram. The following table is intended to help you select the appropriate diagram for your data by identifying the diagrams most commonly associated with different types of data.

Data type		*Type of display*
Discrete data	Few different values	Tally chart for counting, bar chart for display
	Many different values	Stem-and-leaf diagram, or histogram, or boxplot (Section 13.9)
Qualitative data	To show frequencies	Bar chart
	To show proportions	Pie chart
Grouped data	Unequal intervals	Histogram
	Equal intervals	Histogram, or frequency polygon
Cumulative data	Continuous variable	Cumulative frequency diagrams, or cumulative proportion diagrams
	Discrete variable	Step diagram

[1] Anyone interestied in graphical displays is strongly advised to study one of the books by Edward Tufte, particularly *The Visual Display of Quantitative Information*.

Data type		*Type of display*
Two samples	Same categories	Multiple bar chart, or composite bar chart, or side-by-side pie charts
	Discrete	Back-to-back stem-and-leaf, or side-by-side bar charts, or multiple bar chart
	Continuous	Cumulative frequency diagram, or cumulative proportion diagram
Two variables	One variable is time	Time series
	Data in pairs	Scatter diagram

When reporting data, either in a table or as individual values, try to organize the units so that the figures presented are simple integers. (This is the idea that underlies the stem-and-leaf diagram.) Here are some examples:

Context	*A bad choice*	*A good choice*
Masses of flour bags	Report in kg 1.004, 1.032, 1.040, 1.011	Report in g the excess over 1kg 4, 32, 40, 11
Year of millenium	2020, 2005, 2006, 2012	20, 5, 6, 12
Average exam marks	With irrelevant accuracy 53.377, 62.401, 15.822	To nearest integer 53, 62, 16

12.5 Dirty data

It is almost certain that a large set of data will contain errors! This may seem rather pessimistic but remember that 'To err is human ...'. Here are some examples of the way that errors arise.

- **Mistype** The correct value was obtained, but it was written down or typed incorrectly. The most common errors are digits interchanged (1183 instead of 1138) and a digit double-typed (993 instead of 93).
- **Mistaken answer** An interviewee misunderstands a question and gives the wrong answer. For example, an individual earning £24,000 a year replies £24,000 when asked to state his monthly earnings.
- **Mistaken measurement** There is an innate preference for 'nice' round numbers. Suppose that a large collection of pebbles with masses in the range 20g to 60g are weighed. The values reported are likely to show pronounced peaks at 30g, 40g, 50g as well as lesser peaks at 35g, 45g, etc.
- **Mistaken rounding** Values are frequently rounded. This causes a problem with 'halves'. For example, how do you round 3.465 to two decimal places, or round 4.5 to the 'nearest' integer? A bad rule is to always round in the same direction, since this will bias totals and averages. A better rule is to always round to an even digit—thus 3.465 becomes 3.46 and 4.5 becomes 4.

- **Misreporting** In one famous example of misreporting, the specific gravity of beer barrels was being determined. Barrels with a specific gravity greater than g had an acceptable alcohol quantity and were rolled down hill. However, barrels with a specific gravity less than g had to be rolled uphill for further treatment. When the data were examined it was found that there were remarkably few barrels with specific gravity just below g! The measurer had taken a 'favourable view' of his measuring instrument.

- **Biased sampling** The sample may misrepresent the population as a consequence of a poor sampling procedure (such as only interviewing customers leaving a particular supermarket concerning their shopping preferences, as opposed to a wider survey).

The first job of a statistician, with any data set, is to examine it using pictures and summary statistics (Chapter 13) in a search for mistaken data items. Even when there are no errors in the data there are often unusual values. These are called **outliers** (Section 13.8) and their presence can cause problems for the subsequent data analysis.

R

With the data in the vector `x`:
Frequency table: `table(x)`.

The figures in this book have been created using ggplot2, which is an open-source data visualization package that gives publication-quality graphics. The commands given below are the relatively simple commands that enable the data analyst to study the data, without ornamentation.

Stem-and-leaf plot: `stem(x)`. Experiment! You may need to add the option `(scale=n)` (where $n = 2$ was needed for the cricket scores).
Bar chart: (including multiple and compound bars) Use `barplot(x)`. There are many options available—see the examples provided in *RStudio*.
Pie chart: Of course: `pie(x)`.
Histogram: The basic command is `hist(x)`. Experiment with options.
Population pyramid: First, download the `DescTools` library. Now type `library(DescTools)`. Then use `PlotPyramid(lx,rx)` where `lx` and `rx` are vectors containing the data for the left and right sides of the pyramid.

13

General Summary Statistics

The main purpose of the subject statistics is to draw conclusions about a (usually large) **population** from a (usually small) **sample** of **observed values**: the **observations**. In this chapter we study various ways of providing numerical summaries of the observations. For **univariate** data (i.e. data concerned with a single quantity) there are two main classes of summary statistic:

- **Measures of location**
  These answer the question 'What size are the values that we are talking about?'.
- **Measures of spread**
  These answer the question 'To what extent do the values vary?'.

Both are discussed in this chapter. We start with some simple measures of location.

13.1 Measure of location: the mode

The **mode** of a set of discrete data is the single value that occurs most frequently. This is simple but it can be misleading.

Example 13.1

At the supermarket I buy 8 tins of soup. Four tins have mass 400 g, three have mass 425 g, and one has mass 435 g. Determine the mode.

The mode is 400 g because 400 g is the most common value. However, since this is also the smallest value it is rather misleading.

If there are two equally common outcomes, then the data would be described as being **bimodal**; if there were three or more such outcomes, then the data would be called **multimodal**.

13.2 Measure of location: the mean

This measure of location, more formally referred to as the **sample mean**, is often called the **average**. It can be used with both discrete and continuous data. The mean is equal to the sum of all the observed values divided by the total number of observations. The value of the mean will usually not be equal to any one of the individual observed values.

Example 13.1 (cont.)

The mean mass (in g) of the tins of soup is

$$\frac{1}{8}(400 + 400 + 400 + 400 + 425 + 425 + 425 + 435) = 413.75.$$

In general, if the data set consists of n observed values, denoted by $x_1, x_2, \dots, x_n$, then the mean, which is usually denoted by $\bar{x}$, is given by

$$\bar{x} = \frac{1}{n}(x_1 + x_2 + \cdots + x_n). \tag{13.1}$$

One way of thinking about the mean is as the *centre of mass* when the observations are 'balanced' on the x-axis.

Example 13.1 (cont.)

For the soup tins, the mean is 413.75, which is the centre of gravity or balance point for the arrangement in Figure 13.1.

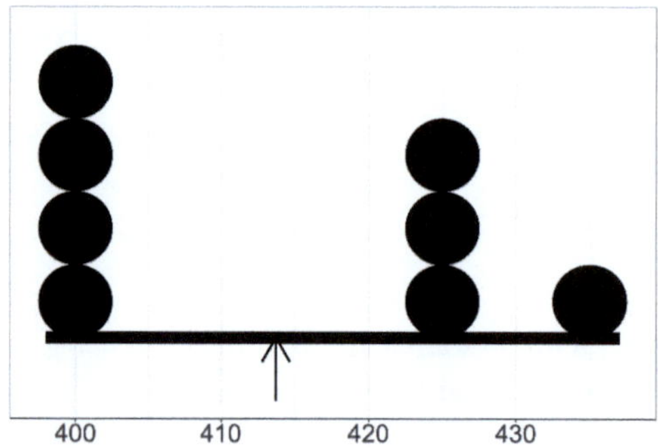

Figure 13.1 The mean is the balance point.

Exercises 13a

1. Find the mode and mean of the following numbers:

$$1, 1, 1, 3, 5, 6, 7, 9, 9, 11, 15, 24, 25.$$

2. Explain why most people have more than the mean number of legs.

Group project: How many four-legged pets does the typical family have? Use a tally chart to record the combined number of dogs, cats, hamsters, etc., for each member of your group. Determine the mean, median, and mode of these values. Which was easiest to calculate? An organization wishes to estimate the total number of four-legged pets in your area. Which of your three statistics is likely to be most useful to them?

13.3 Measure of location: the mean of a frequency distribution

We have seen in Chapter 12 that data are often represented by a frequency distribution. For example, for the soup tins we have

Reported mass (g)	x	400	425	435
Observed frequency	f	4	3	1

The sum of the frequencies $(4 + 3 + 1)$ is equal to n, the total number of observations. The sum of the three products 4×400, 3×425 and 1×435 is equal to the sum of the eight observations, and so the mean mass is $(1600 + 1275 + 435)/(4 + 3 + 1) = 413.75$ g, as before. All that we have done is to collect together equal values of x. So, using sigma notation (Section 1.9), an alternative general formula for the mean is

$$\bar{x} = \sum_{j=1}^{m} f_j x_j \bigg/ \sum_{j=1}^{m} f_j \,,$$

where here the summation is over the m different values of x that were recorded.[1]

Since $\sum_{j=1}^{m} f_j$ equals n, the total number of observations, a simpler form for the previous formula is

$$\bar{x} = \frac{1}{n}\Sigma f_j x_j, \tag{13.2}$$

which might be written as $\frac{1}{n}\Sigma fx$.

13.4 Measure of location: the mean of grouped data

The formula for the mean of a frequency distribution can also be used to provide an estimate of the sample mean of a set of grouped data:

$$\bar{x} = \frac{1}{n}\sum f_j x_j.$$

In this case x_j is the **class mid-point** for the jth of m classes, f_j is the frequency for this class, and $n = \sum_{j=1}^{m} f_j$. This is only an estimate of the actual sample mean since we do not know the individual sample values.

[1] In the example, $m = 3$, $x_1 = 400$, $x_2 = 425$, $x_3 = 435$, $f_1 = 4$, $f_2 = 3$, and $f_3 = 1$.

Example 13.2

The following data summarize the distances travelled by a fleet of 190 buses before experiencing a major breakdown.

Distance ('000 miles) (d)	$d \leq 60$	$60 < d \leq 80$	$80 < d \leq 100$
Mid-point (x)	30	70	90
Frequency (f)	32	25	34

Distance ('000 miles) (d)	$100 < d \leq 120$	$120 < d \leq 140$	$140 < d \leq 220$
Mid-point (x)	110	130	180
Frequency (f)	46	33	20

An estimate of the total distance travelled by the 32 buses in the first category is $32 \times 30,000 = 960,000$ miles. Repeating for each of the classes, the overall estimate of the total mileage is $\sum f_j x_j = 18,720,000$, and hence the grouped mean, $\frac{1}{n} \sum f_j x_j$, is about 98,500 miles.

13.5 Simplifying calculations

Sometimes, by applying a simple transformation to every data item, calculations can be simplified making it is easier to 'see' the data. It also implies fewer buttons to push whether using a calculator or computer.

Example 13.3

Suppose the data are

$$3001, 3003, 3005, 3005, 3007, 3007, 3007, 3009.$$

We could calculate

$$\{3001 + 3003 + (2 \times 3005) + (3 \times 3007) + 3009\}/8 = 3005.5,$$

but, since every observation includes 3000, it is much easier to think of the data as

$$1, 3, 5, 5, 7, 7, 7, 9.$$

The mean of these numbers is $44/8 = 5.5$. So, adding back the 3000, the mean of the original data is 3005.5.

Example 13.4

Consider the problem of finding the mean of

$$0.00001, 0.00003, 0.00005, 0.00005, 0.00007, 0.00007, 0.00007, 0.00009.$$

We could calculate

$$\{0.00001 + 0.00003 + (2 \times 0.00005) + (3 \times 0.00007) + 0.00009\}/8 = 0.000055,$$

but it is much easier to calculate:

$$0.00001 \times \{1 + 3 + (2 \times 5) + (3 \times 7) + 9\}/8 = 0.000055.$$

Example 13.5

A collection of items have the following prices (in £):

$$49.95, 79.95, 79.95, 99.95, 139.95.$$

Adding 5p to each price makes them much easier to deal with:

$$50, 80, 80, 100, 140.$$

These revised prices sum to £450, so their average is £90. Since every original price was 5 pence less, the average of the original prices is £90 − 0.05 = £89.95

Exercises 13b

1. One day an amateur meteorologist records the atmospheric pressure on each hour from 6 a. m. to 9 p. m.. The results were

$$1019, 1020, 1016, 1016, 1015, 1015, 1010, 1008,$$

$$1007, 1007, 1007, 1005, 1005, 1004, 1003, 1003$$

Find the mean of these observations.

2. The gaps, x mm, in a sample of 13 spark plugs were as follows:

$$0.81, 0.83, 0.81, 0.81, 0.82, 0.81, 0.83, 0.84, 0.81, 0.82, 0.84, 0.81, 0.82$$

Find the mean gap.

13.6 Measure of location: the median

The mean can be misleading if the data contains one or more **outliers** (values that are much larger or smaller than the majority of values in the data). For example, if the data consists of the values

$$5, 7, 8, 9, 11, 13, 15, 17, 95,$$

then the mean is 20, which is greater than all but one of the nine values. In this case (assuming that the value 95 is a correct value), then it might be appropriate to report the median, which is the value that subdivides the ordered data into two halves.

When there is an odd number of observations, the median is not included in either half.

When there is an even number of observations, the median is the average of the two central observations.

Example 13.6

The numbers of words in the 18 sentences of Chapter 1 of *A Tale of Two Cities* by Charles Dickens are as follows:

118, 39, 27, 13, 49, 35, 51, 29, 68, 54, 58, 42, 16, 221, 80, 25, 41, 33,

whilst the numbers of words in the first 17 sentences of Chapter 1 of *Not a Penny More, Not a Penny Less* by Jeffrey Archer are as follows:

8, 10, 15, 13, 32, 25, 14, 16, 32, 25, 5, 34, 36, 19, 20, 37, 19.

Rearranging the Dickens data in order of magnitude we get:

(13 16 25 27 29 33 35 39 41) (42 49 51 54 58 68 80 118 221)

M

The median, denoted here by M, is 41.5.

Rearranging the Archer data in order we get:

(5 8 10 13 14 15 16 19) 19 (20 25 25 32 32 34 36 37)

M

The median for the much shorter Archer sentences is 19.

13.7 Quantiles

Quantiles are cut points dividing the ordered observations in a sample, or the values in a population, into equal fractions of the whole. The number of cut points is always one less than the number of sections created.

The simplest example is the median, which divides the ordered sample or population into halves. Other examples are the three **quartiles** which divide into quarters, the nine **deciles** which divide into tenths, and the ninety-nine **percentiles** which divide into hundredths.

> The lower quartile (Q_1) is the median of the lower half.
>
> The upper quartile (Q_3) is the median of the upper half.

The median itself is therefore the 2nd quartile, the 5th decile and the 50th percentile.

Example 13.6 (cont.)

For the Dickens data

(13 16 25 27 29 33 35 39 41) (42 49 51 54 58 68 80 118 221)

Q_1 (29), Q_2 (between 41 and 42), Q_3 (58)

the lower quartile is 29 and the upper quartile is 58.

For the Archer data

(5 8 10 13 14 15 16 19) 19 (20 25 25 32 32 34 36 37)

Q_1 (between 13 and 14), Q_2 (19), Q_3 (between 32 and 32)

the quartiles are 13.5 and 32.

13.8 Measures of spread: the range and the inter-quartile range

The **range** of a set of numerical data is the difference between the highest and lowest values. The range can be misleading if there is an outlier.

The **interquartile range (IQR)** is the difference between the upper and lower quartiles: $Q_U - Q_L$.

Example 13.6 (cont.)

For the Dickens data the range is enormous, ($221 - 13 = 208$), because of the two outlier values (118 and 221). The interquartile range is $Q_U - Q_L = 58 - 29 = 29$.

Exercises 13c

1. One day an amateur meteorologist records the atmospheric pressure on each hour from 6 a. m. to 9 p. m.. The results were

$$1019, 1020, 1016, 1016, 1015, 1015, 1010, 1008,$$

$$1007, 1007, 1007, 1005, 1005, 1004, 1003, 1003$$

Find the median and upper and lower quartiles.

2. The gaps, x mm, in a sample of 13 spark plugs were as follows:

$$0.81, 0.83, 0.81, 0.81, 0.82, 0.81, 0.83, 0.84, 0.81, 0.82, 0.84, 0.81, 0.82$$

Find the median and upper and lower quartiles.

13.9 Boxplot

A boxplot (also known as a **box-whisker diagram**) provides an illustration of the data based on the values of the quartiles.

The diagram consists of a central box with a 'whisker' at each end. The ends of the central box are the lower and upper quartiles, with either a central line or a notch indicating the value of the median. The simplest form of the diagram has two lines (the whiskers): one joining the lowest value to the lower quartile, and the other joining the highest value to the upper quartile. An alternative presentation indicates outlier values separately. Boxplots provide a convenient way of comparing two or more sets of values.

Example 13.6 (cont.)

Figure 13.2 shows boxplots comparing the sentence lengths of the two authors. In this version the whisker lengths can be no more than 1.5 times the inter-quartile range, with outliers indicated individually.

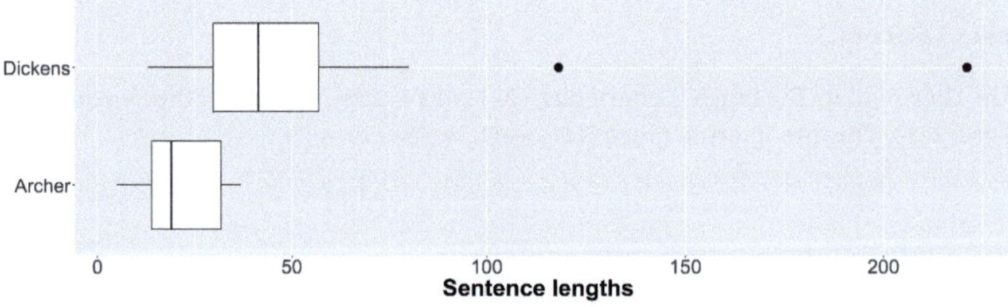

Figure 13.2 Boxplots comparing the sentence lengths of sentences in the first chapters of works by Charles Dickens and Jeffrey Archer.

Put theory into practice: Choose two of your own favourite authors and repeat the Dickens/Archer experiment. Try and choose authors whose styles you think may be different. Choose descriptive passages rather than passages of dialogue, but don't choose them because they seem to have particularly long (or short) sentences, or you will bias the results! Construct box-whisker diagrams or refined boxplots for each author. Do there seem to be differences in the two distributions of sentence length?

Put theory into practice: What do you suppose affects the price of a second-hand car? Three obvious variables are the size of the car, its age, and how far it has travelled. Choose two different makes of cars that offer models of similar size. Use an on-line list of second-hand cars and, choosing models of similar age, find the prices of at least 10 examples of these models. Display the data on side-by-side boxplots.

13.10 Deviations from the mean

Compare these two sets of data:

Set 1:	0	99	99	100	100	100	100	100	101	101	200
Set 2:	0	0	99	99	100	100	100	101	101	200	200.

The two sets have the same minimum and maximum, the same quartiles (99 and 101) and both sets have mean, median, and mode all equal to 100.

However, the second set of data has four extreme observations, compared to only two in the first set. We can make the difference between the two sets more apparent by calculating the differences between each observations and the mean of the set to which it belongs (in this case, 100 for each set). These deviations from the mean are

Set 1 :	−100	−1	−1	0	0	0	0	0	1	1	100
Set 2 :	−100	−100	−1	−1	0	0	0	1	1	100	100

In each case the differences sum to 0. For a set of n observations $x_1, \dots, x_n$ with sample mean $\bar{x}$, given by $n\bar{x} = \sum x_i$, this always happens, since

$$\begin{aligned}\sum_{i=1}^{n}(x_i - \bar{x}) &= (x_1 - \bar{x}) + \cdots + (x_n - \bar{x}) \\ &= (x_1 + \cdots + x_n) - (\bar{x} + \cdots + \bar{x}) \\ &= \sum_{i=1}^{n} x_i - n\bar{x} \\ &= n\bar{x} - n\bar{x} \\ &= 0.\end{aligned}$$

13.11 The mean deviation

If we ignore the signs of the differences between the observations and their mean, and work with absolute values (moduli), then a natural measure of spread is provided by the average value of the differences. This is called the **mean deviation** or **mean absolute deviation** (MAD):

$$\frac{1}{n}\sum |x_i - \bar{x}|.$$

For the two sets of data in Section 13.10, the mean deviations are $204/11 = 18.5$ and $404/11 = 36.7$. The extra variability of the second set leads to a larger value of the mean deviation.

Despite its apparent simplicity the mean deviation is little used because its use of absolute values makes the subsequent mathematics difficult.

13.12 Measure of spread: the variance

The preferable alternative is to work with the sum of the squares of the deviations from the mean:

$$\sum(x_i - \bar{x})^2 = (x_1 - \bar{x})^2 + \cdots + (x_n - \bar{x})^2.$$

The more variation there is in the x-values, the larger will be the value of $\sum(x_i - \bar{x})^2$. However, since the sum might be large only because of the number of x-values, we need to take that number into account by using s^2 given by

$$s^2 = \frac{1}{n-1}\sum_{i=1}^{n}(x_i - \bar{x})^2. \tag{13.3}$$

The $(n-1)$ divisor is appropriate[2] when the values $x_1, \dots, x_n$ are a sample from a population with unknown variance, σ^2. We describe s^2 as the **sample variance** and σ^2 as the **population variance**. When there is no chance of confusion, either may be referred to simply as the **variance**.

Since s^2 is a positive multiple of a sum of squares:

- it cannot be negative,
- it has units which are the square of the units of x.

13.13 Calculating the variance by hand

Usually $\bar{x}$ will be an awkward decimal. Therefore, when calculating a variance by hand, it is both simpler and more accurate to use the equivalent form

$$s^2 = \frac{1}{n-1}\left\{\sum x_i^2 - \frac{1}{n}\left(\sum x_i\right)^2\right\}. \tag{13.4}$$

The proof of the equivalence requires some messy algebra:

[2] This is discussed at length in Section 14.1.1.

$$\begin{aligned}\sum(x_i - \bar{x})^2 &= \sum(x_i^2 - 2x_i\bar{x} + \bar{x}^2)\\ &= \sum x_i^2 - 2\bar{x}\sum x_i + n\bar{x}^2\\ &= \sum x_i^2 - \frac{2}{n}\sum x_i \times \sum x_i + n\left(\frac{1}{n}\sum x_i\right)^2\\ &= \sum x_i^2 - \frac{1}{n}\left(\sum x_i\right)^2.\end{aligned}$$

In all important cases the quantities $\sum x_i^2$ and $\left(\sum x_i\right)^2$ are *not* equal, since

$$\sum x_i^2 = x_1^2 + x_2^2 + \cdots + x_n^2,$$

whereas

$$\begin{aligned}\left(\sum x_i\right)^2 &= (x_1 + x_2 + \cdots + x_n)^2\\ &= (x_1^2 + x_2^2 + \cdots + x_n^2) + 2(x_1x_2 + x_1x_3 + \cdots + x_{n-1}x_n).\end{aligned}$$

13.14 Measure of spread: the standard deviation

The standard deviation is the square root of the variance. It is often quoted instead of the variance, because its units are the same as the units of the original observations. For example, if x is a weight in kg, then the standard deviation of the values of x is also a weight in kg.

Example 13.7

The nine planets of the Solar System have approximate equatorial diameters (in thousands of kms) as follows:

4.9, 12.1, 12.8, 6.8, 143.0, 120.5, 51.1, 49.5, 2.4

To determine the standard deviation of these diameters by hand we first calculate

$$\sum x_i = (4.9 + \cdots + 2.4) = 403.1 \text{ and } \sum x_i^2 = (4.9^2 + \cdots + 2.4^2) = 40,416.97.$$

Notice that this is not a sample; it is the entire population. When calculating σ^2, the variance of an entire population, we are working with the population mean rather than a sample mean. A consequence is that the fraction $1/(n-1)$ must be replaced by $1/n$. In this case, therefore,

$$\sigma = \sqrt{\frac{1}{n}\left\{\sum x_i^2 - \frac{1}{n}\left(\sum x_i\right)^2\right\}} = \sqrt{\frac{1}{9}\left(40,416.97 - \frac{403.1^2}{9}\right)} = 49.85 \text{ (thousand km)}.$$

We have found that the standard deviation of the diameters for the population of nine planets of the Solar System is about 50 thousand km.

When reporting results, a reasonably accurate value (e.g. 49.85) should be available, but the description (50) of the result should be as simple as possible.

On no account quote the result from your computer or calculator (49.84706), since the original data are nowhere near that accurate.

Example 13.8

An office manager wishes to get an idea of the number of phone calls received by the office during a typical day. A week is chosen at random and the numbers of calls on each day of the (5-day) week are recorded. They were as follows:

$$15, 23, 19, 31, 22.$$

In this case the divisor for the variance is $(n - 1) = 4$, since it is not the variance of this week's data that is of interest. This week is being used as an example of what might be expected in other weeks.

Here $\sum x = 110$ and $\sum x^2 = 2560$. The mean is $110/5 = 22$. Using the divisor $(5 - 1) = 4$, the standard deviation s is 5.92 (for simplicity in later calculations we might report this as 6).

13.14.1 Approximate properties of the standard deviation

Providing the sample size is reasonably large, and the data are not too skewed (i.e. there is not a long 'tail' of very large or very small values), it is possible to make some approximate statements which are based on the theory covered in Chapter 6:

About two-thirds of the individual observations will lie within one standard deviation of the sample mean.

About 95% of the individual observations will lie within two standard deviations of the sample mean.

Almost all the data will lie within three standard deviations of the sample mean.

Example 13.8 (cont.)

The mean number of calls per day was 22 and the standard deviation was 5.92.

The office manager can conclude that on about two-thirds of days the office will receive between $22 - 6 = 16$ and $22 + 6 = 28$ calls (there is no point in using great precision since these are only very crude approximations).

Assuming that the week sampled was typical, the office is unlikely ever to receive less than $22 - (3 \times 6) = 4$ calls, or more than $22 + (3 \times 6) = 40$ calls.

Here the sample is very small, so we cannot place too much reliance on our approximations.

Exercises 13d

1. The numbers of books of stamps bought at a post office on five randomly chosen weekdays were 15, 9, 23, 12, 17. Find the mean and standard deviation of this sample. Suggest bounds (based on your results) for the number of books of stamps bought on another weekday.
2. One day Mr I. Walton, an angler, was fishing in his favourite river. He caught six fish. Their masses, in kilos, were 1.35, 0.87, 1.61, 1.24, 0.95, 1.87. Assuming that these fish constituted a random sample from the fish in the river, deduce bounds for the mass of the next fish that he might catch from that river.

13.15 Variance and standard deviation for frequency distributions

When data have been summarized in the form

'the value x_i occurs with frequency f_i,'

the formula for the variance needs rewriting. With m distinct values of x, the formula becomes

$$s^2 = \frac{1}{n-1}\left\{\sum_{j=1}^{m} f_j x_j^2 - \frac{1}{n}\left(\sum_{j=1}^{m} f_j x_j\right)^2\right\}, \tag{13.5}$$

where n is the total of the individual frequencies:

$$n = \sum_{j=1}^{m} f_j.$$

. As before, the sample standard deviation is simply the square root of the sample variance. The divisor $n - 1$ is replaced by n if the values represent the entire population.

The same revised formulae are used when working with grouped data. In this case the x-values are the mid-points of the class intervals and the f-values are the class frequencies. The value obtained will usually be a slight underestimate of the true sample variance.

13.16 Symmetric and skewed data

If a population is approximately **symmetric**, then a reasonable-sized sample will have mean and median having similar values. Typically their values will also be close to that of the mode of the population (if there is one!).

A population that is not symmetric is said to be **skewed**. A distribution with a long 'tail' of high values is said to be **positively skewed**, in which case the mean is usually greater than the mode or the median. If there is a long tail of low values then the mean is likely to be the lowest of the three location measures and the distribution is said to be **negatively skewed**

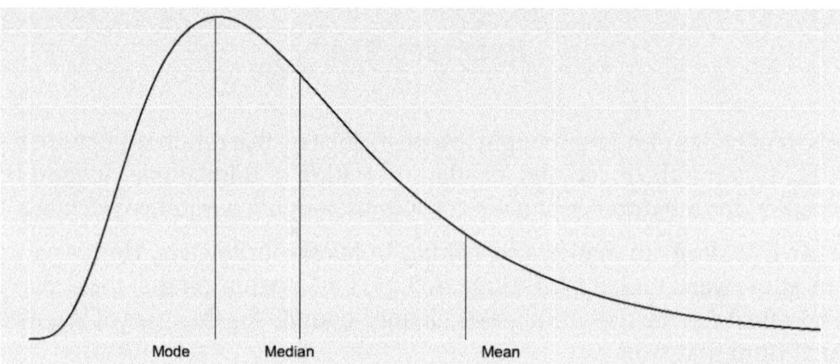

13.17 Standardizing to a prescribed mean and standard deviation

By careful choice of the constants a and b, where $b > 0$, it is always possible to use the coding $y = (x - a)/b$, to transform the original x-values to new y-values having some predetermined mean and standard deviation (which we will denote by $\bar{y}$ and σ_y). Let the mean and standard deviation of the x-values be $\bar{x}$ and σ_x, respectively. The required values are

$$b = \sigma_x/\sigma_y,$$

and

$$a = \bar{x} - b\bar{y},$$

so that, the revised value y corresponding to an original value x is given by the equation

$$y = \bar{y} + \frac{\sigma_y}{\sigma_x}(x - \bar{x}).$$

An equivalent expression, which presents the original and standardized values in a pleasingly symmetric form, is

$$\frac{y - \bar{y}}{\sigma_y} = \frac{x - \bar{x}}{\sigma_x}.$$

These results may be easily obtained by using the formulae $\bar{x} = a + b\bar{y}$ and $\sigma_x^2 = b^2\sigma_y^2$.

Example 13.9

The mean and standard deviation of a set of exam marks were found to be 40.06 and 15.32, respectively. There is a policy that all exams should have mean 50 and standard deviation 12. Determine the necessary transformation.

Here x is the original exam mark and y is the required mark. The transformation required therefore has $b = 15.32/12 = 1.277$ and $a = 40.06 - (1.277 \times 50) = -23.77$. An original mark of 80 is transformed to a new mark of $50 + (80 - 40.06)/1.277$ which is equal to 81 (to the nearest whole number).

13.18 *Calculating the combined mean and variance of several samples

Sometimes we have information in the form of the sample size, the sample mean, and the sample variance for each of several independent samples. We wish to amalgamate the information so as to discover the overall mean and variance of the combined set of data. We illustrate the calculations for the case of two samples having sample sizes n_1 and n_2, sample means $\bar{x}_1$ and $\bar{x}_2$, and sample variances $s_{n_1}^2$ and $s_{n_2}^2$.

The sum of the n_1 observed values in the first sample is $n_1\bar{x}_1$ and the sum of the n_2 observed values in the second sample is $n_2\bar{x}_2$, so that the overall sum of the two sets of observed values is $n_1\bar{x}_1 + n_2\bar{x}_2$. If we denote the overall mean by $\bar{x}$ and the combined sample size by n, then the overall mean is given by

$$\bar{x} = (n_1\bar{x}_1 + n_2\bar{x}_2)/n.$$

With k samples this formula generalizes to

$$\bar{x} = \sum_{j=1}^{k} n_j\bar{x}_j/n, \quad (13.6)$$

where $\bar{x}_j$ is the mean of the jth sample, n_j is the size of the jth sample, and $n = \Sigma n_j$.

In order to calculate s_n^2, the variance of the combined sample, it is necessary first to calculate the combined sum of squares of the observed values. The general formula for a sample variance for a single sample of size n having observations $x_1, \ldots, x_n$ and sample mean $\bar{x}$ is given by the equation

$$s_n^2 = \frac{1}{n-1}(\Sigma x_i^2 - n\bar{x}^2),$$

which can be rearranged in the form

$$\Sigma x_i^2 = (n-1)s_n^2 + n\bar{x}^2.$$

Thus, in the case of two samples, with sample variances $s_{n_1}^2$ and $s_{n_2}^2$, the total sum of squares, T, is given by

$$T = \frac{1}{n_1-1}(\Sigma x_i^2 - n_1\bar{x}^2) + \frac{1}{n_2-1}(\Sigma x_i^2 - n_2\bar{x}^2).$$

For the case of k samples this generalizes to

$$T = \sum_{j=1}^{k} \{(n_j - 1)s_{n_j}^2 + n_j \bar{x}_j^2)\}.$$

The variance of the combined sample is therefore

$$s_n^2 = \frac{1}{n-1}\left\{T - \frac{1}{n}\left(\sum_{j=1}^{k} n_j \bar{x}_j\right)^2\right\}.$$

13.19 Combining proportions

Suppose that we are told that, in a certain population (consisting only of middle class and working class families), 54% of middle class families have a video recorder, whereas the proportion in working class families is just 14%. Without knowledge of the relative sizes of the two classes, all that we can say about the overall proportion of families that have a video recorder is that it lies in the range 14% to 54%.

If we are also told that 63% of all families are middle class with the remainder being working class, then we can be more precise! Suppose there are n families in the population:

Class	*Number of families*		*Proportion with a video recorder*		*Number with a video recorder*
Middle	$0.63n$	×	0.54	=	$0.3402n$
Working	$0.37n$	×	0.14	=	$0.0518n$
Total	n	×	?	=	$0.3920n$

The overall proportion with a video recorder is therefore just short of 40%.

Key facts

- **Mode** The most frequent value.
- **Median** The central point of values ordered in magnitude.
- **Mean** The average value.
- **Quartiles** The median values of the lower and upper halves of ordered values.
- **Range** The greatest value minus the smallest value.
- **Boxplot** The central box bounded by the quartiles with the median indicated. Whiskers to outermost values (or limited, with outliers individually indicated).
- **Variance** $s^2 = \sum(x - \bar{x})^2 / (n - 1)$.
- **Standard deviation** s.

R

With the data in the vector `x`:
For the **mode** a simple approach could begin with `sort(table(x))`.
For the **median** the command `median(x)` can be used, but if **quartiles** are also required then use `quantile(x,c(.25,.50,.75))`.
For the **mean** use `mean(x)`.

Note that all of the mean, median, and quartiles, together with the minimum and maximum values, are provided by using the command `summary(x)`.

For the **inter-quartile range** use `IQR(x)`.
For a **boxplot** use `boxplot(x)`. There are many options available to tweak the output.
For the **variance** and **standard deviation** use `var(x)` and `sd(x)`.

14

Point and interval estimation

14.1 Point estimates

A **point estimate** is a numerical value, calculated from a set of data, which is used as an estimate of an unknown parameter in a population. The random variable corresponding to an estimate is known as the **estimator**. The most familiar examples of point estimates are

the sample mean, $\bar{x}$,	used as an estimate of the population mean, μ;
the sample proportion, r/n,	used as an estimate of the population proportion, p;
$s^2 = \sum(x_i - \bar{x})^2/(n-1)$	used as an estimate of the population variance, σ^2.

These three estimates are very natural ones and they also have a desirable property that is the subject of Section 14.1.1: the expected value of the sample value is exactly equal to the population value.

14.1.1 Unbiasedness

Suppose that θ is some population parameter (e.g. μ) with an unknown value. Let U be an estimator of this parameter. Ideally $\text{E}(U) = \theta$. If this is true, then U is said to be an **unbiased** estimator. If instead, $\text{E}(U) = \theta + b$, then b would be termed the **bias** of the estimator.

All three of the estimators in this chapter's introduction are unbiased: $\text{E}(\bar{X}) = \mu$, $\text{E}(R/n) = p$, and $\text{E}(S^2) = \sigma^2$, where $\bar{X}$, R, and S^2 are the random variables corresponding to $\bar{x}$, r, and s^2, respectively.

Sometimes the bias, b, is a function of the sample size, n, and often it may reduce to 0 as the sample size increases to infinity. In this case the estimator is described as being **asymptotically unbiased**.

Example 14.1

> *A random sample of* n *observations,* x_1, x_2, ..., x_n, *has mean* $\bar{x}$. *The sample has been taken from a population with mean* μ. *Show that* $\bar{x}$ *is an unbiased estimate of* μ.

Corresponding to the n observations are the random variables $X_1, X_2, \ldots, X_n$, with $\bar{X} = \frac{1}{n}\sum_i X_i$. Each X-variable has expectation μ, so that

$$\mathrm{E}(\bar{X}) = \frac{1}{n}\sum_i \mu = \frac{1}{n} \times n\mu = \mu.$$

Thus $\bar{x}$ is an unbiased estimate of μ.

Example 14.2

> *A random sample of* n *observations,* x_1, x_2, ..., x_n, *has mean* $\bar{x}$ *and variance* $s^2 = \frac{1}{n-1}\sum_i(x_i-\bar{x})^2$. *The sample has been taken from a population with mean* μ *and variance* σ^2. *Show that* s^2 *is an unbiased estimate of* σ^2.

This is not as straightforward as the previous answer!
Interest here focuses on the random variable S^2 given by

$$S^2 = \frac{1}{n-1}\sum_{i=1}^{n}(X_i - \bar{X})^2.$$

We begin by looking at the first term in the summation:

$$\begin{aligned} X_1 - \bar{X} &= X_1 - \frac{1}{n}(X_1 + X_2 + \cdots + X_n) \\ &= \left(1 - \frac{1}{n}\right)X_1 - \frac{1}{n}(X_2 + \cdots + X_n). \end{aligned}$$

Since the X-variables are mutually independent, the covariance between any pair is 0. This means that

$$\begin{aligned} \mathrm{Var}(X_1 - \bar{X}) &= \left(1 - \frac{1}{n}\right)^2 \mathrm{Var}(X_1) + \left(-\frac{1}{n}\right)^2 \{\mathrm{Var}(X_2) + \cdots + \mathrm{Var}(X_n)\} \\ &= \frac{(n-1)(n-1)}{n^2}\sigma^2 + \frac{1}{n^2}(n-1)\sigma^2 \\ &= \frac{\{(n-1)+1\}(n-1)}{n^2}\sigma^2 \\ &= \frac{n-1}{n}\sigma^2. \end{aligned}$$

Using Equation (4.8), since the expectation of $(X_1 - \bar{X})$ is 0, we have

$$E[(X_1 - \bar{X})^2] = \frac{n-1}{n}\sigma^2. \tag{14.1}$$

The same result will apply to all n terms in the summation. Since the expectation of any sum is the sum of the separate expectations, we have proved that

$$E\left(\frac{1}{n-1}\sum_{i=1}^{n}(X_i - \bar{X})^2\right) = \frac{1}{n-1} \times n \times \frac{n-1}{n}\sigma^2 = \sigma^2.$$

This proves that the sample variance is an unbiased estimate of the population variance.

The sample mean, $\bar{x}$ and the sample variance, s^2, are unbiased estimates of μ and σ^2, respectively.

14.1.2 Efficiency

Obviously we would like Var(U) to be as small as possible. Thus, if U and V are two unbiased estimators of θ with

$$\text{Var}(U) < \text{Var}(V),$$

then we naturally prefer U, because it seems likely that the estimate u will be closer than v to the unknown θ. In this case U is said to be more **efficient** than V.

Example 14.3

A random sample of two observations (X_1, X_2) *is to be taken from a population with unknown mean* μ *and variance* σ^2*. Three estimators have been proposed for* μ*. These are* U_1, U_2, *and* U_3, *defined by*

$$U_1 = X_1, \qquad U_2 = (X_1 + X_2)/2, \qquad U_3 = 2X_1 - X_2.$$

Show that all three estimators are unbiased and determine which is the most efficient, and which is the least efficient.

Since $E(U_1) = \mu$, U_1 is an unbiased estimator of μ, with variance σ^2.

For U_2 we have

$$\begin{aligned}
E(U_2) &= \frac{1}{2}E(X_1 + X_2) \\
&= \frac{1}{2}\{E(X_1) + E(X_2)\} \\
&= \frac{1}{2}(\mu + \mu) = \mu,
\end{aligned}$$

showing that U_2 (which is the sample mean) is unbiased. It has variance given by

$$\begin{aligned}
\text{Var}(U_2) &= \text{Var}\left\{\frac{1}{2}(X_1 + X_2)\right\} \\
&= \left(\frac{1}{2}\right)^2 \{\text{Var}(X_1) + \text{Var}(X_2)\} \\
&= \frac{1}{4}(\sigma^2 + \sigma^2) \\
&= \frac{1}{2}\sigma^2,
\end{aligned}$$

which is less than $\text{Var}(U_1)$ and implies that U_2 is more efficient than U_1.

Discovering that U_2, which uses information from both observations, is more efficient than U_1, which uses one observation only, should not be a surprise. However, how does U_3 fare?

$$\begin{aligned}
E(U_3) &= E(2X_1 - X_2) \\
&= 2E(X_1) - E(X_2) \\
&= 2\mu - \mu \\
&= \mu,
\end{aligned}$$

confirming that all three estimators are unbiased. Finally,

$$\begin{aligned}
\text{Var}(U_3) &= \text{Var}(2X_1 - X_2) \\
&= 2^2\text{Var}(X_1) + (-1)^2\text{Var}(X_2) \\
&= 4\sigma^2 + \sigma^2 \\
&= 5\sigma^2.
\end{aligned}$$

The variance of U_3 is ten times that of U_2 and five times that of U_1. The most efficient estimator is U_2 and the least efficient is U_3.

Example 14.4

A random sample of two observations, X_1 *and* X_2 *is taken from a distribution with mean* μ *and variance* σ^2*. Determine the most efficient unbiased estimator of* μ *of the form* $aX_1 + bX_2$*, where* a *and* b *are constants whose values are to be determined.*

Let $Y = aX_1 + bX_2$. For the estimator to be unbiased, we require $E(Y) = \mu$. Now, since

$$E(Y) = E(aX_1 + bX_2) = a\mu + b\mu = (a + b)\mu,$$

it is evident that a and b must satisfy

$$a + b = 1.$$

Also

$$\begin{aligned} \text{Var}(Y) &= a^2\text{Var}(X_1) + b^2\text{Var}(X_2) \\ &= (a^2 + b^2)\sigma^2 \\ &= \{a^2 + (1 - a)^2\}\sigma^2 \qquad [\text{using } a + b = 1\,] \\ &= (2a^2 - 2a + 1)\sigma^2. \end{aligned}$$

To maximize efficiency, the variance must be as small as possible. By completing the square and writing

$$2a^2 - 2a + 1 = 2\left(a - \frac{1}{2}\right)^2 + \frac{1}{2},$$

we see that the minimum occurs when $a = 1/2$. Since $a + b = 1$, this implies that b is also equal to $1/2$.

We have shown that, of all possible linear combinations, the unbiased estimator of μ that has the minimum variance is, in fact, $\bar{X}$. This result extends to samples of any size.

14.1.3 Mean squared error (MSE)

Suppose we have two alternative estimators, one is unbiased but inefficient, whereas the other is biased, but efficient. Which should we choose? Clearly that depends on just how large is the variance of the first and the extent of the bias of the second. The mean squared error is a convenient measure that combines variance and bias into a single statistic, R, given by

$$R = E\{(U - \theta)^2\},$$

where U is an estimator of the unknown θ. On the face of it, this expression does not refer to either the mean or the variance of U. However, a little algebra reveals the connection. Rearranging the definition of variance given by Equation (4.5) we have $\mathrm{E}(U^2) = \mathrm{Var}(U) + \mathrm{E}(U)^2$. Therefore,

$$\begin{aligned} R &= \mathrm{E}(U^2) - 2\theta \mathrm{E}(U) + \theta^2 \\ &= \mathrm{Var}(U) + \mathrm{E}(U)^2 - 2\theta \mathrm{E}(U) + \theta^2 \\ &= \mathrm{Var}(U) + \{\mathrm{E}(U) - \theta\}^2. \end{aligned}$$

Thus,

$$R = \mathrm{Var}(U) + \mathrm{Bias}(U)^2. \tag{14.2}$$

In practice, the more commonly quoted term is $\sqrt{R}$, the **root mean squared error** (RMSE). Comparing estimators we would usually prefer the one with the smaller RMSE. For unbiased estimators this implies choosing the most efficient.

14.1.4 *Consistency

If U is unbiased (or asymptotically unbiased), and if $\mathrm{Var}(U)$ reduces to 0 as the sample size increases, then U is said to be **consistent**.

Example 14.5

A random sample of n *observations are taken from a distribution with mean* μ *and variance* σ^2. *Show that the sample mean* $\bar{\mathrm{X}}$ *is a consistent estimator of* μ.

An independent random sample of $(\mathrm{n}+1)$ *observations is taken from the same distribution. The sample mean is denoted by* $\bar{\mathrm{X}}'$. *Show that* $\bar{\mathrm{X}}'$ *is more efficient than* $\bar{\mathrm{X}}$ *as an estimator of* μ.

We denote the n individual observations by $X_1, \dots, X_n$ and begin by calculating $\mathrm{E}(\bar{X})$:

$$\begin{aligned} \mathrm{E}(\bar{X}) &= \mathrm{E}\left(\frac{1}{n}X_1 + \cdots + \frac{1}{n}X_n\right) \\ &= \frac{1}{n}\mu + \cdots + \frac{1}{n}\mu = \mu. \end{aligned}$$

Thus $\bar{X}$ is an unbiased estimator of μ. To show that it is consistent we also need to show that its variance approaches 0 as the sample size increases. Now,

$$\begin{aligned} \mathrm{Var}(\bar{X}) &= \mathrm{Var}\left(\frac{1}{n}X_1 + \cdots + \frac{1}{n}X_n\right) \\ &= \left(\frac{1}{n}\right)^2\sigma^2 + \cdots + \left(\frac{1}{n}\right)^2\sigma^2 \\ &= \frac{\sigma^2}{n}. \end{aligned}$$

As n increases, so σ^2/n approaches 0 as required. We have therefore shown that $\bar{X}$ is a consistent estimator of μ.

For the second sample, replacing n by $n+1$, we obtain $\mathrm{E}(\bar{X}') = \mu$, and $\mathrm{Var}(\bar{X}') = \sigma^2/(n+1)$. Both $\bar{X}$ and $\bar{X}'$ are unbiased estimators of μ. Since $\bar{X}'$ has the smaller variance it is the more efficient.

*14.1.5 *Sufficiency*

An estimator is said to be sufficient if the sample contains no extra information that would be helpful in the estimation of the unknown parameter.

For example, suppose, our aim is to estimate the mean of a distribution. We take a sample and use the sample mean as an estimate of the population mean. If we are now told the value of the sample median, or the sample variance, does this help with the estimation process? No! Knowing the value of the sample mean is **sufficient**.

14.2 Estimation methods

Bearing in mind the desirable properties listed in Section 14.1, we now turn to the general question of how best to estimate the parameter(s) of a distribution? We consider two methods here: the method of moments and the method of maximum likelihood. A third method, that of least squares, will be introduced in Section 19.3.

14.2.1 The method of moments

This is an intuitive method that we have already used: take the value obtained from a sample as an estimate of the corresponding population quantity. For example, we have used the sample mean as an estimate of the population mean.

In Section 13.2 it was observed that the sample mean marked the centre of gravity of a set of data. Since applied mathematicians might refer to the centre of gravity as the first **moment**, Karl Pearson[1] used the same term in the statistics context. The first moment of a population is $\mathrm{E}(X)$, with the second population moment being $\mathrm{E}(X^2)$, and so forth.

Example 14.6

> *A sample of observations is taken from the exponential distribution* $f(\mathrm{x}) = \lambda e^{-\lambda \mathrm{x}}$.
> *We are asked to find an estimate of the unknown* λ *based on the observations* $\mathrm{x}_1, \mathrm{x}_2, \ldots, \mathrm{x}_\mathrm{n}$ *which have mean* $\bar{\mathrm{x}}$.

Since the exponential distribution has mean $1/\lambda$ (see Section 5.6), we equate $\bar{x}$ to $1/\lambda$. The method of moments estimate of λ is therefore $1/\bar{x}$.

[1] Karl Pearson (1857–1936), an English biometrician, founded the first university department of statistics (at University College London). In addition to the method of moments (a phrase he introduced in 1894), he wrote many papers on correlation (Chapter 18) and the chi-squared goodness-of-fit test (Chapter 17).

Example 14.7

The random variable X has the negative binomial distribution $P(X = n) = \binom{n-1}{r-1} p^r (1-p)^{n-r}$*, where* r *is specified, but* p *is unknown and requires estimation. The sample mean ihas been found to be* $\bar{x}$.

The mean of the negative binomial distribution was found in Section 3.14 to be $\frac{r}{p}$. We therefore set $\frac{r}{p} = \bar{x}$ which implies that the estimate of p is $\frac{r}{\bar{x}}$.

14.2.2 *The method of maximum likelihood*

Suppose we have observations $x_1, x_2, ..., x_n$ that come from the probability density function $f(x; \theta)$, where θ is an unknown parameter of the distribution. The joint **likelihood** of obtaining these values is L, given by

$$L = f(x_1; \theta) \times f(x_2; \theta) \times \cdots \times f(x_n; \theta) = \prod_{i=1}^{n} f(x_i; \theta).$$

An intuitively reasonable solution to the problem of estimating the unknown θ is to choose that value that makes the next sample of n observations most likely to resemble the data in the current sample. This is simply the value that maximizes L. The idea was formalized by Fisher[2] in the early 1920s.

In order to find the maximum, we can differentiate L with respect to θ, set the differential equal to 0, and solve. However, in practice, it is usually easier to work with $\log(L)$ rather than L, and then differentiate that. The two functions L and $\log(L)$ have their maxima at the same values of any parameters.

Example 14.6 (cont.)

In this case the likelihood is

$$L = \lambda e^{-\lambda x_1} \times \lambda e^{-\lambda x_2} \times \cdots \times \lambda e^{-\lambda x_n} = \lambda^n e^{-\lambda \sum x_i},$$

so that the log(likelihood) is

$$\log(L) = n \log(\lambda) - \lambda \sum_{i=1}^{n} x_i.$$

Differentiating with respect to λ gives

$$\frac{n}{\lambda} - \sum_{i=1}^{n} x_i.$$

[2] See footnote in Section 7.3.

Setting this equal to 0, and rearranging the terms, gives the maximum likelihood estimate of λ as $n/\sum x_i$, which is more conveniently written as $1/\bar{x}$. In this case the maximum likelihood estimate is the same as the method of moments estimate.

Example 14.9

The random variable X has a geometric distribution with parameter p. *A sample of* n *observations* $x_1, x_2, ..., x_n$ *is available. Obtain the maximum likelihood estimate of* p.

The likelihood is given by

$$L = (1-p)^{x_1-1}p \times (1-p)^{x_2-1}p \cdots \times (1-p)^{x_n-1}p = (1-p)^{\sum(x_i-1)}p^n.$$

Taking logarithms we have

$$\log(L) = \left(\sum_{i=1}^{n} x_i - n\right)\log(1-p) + n\log(p).$$

Differentiating with respect to p gives

$$\left(\sum_{i=1}^{n} x_i - n\right) \times \frac{-1}{1-p} + \left(n \times \frac{1}{p}\right).$$

Setting this equal to 0 and multiplying by $p(1-p)$ gives

$$-p\left(\sum_{i=1}^{n} x_i - n\right) + n(1-p) = 0.$$

The np terms conveniently cancel, to give,

$$-p\sum_{i=1}^{n} x_i + n = 0,$$

from which we get the maximum likelihood estimate of p as being $\hat{p} = 1/\bar{x}$, where $\bar{x}$ is the sample mean.

Exercises 14a

1. Find an expression for the method of moments estimate of the parameter p of the geometric distribution $P(X = x) = (1-p)^{x-1}p$ using the observations $x_1, x_2, ..., x_n$.
2. A sample of observations, $x_1, x_2, ..., x_n$, is taken from the distribution

$$f(x) = \frac{1}{\sqrt{2\pi}} \exp^{-(x-\mu)^2/2},$$

which is a normal distribution with unit variance and unknown mean, μ. Obtain the maximum likelihood estimate of μ.

14.3 Confidence intervals

We hope that our point estimate, $\hat{\theta}$, of some unknown parameter, θ, will be close to the true population value, but we cannot know for certain how close it is. A **confidence interval** (an **interval estimate**), is a statement that the true value of θ is likely to lie in some interval $(\hat{\theta} - \delta, \hat{\theta} + \delta)$, where the quantity δ is chosen in a way that allows us to assign a probability to the interval.

14.3.1 Confidence interval for a mean

There are four cases that we need to consider:

1. A sample (large or small) is taken from a normally distributed population, with known variance.
2. A large sample is taken from a population with known variance.
3. A sample (large or small) is taken from a normally distributed population with unknown variance.
4. A large sample is taken from a population with unknown variance.

The distinction between 'large' and 'small' sample sizes is arbitrary, but, typically, 'large' is taken in this context to mean 30 or more observations.

14.3.1.1 Normal distribution with known variance

This is a rather unlikely situation, but it explains the general approach.

A sample of n observations is taken from a $\mathrm{N}(\mu, \sigma^2)$ distribution. We denote the random variable corresponding to the sample mean by $\bar{X}$. Since $\bar{X}$ is a linear combination of independent normal random variables, it too has a normal distribution. From Section 4.5 we know that $\bar{X}$ has expectation μ and variance σ^2/n, hence

$$\bar{X} \sim \mathrm{N}\left(\mu, \frac{\sigma^2}{n}\right).$$

Supposing, for the moment, that μ was known, we could work with the random variable Z, given by

$$Z = \frac{\bar{X} - \mu}{\sigma/\sqrt{n}},$$

where the quantity $\sigma/\sqrt{n}$ is often called the **standard error** of the mean.

Since the distribution of Z is known to be $\mathrm{N}(0,1)$, we find, by looking at the table of percentage points for a standard normal distribution, that

$$\mathrm{P}(Z > 1.96) = 0.025,$$

from which it follows that

$$\mathrm{P}(Z < -1.96) = 0.025,$$

and hence that

$$\mathrm{P}(|Z| < 1.96) = 0.95.$$

Substituting for Z, this implies that

$$\mathrm{P}\left(\frac{|\bar{X}-\mu|}{\sigma/\sqrt{n}} < 1.96\right) = 0.95.$$

Multiplying through by $\sigma/\sqrt{n}$, this statement becomes

$$\mathrm{P}\left(|\bar{X}-\mu| < 1.96\frac{\sigma}{\sqrt{n}}\right) = 0.95.$$

In words, this states that the probability that the distance between μ and $\bar{X}$ is less than $1.96\sigma/\sqrt{n}$ is 0.95. We can conveniently rewrite this as

$$\mathrm{P}\left(\bar{X} - 1.96\frac{\sigma}{\sqrt{n}} < \mu < \bar{X} + 1.96\frac{\sigma}{\sqrt{n}}\right) = 0.95.$$

Note that, despite its present appearance, this is still a probability statement concerning the random variable $\bar{X}$. It is *not* a probability statement about μ which is a constant (albeit an unknown constant).

Suppose we now collect our n observations on X, and compute the sample mean, $\bar{x}$. The interval

$$\left(\bar{x} - 1.96\frac{\sigma}{\sqrt{n}},\ \bar{x} + 1.96\frac{\sigma}{\sqrt{n}}\right) \tag{14.3}$$

is called a **95% symmetric confidence interval** for μ. Often the adjective 'symmetric' is omitted and we just write **confidence interval**. The two limiting values that define the interval are known as the **confidence limits**.

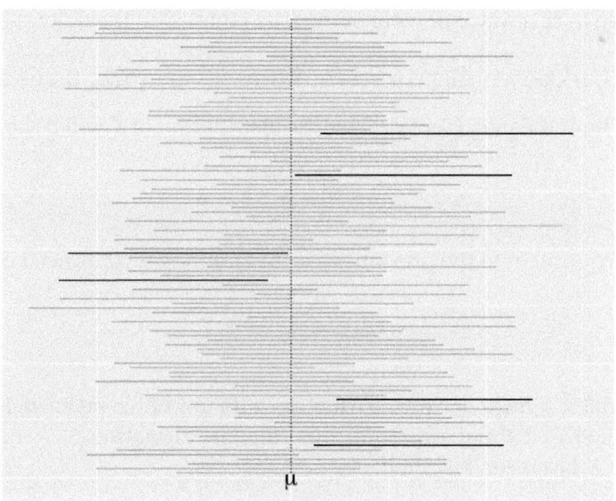

Figure 14.1 The 95% confidence intervals for the mean that result from 100 random samples from a normal distribution: in this case six of the intervals fail to include the true value μ.

As Figure 14.1 shows, different samples will lead to different values of $\bar{x}$ and hence to different 95% confidence intervals: on average, 95% will include the true population value.

If we wish to be more confident that our interval includes the true value of μ, all we need do is to replace 1.96 by a larger value. This will make the intervals wider. If we wish to have a smaller interval, then we must either take a larger sample or be less confident that the interval includes μ.

The most common values used in the construction of symmetric confidence intervals based on the normal distribution are given in the following table; we refer to these values as **critical values**.

Two-sided confidence interval	90%	95%	98%	99%
Critical value	1.645	1.960	2.326	2.576

Example 14.10

A machine cuts metal tubing into pieces. It is known that the lengths of the pieces have a normal distribution with standard deviation 4 mm. After the machine has undergone a routine overhaul, a random sample of 25 pieces are found to have a mean length of 146 cm. Assuming the overhaul has not affected the variance of the tube lengths, determine a 99% symmetric confidence interval for the population mean length.

Working in centimetres, the confidence interval is

$$\left(146 - 2.576\frac{4}{\sqrt{25}}, 146 + 2.576\frac{4}{\sqrt{25}}\right),$$

which simplifies to (143.94, 148.06). This particular interval either does or does not include the true population mean length—we cannot say which is true! What we *can* say is that 99% of the intervals constructed in this way will include the true population mean length.

14.3.1.2 Unknown distribution, known variance, large sample

Because of the central limit theorem (Section 6.4), the distribution of $\bar{X}$ will be approximately normal:

$$\bar{X} \sim \mathrm{N}\left(\mu, \frac{\sigma^2}{n}\right).$$

This case is therefore equivalent to the previous case and nothing further need be added.

Exercises 14b

1. The random variable X has a normal distribution with unknown mean, μ. From past experience it can be assumed that the distribution has standard deviation 3. A random sample of 10 observations on X has mean 8.2. Find:

   (a) a 95% symmetric confidence interval for μ,

   (b) a 99% symmetric confidence interval for μ,

2. The weights of 4-month-old pigs are known to be normally distributed with variance 16 kg^2. A new diet is suggested and a random sample of 25 pigs given this diet has a mean weight of 30.42 kg. Determine a 99% confidence interval for the mean weight of 4-month-old pigs that are fed on this diet.

3. A supplier provides pre-packaged goods to the supermarket. Each package has a nominal weight of 1 kg. To verify the accuracy of the supplier, 100 packages are weighed. Their mean weight is found to be 1020 g. Taking the population variance to be 144 g^2, determine a 95% confidence interval for the mean package weight.

14.3.1.3 Normal distribution, unknown variance

When σ^2 is unknown, it is natural to use the sample estimate, s^2 in its place. However, s^2 is itself an observation on a random variable (S^2), since different samples from the same population would have both different means and different variances. The random variable T, defined by

$$T = \frac{\bar{X} - \mu}{S/\sqrt{n}}, \tag{14.4}$$

therefore involves *two* random variables: $\bar{X}$ in the numerator, and $\sqrt{S^2}$ in the denominator. Rather than a normal distribution, it has a t_{n-1}-distribution (Section 7.1).

Suppose that the relevant critical value of this distribution is c. Then an argument exactly parallel to that used for the case of known σ would lead to

$$\mathrm{P}\left(\bar{X} - cS/\sqrt{n} < \mu < \bar{X} + cS/\sqrt{n}\right) = 0.95.$$

The symmetric confidence interval for μ therefore becomes

$$\left(\bar{x} - cs/\sqrt{n},\ \bar{x} + cs/\sqrt{n}\right). \tag{14.5}$$

Example 14.11

A random sample of 31 sweets are chosen from a box and each is weighed. The sample is found to have mean 1g and standard deviation 0.2g. Assuming that the masses have a normal distribution, determine a 99% symmetric confidence interval for the population. mean.

Since the population is assumed to have a normal distribution, it is appropriate to base a confidence interval on a t-distribution—in this case the t_{30}-distribution. For a 99% symmetric confidence interval, the relevant value of c is 2.750. The interval is therefore

$$\left(1 - 2.750 \times \frac{0.2}{\sqrt{31}},\ 1 + 2.750 \times \frac{0.2}{\sqrt{31}}\right),$$

which, with appropriate rounding, simplifies to

$$(0.90,\ 1.10).$$

Exercises 14c

1. The random variable X has a normal distribution with mean μ. A random sample of 11 observations of X is taken and gives $\sum x = 83.6$ and $\sum x^2 = 742.84$. Obtain a symmetric 95% confidence interval for μ.
2. Thirteen students independently performed an experiment to estimate the value of π. Their results were summarized by $\sum x = 40.42$ and $\sum x^2 = 126.4111$.

   (a) Stating any necessary assumption that you make, calculate a 99% symmetric confidence interval for π based on these data, giving the confidence limits correct to two decimal places.

   (b) Estimate the minimum number of results that would be needed if it is required that the width of the resulting 99% symmetric confidence interval should be at most 0.02.
3. A supplier provides pre-packaged goods to the supermarket. Each package has a nominal weight of 1 kg. To verify the accuracy of the supplier, a random sample of 100 packages are weighed. Their mean weight is found to be 1020 g with a sample variance of 144 g^2. Determine a 95% confidence interval for the mean package weight.

Put theory into practice: How many words are there in this book?

- Choose 20 pages at random.
- For each page, count the number of lines on that page.
- For each page, choose five lines at random and count the number of words on each line.
- Multiply the number of lines on the page by the average number of words on a line to give an estimate of the number of words on that page.
- Calculate the mean and variance of the 20 estimates.
- Assume that the estimates may be regarded as arising from a (discretized) normal distribution; obtain a 95% symmetric confidence interval for the number of words on a page.
- Scale up this interval by multiplying by the number of pages in the book to arrive at a 95% symmetric confidence interval for the number of words in the book.

14.4 Confidence intervals with discrete distributions

14.4.1 Confidence interval for a Poisson mean

When the mean of a Poisson distribution is large, the normal distribution provides a reasonable approximation (see Section 6.5.2). Since the variance of a Poisson distribution is equal to its mean, there is no need to estimate its value from the data. We simply use the value of the sample mean as its estimate. In this case, therefore, the 95% confidence interval for the population mean is given (approximately) by

$$\left(\bar{x} - 1.96\sqrt{\frac{\bar{x}}{n}},\quad \bar{x} + 1.96\sqrt{\frac{\bar{x}}{n}}\right). \tag{14.6}$$

This is known as the **Wald interval** for the Poisson mean.[3]

Example 14.12

> *An environmentalist takes a random sample of water from a river. She discovers that her 100 ml sample contains 64 individuals of a particular type. Give a 99% confidence interval for the mean number of these organisms in a litre of this river water.*

We must first obtain a confidence interval for a water sample of the size obtained (since the size of our sample reflects its precision). We can then scale this to the required size. The 99% confidence interval for 100 ml (with $n = 1$ and $\bar{x} = 64$) is

$$(64 - 2.576\sqrt{64},\ 64 + 2.576\sqrt{64}).$$

This interval simplifies to (43.4, 84.6). The required confidence interval for a litre of the river water is therefore (434, 846).

A recent study by the Indian statisticians Patil and Kulkarni [4] compared 19 methods for constructing confidence intervals for a Poisson mean. They concluded that the probability associated with the Wald interval consistently exceeded the nominal confidence level. For cases where the observed mean is 4 or more, they recommended a number of methods of which the most straightforward is that due to Garwood.[5] With $\bar{x}$ denoting the observed mean, the **Garwood limits** are (for a two-sided 95% confidence interval) x_L and x_U, the solutions of

$$\mathrm{P}(2x_L < \chi^2_{2\bar{x}}) = 0.025 \qquad \text{and} \qquad \mathrm{P}(2x_U > \chi^2_{2(\bar{x}+1)}) = 0.025. \tag{14.7}$$

[3] Abraham Wald (1902–50) was a Hungarian mathematician who studied at the University of Vienna. On the Nazi seizure of Austria, he emigrated to the United States. He was elected President of the Institute of Mathematical Statistics in 1948, but died in a plane crash in India two years later.

[4] Patil, V. V., and Kulkarni, H. V. (2012) Comparison of confidence intervals for the Poisson mean: some new aspects, *REVSTAT-Statistical Journal*, **10**, 211–227. There are no 'best buys' and the methods mentioned here are chosen for their simplicity.

[5] Frank Garwood (1911–88) was a student at London University (now called University College, London). This result formed part of his 1934 PhD thesis. The method is described as 'exact' by R.

Example 14.12 (cont.)

For 100 ml, the Garwood limits are (45.3, 87.6). Both limits are greater than their Wald counterparts, but they are no longer symmetric about the observed 64. This correctly reflects the asymmetry inherent in a Poisson distribution.

For cases where the count is less than 4, one simple method recommended by Patil and Kulkarni is to apply the continuity correction, given by Equation (6.2), to the Wald interval. With a total count of x events from n observations this becomes

$$\left\{\left(\frac{x-0.5}{n}\right)-\frac{1.96}{n}\sqrt{x-0.5},\quad \left(\frac{x+0.5}{n}\right)+\frac{1.96}{n}\sqrt{x+0.5}\right\}. \tag{14.8}$$

If the lower limit is negative, then it is replaced by 0.

Example 14.13

A court stenographer made 12 transcription errors on a total of 20 pages of transcription. Determine a 95% confidence interval for the number of errors per page.

Since the mean is 0.6, we use the continuity-corrected version of the Wald limits:

$$\left(\frac{11.5}{20}-\frac{1.96}{20}\sqrt{11.5},\quad \frac{12.5}{20}+\frac{1.96}{20}\sqrt{12.5}\right)=(0.24, 0.97).$$

Exercises 14d

1. The number of telephone calls arriving at a school was monitored on 10 randomly chosen days, with the total number of calls being 1053. Assuming a Poisson distribution, find an approximate 95% symmetric confidence interval for the mean number of calls per day,

   (a) using the Wald interval,

   (b) using the Garwood limits.

2. In spring a field of area 7000 m^2 is sown with a wild flower mixture. In summer 16 non-overlapping squares, each of side 0.1 m, are chosen at random. These squares contained a total of 25 flowers of a particular variety. Assuming that the seeds have been randomly distributed, find an approximate 95% confidence interval for the mean number of these flowers per square metre.

14.4.2 Confidence interval for a binomial proportion

Suppose that a random sample of n observations is taken from a population in which the proportion of successes is p and the proportion of failures is q $(= 1 - p)$. Suppose the number of successes in the sample is denoted by r (an observation on the random variable R). Define $\widehat{p}$ by $\widehat{p} = r/n$, with the corresponding random variable being given by $\widehat{P} = R/n$.

The random variable R has a binomial distribution with parameters n and p and therefore $\mathrm{E}(R) = np$ and $\mathrm{Var}(R) = npq$. Hence

$$\mathrm{E}(\widehat{P}) = \mathrm{E}\left(\frac{R}{n}\right) = \frac{1}{n}\mathrm{E}(R) = \frac{1}{n}np = p,$$

which shows that $\widehat{P}$ is an unbiased estimator of p. Its variance is given by

$$\mathrm{Var}(\widehat{P}) = \mathrm{Var}\left(\frac{R}{n}\right) = \left(\frac{1}{n}\right)^2 \mathrm{Var}(R) = \left(\frac{1}{n}\right)^2 npq = \frac{pq}{n}.$$

We begin with cases where n is sufficiently large that the normal approximation to the binomial distribution may be used, so that then

$$\widehat{P} \sim \mathrm{N}\left(p, \frac{pq}{n}\right).$$

The standardized variable, Z, given by

$$Z = \frac{\widehat{P} - p}{\sqrt{pq/n}},$$

will have an approximate N(0,1) distribution.

As before, we note that

$$\mathrm{P}(|Z| < 1.96) = 0.95.$$

Substituting for Z, this implies that

$$\mathrm{P}\left(\frac{|\widehat{P} - p|}{\sqrt{pq/n}} < 1.96\right) = 0.95.$$

Multiplying the inequality through by $\sqrt{pq/n}$, and rearranging, we obtain

$$\mathrm{P}\left\{(\widehat{P} - 1.96\sqrt{pq/n}) < p < (\widehat{P} + 1.96\sqrt{pq/n})\right\} = 0.95.$$

Replacing $\widehat{P}$ by its observed value $\widehat{p}$, and replacing the unknown pq by the approximation $\widehat{p}\widehat{q}$, where $\widehat{q} = 1 - \widehat{p}$, an approximate 95% symmetric confidence interval for p is given by

$$\left(\widehat{p} - 1.96\sqrt{\widehat{p}\widehat{q}/n},\ \widehat{p} + 1.96\sqrt{\widehat{p}\widehat{q}/n}\right). \tag{14.9}$$

There are three approximations involved in the production of this confidence interval (which, as in the case of the interval for the Poisson mean, is usually referred to as a Wald interval). The approximations are:

(i) the normal approximation to the binomial,

(ii) the replacement of pq by $\widehat{p}\widehat{q}$,

(iii) the omission of a continuity correction for the normal approximation.

Unfortunately, using Equation (14.9), the actual size of the confidence interval can be much smaller than its intended value. As an example, when $n = 12$ and the true p-value is 0.5, the average size of a supposed 95% confidence interval is 0.85. Recognizing this feature, Wilson[6] suggested that an improvement was provided by

$$\left\{(2r + z_0^2) \pm z\sqrt{z_0^2 + 4r - 4r^2/n}\right\}/2(n + z_0^2)\,, \tag{14.10}$$

where z_0 is the desired critical value (in this case 1.96).

There is no perfect procedure. The R package currently lists no fewer than 15 alternatives and, tellingly, defaults to Wilson's unless the user specifies otherwise.

Example 14.14

An importer has ordered a large consignment of tomatoes. When it arrives he examines a randomly chosen sample of 50 boxes and finds that 12 contain at least one bad tomato. Assuming that these boxes may be regarded as being a random sample from the tomatoes in the consignment, obtain an approximate 95% confidence interval for the proportion of boxes containing bad tomatoes, giving your confidence limits correct to three decimal places.

We have $\widehat{p} = 0.24$, $\widehat{q} = 0.76$, and a critical value of 2.576. Using Laplace's suggestion, Equation (14.9), the confidence interval is

$$\left(0.24 - 1.96\sqrt{\frac{0.24 \times 0.76}{50}},\ 0.24 + 1.96\sqrt{\frac{0.24 \times 0.76}{50}}\right),$$

which simplifies to

$$(0.122,\ 0.358).$$

Using Wilson's version, Equation (14.10), we have

$$\left\{24 + 1.96^2 \pm \sqrt{1.96^2 + 48 - 576/50}\right\}/2(50 + 1.96^2),$$

which simplifies to

$$(0.143,\ 0.374).$$

There is a notable difference between the two intervals. Indeed the 15 alternatives provided by the R program give lower bounds varying between 0.112 and 0.148 with upper bounds varying between 0.345 and 0.385.

[6] Edwin Bidwell Wilson (1879–1964) was an American polymath who suggested his alternative in the publication 'Probable Inference, the Law of Succession, and Statistical Inference' , which appeared in issue 158 of the *Journal of the American Statistical Association.*

Evidently, in the case of a confidence interval for a binomial proportion, when the sample size is not large, it would be unwise to assign too much precision to the answer obtained by the method that you choose to use.

Exercises 14e

1. A random sample of 1000 voters are interviewed. Of these, 349 state that they would support the Conservative Party. Determine an approximate 95% symmetric confidence interval for the proportion of Conservative supporters in the population.
2. A coin, which is possibly biased, is thrown 400 times. The number of heads obtained is 217. Find an approximate 95% confidence interval for the probability of obtaining a head.

Put theory into practice: This practical answers a question first posed by the Comte de Buffon in 1777. If a needle of length l is dropped randomly (i.e. without looking!) onto a grid of equi-spaced parallel lines (separated by a distance d), then, the Comte enquired, what is the probability that the needle crosses a line? For the case $l < d$ the Comte showed that the answer is $2l/d\pi$. Using a matchstick rather than a needle, and using a grid in which d is chosen to be about $4l/3$, perform Buffon's experiment 100 times. Show that, with this number of tosses and this choice of d, a result of r successes (the crossing of a line) corresponds to an estimate of π as $150/r$.

Obtain a 99% symmetric confidence interval for the proportion crossing a line and deduce a 99% confidence interval for π.

Group project: Compare your results with those of the rest of your group. Obtain a narrower confidence interval based on the pooled information from the entire class.

14.5 One-sided confidence intervals

In Section 12.3 we implicitly assumed that small values and large values were of equal interest. Writing Z as a random variable having a N(0,1) distribution, our confidence intervals were based on the probability statement

$$P(-1.96 \le Z \le 1.96) = 0.95,$$

which led to

$$\left(\bar{x} - 1.96\frac{\sigma}{\sqrt{n}}, \bar{x} + 1.96\frac{\sigma}{\sqrt{n}}\right).$$

We called this interval a symmetric interval for μ, but we could also have called it a **two-sided confidence interval** for μ, since equal attention was paid to both tails of the distribution.

Suppose instead we consider a one-sided probability statement, such as

$$P(-1.645 \leq Z) = 0.95.$$

If we now substitute for Z using, for example,

$$Z = \frac{\bar{X} - \mu}{\sigma/\sqrt{n}},$$

then, with a simple rearrangement, we arrive at

$$P\left(\bar{X} + 1.645\frac{\sigma}{\sqrt{n}} \geq \mu\right) = 0.95.$$

Replacing the random variable $\bar{X}$ by the sample value $\bar{x}$, we obtain the following **one-sided confidence interval** for μ:

$$\left(-\infty,\ \bar{x} + 1.645\frac{\sigma}{\sqrt{n}}\right).$$

Alternatively, using the opposite tail, we would get

$$\left(\bar{x} - 1.645\frac{\sigma}{\sqrt{n}},\ \infty\right).$$

With a large sample and an unknown variance, approximate one-sided confidence intervals are obtained by replacing the population standard deviation σ by its sample counterpart s and using the t-distribution.

Equivalent arguments lead to one-sided confidence intervals for a Poisson mean,

$$\left(0,\ \bar{x} + 1.645\sqrt{\frac{\bar{x}}{n}}\right) \text{ and } \left(\bar{x} - 1.645\sqrt{\frac{\bar{x}}{n}},\ \infty\right),$$

and for binomial proportions,

$$\left(0,\ \hat{p} + 1.96\sqrt{\frac{\hat{p}\hat{q}}{n}}\right) \text{ and } \left(\hat{p} - 1.96\sqrt{\frac{\hat{p}\hat{q}}{n}},\ 1\right).$$

The only noticeable difference with these latter situations is the restriction to values in $(0, \infty)$ for the Poisson mean, and to $(0, 1)$ for the binomial proportion.

For an observed count x, an exact solution is provided by using the interval (μ_1, ∞) or $(0, \mu_2)$, where μ_1 and μ_2 are the solutions of (for the case of a 95% interval)

$$P(X < x|\mu = \mu_1) = 0.05 \quad \text{and} \quad P(X > x|\mu = \mu_2) = 0.05. \tag{14.11}$$

Example 14.15

A vet is called to examine a large herd of cattle. Out of a random sample of 60 cows, the vet finds that eight show signs of a particular disease. Find a 99% one-sided confidence interval of the form (0, θ) for the proportion of cows in the herd that show signs of the disease.

The estimate of the population proportion is $\hat{p} = 2/15$. The upper 1% point of a standard normal distribution is 2.326; hence the required interval is

$$\left(0,\ \frac{2}{15} + 2.326\sqrt{\frac{2}{15} \times \frac{13}{15} \times \frac{1}{60}}\right),$$

which simplifies to

$$(0,\ 0.235),$$

or from 0 to 24%.

Exercises 14f

1. The random variable X has a normal distribution with mean μ and variance 16. A random sample of 10 observations of X has mean 8.2. Find a 95% confidence interval for μ of the form (θ, ∞).
2. A random sample of 90 one-year-old fireworks was tested. It was found that 72 went off successfully. Find a 95% confidence interval of the form $(\theta, 1)$ for the proportion of one-year-old fireworks that go off satisfactorily.

14.6 Confidence intervals for a variance

14.6.1 Assuming a normal distribution

Suppose that we have a sample of n independent observations, $x_1, x_2, \ldots, x_n$, with mean $\bar{x}$. and variance, s^2. The corresponding random variable, S^2, is defined by

$$S^2 = \frac{1}{n-1}\sum_{i=1}^{n}(X_i - \bar{X})^2,$$

where the random variables X_1, X_2, ..., X_n, and $\bar{X}$ all have normal distributions with mean μ and standard deviation, σ.

Thus $\left(\frac{X_i-\mu}{\sigma}\right)^2$ has a χ_1^2-distribution, for all values of i. Because of the additivity of independent chi-squared random variables it follows that

$$\sum_{i=1}^{n}\left(\frac{X_i-\mu}{\sigma^2}\right)^2 = \frac{1}{\sigma^2}\sum_{i=1}^{n}(X_i-\mu)^2 \sim \chi_n^2.$$

We are interested in $\sum_{i=1}^{n}(X_i - \bar{X})^2$ rather than $\sum_{i=1}^{n}(X_i - \mu)^2$, so we write

$$(X_i - \mu) = (X_i - \bar{X} + \bar{X} - \mu),$$

and, squaring, we then have

$$(X_i - \mu)^2 = (X_i - \bar{X})^2 + 2(X_i - \bar{X})(\bar{X} - \mu) + (\bar{X} - \mu)^2.$$

Summing over i gives

$$\sum_{i=1}^{n}(X_i - \mu)^2 = \sum_{i=1}^{n}(X_i - \bar{X})^2 + 2\sum_{i=1}^{n}(X_i - \bar{X})(\bar{X} - \mu) + \sum_{i=1}^{n}(\bar{X} - \mu)^2.$$

The middle term is 0, since $(\bar{X} - \mu)$ is a constant and $\sum_{i=1}^{n}(X_i - \bar{X})$ equals 0. Also $\sum_{i=1}^{n}(\bar{X} - \mu)^2 = n(\bar{X} - \mu)^2$. Dividing through by σ^2 therefore gives

$$\begin{aligned} \frac{1}{\sigma^2}\sum_{i=1}^{n}(X_i - \mu)^2 &= \frac{1}{\sigma^2}\sum_{i=1}^{n}(X_i - \bar{X})^2 + \frac{n}{\sigma^2}(\bar{X} - \mu)^2 \\ &= \frac{1}{\sigma^2}\sum_{i=1}^{n}(X_i - \bar{X})^2 + \left(\frac{\bar{X} - \mu}{\sigma/\sqrt{n}}\right)^2. \end{aligned}$$

We know that the left-hand term has a χ_n^2-distribution. Also, since $\bar{X}$ has a normal distribution with mean μ and variance σ^2/n, the right-hand term has a χ_1^2-distribution. The middle term, which is equal to $(n-1)S^2/\sigma^2$, must therefore have a χ_{n-1}^2-distribution.

$$(n-1)S^2/\sigma^2 \sim \chi_{n-1}^2. \qquad (14.12)$$

To see that this is plausible, examine the expected values. A chi-squared distribution with d degrees of freedom, has mean d, while we know that S^2 is an unbiased estimator of σ^2. Thus both sides of Equation (14.12) have expectation $(n-1)$.

Suppose we denote the lower and upper relevant 2.5% points of a χ_{n-1}^2 distribution by L and U, then we have

$$\mathrm{P}\left[L < \frac{(n-1)S^2}{\sigma^2} < U\right] = 0.95.$$

Taking reciprocals, we need to reverse the inequalities,

$$\mathrm{P}\left[\frac{1}{L} > \frac{\sigma^2}{(n-1)S^2} > \frac{1}{U}\right] = 0.95,$$

giving

$$\mathrm{P}\left[\frac{(n-1)S^2}{U} < \sigma^2 < \frac{(n-1)S^2}{L}\right] = 0.95.$$

The 95% confidence interval for σ^2 is

$$\left(\frac{(n-1)s^2}{U}, \frac{(n-1)s^2}{L}\right), \tag{14.13}$$

where L and U are, respectively, the lower and upper relevant 2.5% points of a χ^2_{n-1} distribution.

Example 14.16

A machine fills containers with orange juice. A random sample of 10 containers are examined in a laboratory, and the amounts of orange juice in each container were determined correct to the nearest 0.05 ml. The results (after subtracting 500ml from each) were as follows:

3.45, 7.80, 1.10, 6.45, 5.85, 4.40, 3.45, 5.50, 2.95, 7.75.

Assuming that the observations have a normal distribution, obtain a 95% symmetric confidence interval for the population variance, explaining the sense in which the confidence interval is symmetric.

We find that s^2 is 42.885, while the lower and upper 2.5% points of the relevant χ^2_9 distribution are $L = 2.70$ and $U = 19.02$. The 95% confidence interval is therefore

$$\left(\frac{42.886}{19.02}, \frac{42.886}{2.70}\right),$$

which simplifies to (2.25, 15.88).

The confidence interval is symmetric in the sense that it uses the central 95% of the χ^2 distribution, with 2.5% being excluded in each tail.

Exercises 14g

1. A random sample of 12 observations is taken from a normal distribution. The sample variance is 3.56. Determine a symmetric 99% confidence interval for the population variance.
2. A random sample of 20 observations from a normal distribution has variance 480. Determine a symmetric 95% confidence interval for the variance of the population from which the sample was drawn.

14.6.2 Assuming a moderately symmetric distribution

For the chi-squared procedure to give an accurate confidence interval, it is critical that the distribution is normal. Even moderate departures from normality can lead to incorrect intervals. A procedure that is robust to mild deviations from normality was suggested by Bonett.[7] However, Bonett's procedure is too complex to be discussed here.

[7] See Bonett, D. G. (2006), 'Approximate confidence interval for standard deviation of nonnormal distributions', *Computational Statistics & Data Analysis*, **50**, 775–782.

Key facts

- **Properties of point estimators**
  - **Bias**
    An estimator U is an unbiased estimator of a population parameter θ if $\mathrm{E}(U) = \theta$.
  - **Efficiency**
    If U and V are two unbiased estimators of θ, with $\mathrm{Var}(U) < \mathrm{Var}(V)$, then U is more efficient than V.
  - **Consistency**
    If U is an unbiased estimator of θ, and if $\mathrm{Var}(U)$ approaches 0 as the sample size increases, then U is a consistent estimator of θ.
  - **Sufficiency**
    U is a sufficient estimator of θ, if there is no further information in the sample that would improve the accuracy of the estimator.

- **Methods of estimation**
  - **Method of moments**
    Set the sample mean equal to the population mean (and, if necessary, the sample variance equal to the population variance, etc.). Solve the equations.
  - **Method of maximum likelihood**
    Form the likelihood: $L = \prod \mathrm{P}(x_i)$ or $\prod \mathrm{f}(x_i)$. Maximize by differentiating, setting equal to 0, and solving. Usually easier when working with $\ln(L)$.

- **Confidence intervals for a population mean**
  For a sample of size n, denote the appropriate critical value from a normal distribution by c_N, and its counterpart from a t-distribution with $n-1$ degrees of freedom by c_t.

Condition	*Confidence interval*	*Notes*
X normal, σ^2 known	$\bar{x} \pm c_N\sigma/\sqrt{n}$	Exact
σ^2 known, n large	$\bar{x} \pm c_N\sigma/\sqrt{n}$	Approximate
X normal	$\bar{x} \pm c_t\sigma\sqrt{n}$	Exact

- **Confidence intervals for a Poisson mean**
  Basic (Wald): $\bar{x} - c_N\sqrt{\bar{x}/n}$
  Better: (Garwood): $\mathrm{P}(2x_L < \chi^2_{2\bar{x}}) = 0.025$ and $\mathrm{P}(2x_U > \chi^2_{2(\bar{x}+1)}) = 0.025$
  Small x (continuity-corrected Wald):
  $\left\{\left(\frac{x-0.5}{n}\right) - \frac{c_N}{n}\sqrt{x-0.5}, \quad \left(\frac{x+0.5}{n}\right) + \frac{c_N}{n}\sqrt{x+0.5}\right\}$

- **Confidence interval for a population proportion**
  Write $\widehat{p} = r/n$ and $\widehat{q} = 1 - \widehat{p}$, where r is the number of successes in a large sample of size n. The approximate (Wald) interval is provided by:

$$\widehat{p} \pm c_N\sqrt{\frac{\widehat{p}\widehat{q}}{n}}.$$

- **One-sided confidence intervals**
  These are typified by :

$$\left(-\infty,\quad \bar{x} + 1.645\frac{\sigma}{\sqrt{n}}\right).$$

- **Confidence interval for the variance of a normal distribution**
  The $100(1-\alpha)\%$ interval using a sample of n observations is $\left(\frac{(n-1)s^2}{U}, \frac{(n-1)s^2}{L}\right)$, where U and L are the upper and lower $\alpha/2$ points, of a chi-squared distribution with $n-1$ degrees of freedom.

R

Distributions

- The probability of a value less than x for a t-distribution with d degrees of freedom is given by `pt(x, d)`.
- The probability of a value less than x for a χ^2-distribution with d degrees of freedom is given by `pchisq(x, d)`.

Confidence intervals

First, download the `DescTools` library. Now type `library(DescTools)`.

- For a CI for a mean, with the data in `x`, use either `MeanCI(x,sd)` where `sd` is the known standard deviation, or, simply `MeanCI(x)`. In the latter case the t-distribution will be used.
- For a CI for the Poisson mean, $\bar{x} = x/n$, use `PoissonCI(x,n)`. There are four alternative methods considered (including the Wald interval), with the default being the Garwood limits provided by Equation (14.7).
- For a CI for a binomial proportion, use `BinomCI(r,n)`. This defaults to Wilson's method.
- For a CI for the variance of a normal distribution, use `VarCI(x)`. A total of seven alternative methods are provided. This defaults to the use of the χ^2-distribution. For a moderately non-normal distribution add `method='bonett'`.

All these intervals default to two-sided, 95% intervals. For a one-sided interval, add either `'left'` or `'right'` as an additional argument. To change the width of the interval to (say) 99%, add `conf.level=0.99`.

15

Single-sample hypothesis tests

Hypothesis tests are also called **significance tests**.

15.1 The null and alternative hypotheses

A hypothesis test compares a specific parameter value believed to be correct, the **null hypothesis** (H_0), with an **alternative hypothesis** (H_1) that specifies *a range of values* for the parameter in question. Here are some examples:

Parameter	*Null hypothesis*	*Alternative hypothesis*
Mean, μ	$\mu = 400$	$\mu \neq 400$
Proportion, p	$p = 1/2$	$p \neq 1/2$
Mean, μ	$\mu = 400$	$\mu < 400$
Proportion, p	$p = 1/2$	$p > 1/2$

Since H_0 specifies a value for the unknown parameter, we can determine the probabilities of events of interest (such as the sample mean being greater than 450, or the sample proportion being less than 0.01). This enables us to develop rules for deciding whether or not H_0 is acceptable.

Example 15.1

Ten independent observations are to be taken from a $\mathbf{N}(\mu,40)$ *distribution. The hypotheses are* $H_0 : \mu = 20$, $H_1 : \mu > 20$. *The following procedure has been proposed: 'Reject* H_0 *(and accept* H_1*) only if* $\bar{X} > 23.29$. *Determine the probability of rejecting* H_0 *when using this procedure.*

We begin by assuming that the value specified by H_0 is correct. In that case we are anticipating that the observations have been taken from a N(20, 40) distribution. The value of $\bar{X}$ is therefore, supposedly, a value from a $\text{N}\left(20, \frac{40}{10}\right)$ distribution. Using a table of percentage points for Z (a standard normal random variable), such as the table given near the start of Chapter 6, we find that

$$P\left(Z < \frac{23.29 - 20}{\sqrt{4}}\right) = P(Z < 1.645) = 0.95.$$

Hence, when H_0 is correct, then using the specified procedure, the probability of (incorrectly) rejecting H_0 is 5%.

With the computer, we can obtain the same result using a simple command.[a]

[a] Using R, the command is pnorm(23.29,20,2).

- In English law the prisoner in the dock is treated as being innocent until 'proved' guilty. In the same way, the null hypothesis is accepted, until strong evidence suggests that it should be rejected. However, acceptance of the null hypothesis does not mean that it it is correct (see below)!
- In practice, the null hypothesis will **never** be exactly true. For example, no coin is ever exactly fair, no die is ever completely unbiased, etc. However, the difference between the true parameter value and the value specified by H_0 ($p = 0.5$ for a coin, or $p = 1/6$ for a die) may make a trivial difference to subsequent calculations.

15.2 Critical regions and significance levels

The set of values that leads to the rejection of H_0 in favour of H_1 is called the **rejection region** or the **critical region**. The set of values that leads to the non-rejection of H_0 is the **acceptance region**.

When the population parameter has the value specified by H_0, the probability that H_0 is nevertheless rejected in favour of H_1 is called the **significance level**. Changing the significance level changes the size of the critical region. In Example 15.1, the significance level was 5% and the critical region was values of $\bar{x}$ greater than 23.29.

Hypothesis tests in which H_1 involves either a '>' sign (as in Example 15.1) or a '<' sign are called **one-tailed tests**. The critical regions in these cases involve values in the corresponding tail of the distribution specified by H_0.

Hypothesis tests in which H_1 involves a '$\neq$' sign are called **two-tailed tests**. In these cases the 'critical region' actually consists of two regions—one in each tail of the distribution specified by H_0.

The quantity with the value that determines the outcome of the test is often referred to as the **test statistic**. In Example 15.1 the test statistic was $\bar{X}$ (or, equivalently, Z, if working with the standardized value).

If the value of the test statistic falls in the critical (rejection) region, then the result is said to be **'significant'**. If the significance level were α%, then the result would be described as being **'significant at the α% level'**.

The default significance level is usually 5%. This was the choice of Fisher[a] who wrote that he 'prefers to set a low standard of significance at the 5 per cent point, and ignore entirely all results which fail to reach this level.'

However, significance at the 5% level should be regarded only as an indicator that further sampling (or other investigation) should take place.

[a] See Section 14.2.2

15.3 The test procedure

Theoretically the test procedure would be set out before the data are collected. It might look like this:

1. Write down H_0 and H_1.
2. Determine the appropriate test statistic and the distribution of the corresponding random variable (using the parameter value specified by H_0).
3. Decide on an appropriate significance level.
4. Determine the resulting acceptance and rejection regions.
5. Calculate the value of the test statistic.
6. Determine the outcome of the test.

In practice the data are often collected before the statistician is involved. When this is the case, it is important not to deliberately choose the significance level so that the outcome of the test satisfies the hope of some interested party.

15.4 Identifying the two hypotheses

15.4.1 The null hypothesis

This states that a parameter has some precise value. This might be:

- The value that occurred in the past,
- The value claimed by some person,
- The (target) value that is supposed to occur.

Sometimes the null hypothesis may not appear to refer to a precise value:

> 'The mean breaking strengths of types of climbing rope have never exceeded 200kg, and have sometimes been considerably less. It is claimed that a new rope brought onto the market has a breaking strength in excess of this figure. A random sample of 12 pieces of the new rope are tested.'

Here it appears that the hypotheses are

$$\begin{aligned} H_0&: \quad \mu \le 200 \text{ kg}, \\ H_1&: \quad \mu > 200 \text{ kg}. \end{aligned}$$

In order to see how to proceed, consider two alternative specific null hypotheses such as H_0': $\mu = 200$ kg and H_0'': $\mu = 190$ kg. Suppose we use H_0' and suppose that the outcome of the test is that H_0' is rejected in favour of H_1. Can we say what would have happened if we had used H_0''? The answer is that it too would have been rejected—if the mean of the sample values is so large that $\mu = 200$kg is rejected, then $(\bar{x} - 200)$ must be unacceptably large. Since $(\bar{x} - 190) > (\bar{x} - 200)$ this too must be unacceptably large. The same argument would apply for any value of μ less than 200 kg. Hence we can cover all the cases where μ is less than 200 kg by using

$$\begin{aligned} H_0&: \quad \mu = 200 \text{ kg}, \\ H_1&: \quad \mu > 200 \text{ kg}. \end{aligned}$$

15.4.2 The alternative hypothesis

The alternative hypothesis involves the use of one of the signs >, <, or ≠. A decision has to be made as to which is appropriate. Generally, exam questions attempt to signal which sign is to be used by means of suitable phrases:

$\neq$	'change', 'different', 'affected';
$>$ or $<$	'less than', 'better', 'increased', 'overweight'.

In real life the choice is not usually so clear cut! Suppose, for example, that we have a situation such as the following:

> "The mean breaking strength of a type of climbing rope is 200 kg. Scientists adjust the method of construction so that, they claim, there will be an increase in the breaking strength. A random sample of twelve pieces of the new rope are tested."

This appears very straightforward. We would use the hypotheses

$$\begin{aligned} H_0&: \quad \mu = 200 \text{ kg}, \\ H_1&: \quad \mu > 200 \text{ kg}. \end{aligned}$$

Suppose now that the twelve pieces of new rope have the following breaking strengths (in kg):

187, 196, 193, 187, 194, 193, 197, 194, 191, 195, 194, 199.

We evidently do not reject H_0 in favour of H_1—but do we really wish to believe that $\mu = 200$ kg? The new rope appears to have a mean breaking strength of about 193 kg, and not 200 kg. Some statisticians argue that, because of this type of situation, one-sided tests should never be used.

Exercises 15a

1. Jars of honey are filled by a machine. Samples of these filled jars are taken each week and their weights are found to have a mean of 475 g, with a standard deviation of 3 g. Following an overhaul of the machine, the weights of the next 100 sampled jars are measured in order to determine whether there has been a change in the mean weight. Write down an appropriate pair of hypotheses.
2. Observations of the time taken to inspect a piece of equipment for faults, show that, the average inspection time is 5.82 minutes, with a standard deviation of 0.63 minutes. An incentive scheme is to be introduced, with the next 80 inspections being timed to determine whether there has been a change in the mean inspection time. Write down an appropriate pair of hypotheses.

15.5 Tail probabilities: the *p*-value approach

The procedure set out in Section 15.3 will result in one of two statements: 'Accept H_0 at the chosen significance level', or 'Reject H_0 at the chosen significance level'. Suppose the result has been to reject H_0; whilst we would then know that the result had been an unusual one, the statement does not tell us the extent to which the outcome had been unusual. We would not know whether, if we had chosen a more extreme significance level (say 0.1% rather than 5%), we would still have rejected H_0.

One way of dealing with this dilemma is to quote the **p-value**, which is simply the probability, assuming the parameter value specified by H_0 is correct, of obtaining the observed value or a more extreme one. If we have observed the outcome y, say, we would typically need to report $P(Y \geq y)$ or $P(Y \leq y)$ for a one-sided test or $2P(Y \geq y)$ (or $2P(Y \leq y)$) for a two-sided test.[1]

15.6 Hypothesis tests and confidence intervals

For continuous variables there is a direct link between confidence intervals and significance tests:

> *If an α% symmetric confidence interval excludes the population value of interest, then the null hypothesis that the population parameter takes this value will be rejected at the $100(1 - \alpha)$% level.*

For example, if the symmetric 95% confidence interval for a population mean, μ, is (83.0, 85.1), then the null hypothesis that $\mu = 85.2$ will be rejected at the 5% level since the interval excludes 85.2. Indeed *any* hypothesized value for μ that is either greater than 85.1, or less than 83.0, will be rejected at the 5% level. Conversely, the hypothesis that μ takes *any* specific value in the range (83.0, 85.1) will not, at the 5% level, be rejected.

In the case of a one-sided alternative hypothesis, the link is with the corresponding one-sided confidence interval. Suppose, for example, that H_0 states that $\mu = 15$ with H_1 being that $\mu > 15$. Unusually large values of $\bar{x}$ will lead to rejection of H_0. Since H_1 is concerned with the upper-tail, the relevant confidence interval includes all of that tail. If, for example, the interval has the form $(15.2, \infty)$, which excludes the hypothesized 15, then H_0 would be rejected.

[1] The latter only applies if Y has a continuous distribution; if Y is discrete, then other alternatives may be preferable (because, for a discrete distribution, probability comes in 'chunks').

15.7 Hypothesis tests for a mean

15.7.1 Normal distribution with known variance

Knowing the variance, but being unsure about the mean, might seem an unlikely situation. However, it can occur when there is some adjustment (for example, a change of an ingredient or a change in temperature) that means that a manufacturing process can be expected to remain as variable as past records suggest, though the mean may have changed.

If the population being sampled is known to be normal with variance σ^2, and the null hypothesis specifies a mean μ, then the distribution of $\bar{X}$ is approximately (see Section 6.4)

$$\mathrm{N}\left(\mu, \frac{\sigma^2}{n}\right).$$

Example 15.2

A random sample of 36 observations is to be taken from a normal distribution with variance 100. In the past the distribution has had a mean of 83.0, but it is believed that, while the variance will not have been altered, the mean may recently have changed.

(a) *Using a 5% significance level, determine an appropriate test of the null hypothesis, H_0, that the mean is 83.0. When the sample is actually taken, it is found to have a mean of 86.2. Does this provide significant evidence against H_0?*

(b) *Suppose it is known that, if the population mean has changed, then it can only have increased. How would this knowledge affect the conclusions?*

(a) Since there is no suggestion that any change can only be in one direction, the test is two-sided:

$$\begin{aligned} \mathrm{H}_0&: \quad \mu = 83, \\ \mathrm{H}_1&: \quad \mu \neq 83. \end{aligned}$$

Since it is the population mean that is under investigation, the test statistic is the observed value of $\bar{X}$ (or a function thereof).

We can work directly with $\bar{x}$ or, since $\sigma^2 = 100$ and $n = 36$, with

$$z = \frac{\bar{x} - 83.0}{10/\sqrt{36}}.$$

Assuming H_0, z is an observation from a standard normal distribution.

The test is two-sided with 2.5% in each tail. Since $\mathrm{P}(Z > 1.96) = 0.025$, and $\mathrm{P}(Z < -1.96) = 0.025$, an appropriate procedure is to reject H_0 in favour of H_1 if z does not lie in the interval $(-1.96, 1.96)$. The corresponding acceptance interval for $\bar{X}$ is $(79.7, 86.3)$.

The test statistic is $\bar{x} = 86.2$ corresponding to $z = 1.92$. Since the test statistic lies in the acceptance region, we do not reject the null hypothesis that the mean is still 83.0. Note that this does *not* imply that the mean is unchanged, it is simply that the mean of our particular sample did not happen to fall in the rejection region.

In this case $P(Z > 1.92) \approx 0.0274$, so that the p-value is about 5.5%.[a]

(b) If it is known that the population mean cannot have decreased, then we will only be persuaded to reject H_0 if $\bar{x}$ is unusually large. The test is now one-sided with $H_1\colon \mu > 83.0$. Since $P(Z > 1.645) = 0.05$, H_0 will be rejected if $Z > 1.645$ or, equivalently, $\bar{X} > 85.7$. We therefore have significant evidence, at the 5% level, that the population mean has increased from its previous value. The p-value is about 2.7%.

[a] The doubling is because the test is two-sided.

Failure to reject H_0 does *not* mean that it is true. For any data set there will be an infinite number of null hypotheses that would not be rejected.

When the clock strikes midnight on day D, it is not the case that all foodstuffs with packaging that states 'Eat before day (D+1)' will immediately be uneatable!

Similarly, a p-value of 0.0499 and a p-value of 0.0501 should be viewed in the same way. Both are about 0.05 and both should be treated in the same way. p-values provide no more than a guide. On another day, another sample would give different results.

15.7.2 Unknown distribution, variance known, large sample

If the distribution is known to be reasonably symmetric then, because of the central limit theorem (Section 6.4), we can take the distribution of the sample mean to be normal, and proceed as in Section 15.7.1.

Exercises 15b

1. Jars of honey are filled by a machine. It has been found that the quantity of honey in a jar has mean 460.3 g with standard deviation 3.2 g. However, it is believed that the machine controls have been altered such that the mean may have changed, though the variability should be unaffected. A random sample of 60 jars is now taken and the mean quantity of honey per jar is found to be 461.2 g. State suitable null and alternative hypotheses.

   The computer reports a (2-sided) p-value of 0.02936488. Does this provide significant evidence, at the 5% level, of a change in mean?

2. A lightbulb manufacturer has established that the life of a bulb has mean 95.2 days with standard deviation 10.4 days. Following a change in the manufacturing process which is intended to increase the life of a bulb, a random sample of 96 bulbs has mean life 96.6 days. State suitable null and alternative hypotheses.

   Using a 1% significance level, the computer reports the confidence interval as (94.13071, ∞). Is there significant evidence of a change in mean lifetime?

15.7.3 Normal distribution, variance unknown

When the sample size is small, we use the t-distribution (Section 7.1) rather than the normal distribution. The test statistic, t, is defined by

$$t = \frac{\bar{x} - \mu}{s/\sqrt{n}},$$

where μ is the population mean specified by the null hypothesis. The corresponding random variable, T, has a t_{n-1}-distribution.

Example 15.3

Bottles of wine are supposed to contain 75 cl of wine. An inspector takes a random sample of six bottles of wine and determines the volumes of their contents, correct to the nearest half millilitre. Her results were

747.0, 751.5, 752.0, 747.5, 748.0, 748.0.

Determine whether these results provide significant evidence, at the 5% level, that the population mean is less than 75 cl.

It is simplest to work in millilitres. The target quantity, 75 cl, is 75/100 of a litre, which is the same as 750 millilitres.

The test is one-sided with the hypotheses being

$$\begin{aligned} H_0&: \quad \mu = 750, \\ H_1&: \quad \mu < 750. \end{aligned}$$

The test is one-sided with the t_5-distribution being the reference distribution. The upper 5% point of a t_5-distribution is 2.015, and hence, by symmetry, the lower 5% point is −2.015. An appropriate procedure is therefore to accept H_0 if t is greater than −2.015 and otherwise to reject H_0 in favour of H_1.

Since σ^2 is unknown, we use s^2 which is equal to 4.70. The test statistic is therefore

$$t = \frac{\bar{x} - 750}{\sqrt{4.70/6}} = -1.13,$$

which, assuming H_0, is an observation from a t_5-distribution.

Since $t > -2.015$, we accept H_0: there is no significant evidence, at the 5% level, that the bottles are being underfilled.[a]

[a] In this case the p-value is about 15.5%, which is considerably greater than the target 5%.

15.8 Testing for normality

A critical assumption behind both the z-test and the t-test is that the distribution is normal. The methods used here to determine whether that assumption is plausible, involve **order statistics**. The sample order statistics are simply the sample values arranged in order of increasing magnitude. Using the general suffix '(i)', the n sample order statistics are

$$x_{(1)} \leq x_{(2)} \leq \cdots \leq x_{(n)}.$$

The corresponding distribution order statistics are the n values of x for which

$$\mathrm{F}(x) = \frac{1}{n+1}, \frac{2}{n+1}, \ldots, \frac{n}{n+1}.$$

15.8.1 The Q–Q plot

It consists of a plot of the sample order statistics against the corresponding population order statistics. If the sample has been drawn from the population in question, then the plot will be approximately linear (though there may be some deviations at any extreme where the probability density is low).

> The *Q–Q* plot is a simple visual display that can be used with *any* distribution. It can also be used to compare two samples to observe whether they appear to have come from the same population.

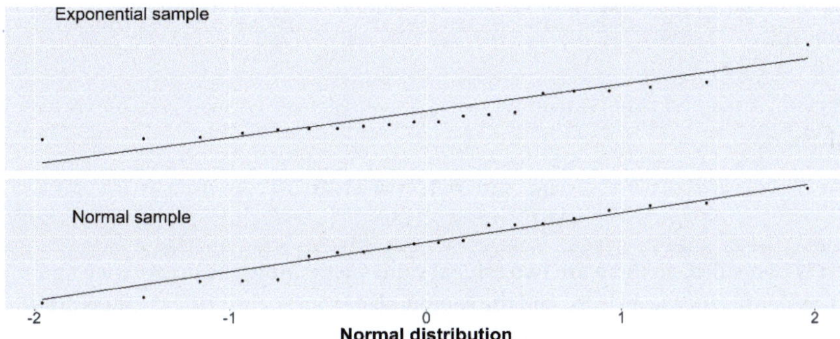

Figure 15.1 Two Q–Q plots comparing the distribution function for a normal distribution (on the x-axis) with 20 observations from an exponential distribution (top) and a normal distribution (bottom).

The Q–Q plot may be suggestive of a departure from normality, but it is not definitive. For that we need a test.

15.8.2 The Shapiro–Wilk test

The test statistic, W, is a complicated function of the order statistics and their means, variances and covariances.[2] The test requires a pre-programmed computer routine. Although the two Q–Q plots shown in Figure 15.1 do not appear greatly different to the eye, they give very different results for the the Shapiro-Wilk test: p-values of 0.44 for the truly normal data, but just 0.01 for the exponential data.

15.9 Hypothesis test for the variance of a normal distribution

Since the sample values, $x_1, x_2, ..., x_n$ come from a normal distribution with mean μ, the same is true for the sample mean, $\bar{x}$. The differences $(x_1 - \bar{x}), (x_2 - \bar{x}), ..., (x_n - \bar{x})$ are therefore observations from a normal distribution with mean 0, while each squared difference is an observation from a chi-squared distribution (Section 7.2) scaled by multiplication by σ^2. The sum of n *independent* χ^2_1 distributions would have a χ^2_n-distribution, but these n differences are *not* independent of one another: they are linked by their sum being 0. If one knows the values of any $(n-1)$ of these values, then one can deduce the value of the nth.[3] The result is that the sample variance $s^2 = \frac{1}{n-1}\sum(x_i - \bar{x})^2$ is an observation from a χ^2_{n-1} distribution scaled by multiplication by σ^2.

Example 15.4

> *Prior to overhaul, a machine is cutting planks of wood with a standard deviation of 4.7 mm. After the overhaul, a random sample of 30 planks is examined and their lengths are found to give* s = 3.84 *mm. Does this provide significant evidence, at the 5% significance level, of a change in population standard deviation? A normal distribution may be assumed.*

Here we have

$$\begin{aligned} \mathrm{H_0}: &\quad \sigma^2 = 4.7^2 = 22.09, \\ \mathrm{H_1}: &\quad \sigma^2 \neq 22.09. \end{aligned}$$

The test is two-sided, so there are two critical values separating the acceptance and rejection regions. The significance level is 5% and the sample size is 30, so the critical values are 16.05 and 45.72, the lower and upper 2.5% points of a χ^2_{29} distribution. The test procedure is therefore to accept $\mathrm{H_0}$ if the observed value of $(n-1)s^2/\sigma_0^2$ lies between 16.05 and 45.72, and otherwise to reject $\mathrm{H_0}$ in favour of $\mathrm{H_1}$.

The observed value of $(n-1)s^2/\sigma_0^2$ is $29 \times 3.84^2/22.09 = 19.36$. This value falls inside the acceptance region[a] and there is therefore no significant evidence, at the 5% level, to suggest that the machine's accuracy has altered.

---

[a] The probability of obtaining a value of 19.36, or less, from a χ^2_{29}-distribution is 0.088.

[2] The test was published in 1965 based on research by the authors (the American Samuel Shapiro (1930–2023) and the Canadian Martin Wilk (1922-2013)) while they were both on the faculty at Rutgers University.

[3] It will be equal to the negative of the sum of the $(n-1)$ that have known values.

15.10 Hypothesis tests with discrete distributions

For both the Poisson and the binomial distributions, hypothesizing the value of a parameter has the result that one has also hypothesized the value of the variance. In the case of the Poisson, the variance equals the mean. In the case of the binomial, specifying a value for p fixes the value of both mean and variance.

15.10.1 Test for a Poisson mean

If X has a Poisson distribution with a large mean, λ, then the distribution can be approximated (see Section 6.5.2) by

$$N(\lambda, \lambda).$$

Using this approximation we need to calculate

$$z = \frac{x - \lambda_0}{\sqrt{\lambda_0}},$$

where x is the observed value and λ_0 is the hypothesized mean. The advantage of the normal approximation is that an approximate value for z requires only some mental arithmetic. That mental arithmetic would continue by comparing the value obtained with 1.96 (the two-sided 5% significance point). The outcome of the test will then often be apparent without the need to use the computer to provide an accurate p-value.

Example 15.5

In a particular river a certain micro-organism occurs at an average rate of 10 per millilitre. A random sample of 0.5 litres of water is taken from a nearby stream and is found to contain 4778 micro-organisms. Does this provide significant evidence, at the 5% level, of a difference in the incidence of the micro-organisms between the stream and the river?

If the incidence in the stream were the same as that in the river, then 0.5 litres (i.e. 500 millilitres) of stream water would contain an average of $10 \times 500 = 5000$ micro-organisms. The question refers to a 'difference' and there is no implication that a low count was anticipated when sampling the stream; we therefore use a two-sided alternative hypothesis:

$$\begin{aligned} H_0: &\quad \lambda = 5000, \\ H_1: &\quad \lambda \neq 5000. \end{aligned}$$

We assume that the micro-organisms are randomly distributed in the pond water, so that a Poisson distribution is appropriate. The single count, 4778, is therefore an observation from a Poisson distribution with mean 5000. The approximating test statistic is

$$z = \frac{4778 - 5000}{\sqrt{5000}} = -3.1.$$

Since the observed -3.1 is well below -1.96, we can confidently state that there is a difference between the incidence of the micro-organisms in the stream and the river.[a]

[a] The computer reports that the probability of obtaining a value of 4778 or less from a Poisson distribution with mean 5000 is approximately 0.0008.

Exercises 15c

1. Rolls of plastic sheeting from the usual manufacturer have minor faults at an average rate of 0.32 per metre. A 100-metre roll is obtained from a new manufacturer. It is found to have 27 minor faults. We wish to use a 5% level to determine whether there is significant evidence that the fault rate for the new manufacturer is significantly different from that for the usual manufacturer.

   Is this a two-sided or a one-sided test? The computer reports that, based on the observed value of 27, the 95% confidence interval for the number per 100 m is (17.79317, 39.28358). Is there significant evidence of a change in fault rate?

2. A traffic survey shows that, between 9 a.m. and 10 a.m., cars pass a particular census point at an average rate of 6.5 per minute. After the opening of a nearby supermarket, the total number of cars passing the census point between 9 a.m. and 10 a.m. is 458.

   State suitable null and alternative hypotheses. The computer reports a p-value of 0.0004279. Interpret this result.

15.10.2 Test for binomial proportion

With a sample of size n that contains r successes, evidence concerning the population success probability, p, is provided by the sample proportion $\widehat{p}$, defined by $\widehat{p} = r/n$, with the corresponding random variable being denoted by $\widehat{P}$. When n is large, the distribution of $\widehat{P}$ is approximately (see Section 6.5.1)

$$\mathrm{N}\left(p, \frac{p(1-p)}{n}\right).$$

As with the Poisson mean, although the computer will provide a more accurate result, the normal approximation is still useful: a rough mental calculation is often sufficient to provide an idea of the test's outcome. Using this approximation, the test statistic is z given by

$$z = \left(\frac{r}{n} - p_0\right) \Big/ \sqrt{\frac{p_0(1-p_0)}{n}},$$

where p_0 is the hypothesized success probability.

Example 15.6

A golf professional sells wooden tees. The type that he usually sells are very brittle, and 25% break on the first occasion that they are used. Hoping for an improvement, the golf professional buys a batch of 'Longlast' tees. From a random sample of 100 of these tees only 18 break on the first occasion that they are used. Does this provide significant evidence, at the 1% level, of the desired improvement?

In this case a success is a breakage! The hypotheses are:

$$\begin{aligned} H_0: &\quad p = 0.25, \\ H_1: &\quad p < 0.25. \end{aligned}$$

The test statistic is

$$z = \frac{\hat{p} - 0.25}{\sqrt{(0.25 \times 0.75)/100}} = -1.50.$$

The lower 1% percentage point of a normal distribution is -2.326. Since $P(Z > 2.326) = 0.01$, $P(Z < -2.326) = 0.01$, and an appropriate (approximate) test procedure is therefore to reject H_0 if $z < -2.326$.

Since -1.50 is greater than -2.326 it lies in the acceptance region. Therefore, at the (approximate) 1% level, the sample result does not give significant evidence that the proportion of 'Longlast' tees which break when first used is any less than that for the previous tees.[a]

[a] The computer reports that the probability of obtaining a value of 18 or less from a binomial distribution with $p = 0.25$ is approximately 0.063, which is appreciably more than 1%.

Exercises 15d

1. A rail company claims that only 4.3% of its trains are late. This is believed to be a false claim, so a passenger association carries out a check on a random sample of 500 trains, finding that 30 trains are late. The computer output from an appropriate test is as follows:

```
data: 30 out of 500, null probability 0.043
X-squared = 3.1105, df = 1, p-value = 0.07779
alternative hypothesis: true p is not equal to 0.043
95 percent confidence interval:
0.04151525 0.08551115
```

   State your conclusions.

2. A survey in a university library reveals that 12% of returned books were overdue. After a big increase in fines for overdue books, a random sample of 80 returned books found that only 6 were overdue. The computer output from an appropriate test is as follows:

```
data: 6 out of 80, null probability 0.12
X-squared = 1.1375, df = 1, p-value = 0.1431
alternative hypothesis: true p is less than 0.12
95 percent confidence interval:
0.0000000 0.1462284
```

   State your conclusions.

15.11 Type I and Type II errors

When conducting hypothesis tests there are two types of error that may occur, as the table below shows.

		Our decision	
		We do not reject H_0	We reject H_0
Reality	H_0 correct	Correct!	TYPE I ERROR
	H_0 incorrect	TYPE II ERROR	Correct!

As the table shows, a **Type I error** is made if a correct null hypothesis is rejected. The probability of this error is under our control since:

P(Type I error) = Significance level.

Calculation of the probability of a **Type II error** is not so straightforward, since the probability depends on the extent to which H_0 is false. If H_0 is only slightly incorrect, then we may not notice that it is wrong, in which case the probability of a Type II error will be large. On the other hand, if H_0 is nothing like correct, then the probability of a Type II error will be low.

In a more positive frame of mind, rather than asking about the probability of making an error, we can ask how good a test is at detecting a false null hypothesis. This is known as the **power** of a test. Formally:

Power = 1 − P(Type II error).

15.11.1 The general procedure

This closely follows that for the construction of hypothesis tests:

1. Write down the two hypotheses, for example, $H_0: \mu = \mu_0$ and $H_1: \mu > \mu_0$.
2. Determine the appropriate test statistic and the distribution of the corresponding random variable (using the parameter value specified by H_0).
3. Determine the significance level. This is P(Type I error).
4. Determine the acceptance and rejection regions.

Now we suppose that the value of the parameter is not that specified by H_0

5. Determine the distribution of the random variable corresponding to the test statistic given the actual parameter value (μ_1, say).
6. Calculate the probability of an outcome falling in the acceptance region (given $\mu = \mu_1$). This is P(Type II error) for the case where $\mu = \mu_1$.

Example 15.7

A machine is supposed to fill bags with 38kg of sand. It is known that the quantities in the bags vary and that the bags have a standard deviation of 500g. When a new employee starts using the machine it is standard practice to determine the masses of a random sample of 20 bags taken from the first batch produced by the employee in order to verify that the mean of the machine has been set correctly. Determine an appropriate test procedure, given that it is desired that the probability of a Type I error should be 5%. Suppose that an employee has set the machine so that it fills bags with an average of μ kg. Determine the probability of a Type II error:

(a) *in the case* $\mu = 38.1$,

(b) *in the case* $\mu = 38.4$.

The test is two-sided with the hypotheses being

$$\begin{aligned} H_0&: \quad \mu = 38.0, \\ H_1&: \quad \mu \neq 38.0. \end{aligned}$$

The appropriate test is one that uses the sample mean, $\bar{x}$. Assuming H_0 is correct, by virtue of the central limit theorem the distribution of $\bar{X}$ is approximately

$$N\left(38.0, \frac{0.250}{20}\right),$$

so that the appropriate test statistic is

$$z = \frac{\bar{x} - 38.0}{\sqrt{0.250/20}}.$$

The acceptance region (in terms of z) is $(-1.96, 1.96)$. In terms of $\bar{x}$ this corresponds to $(37.78, 38.22)$.

(a) If the true mean, μ, is 38.1, then the distribution of $\bar{X}$ is $N\left(38.1, \frac{0.250}{20}\right)$.
The probability of failing to reject H_0 is $P(37.78 < \bar{X} < 38.22) = 0.855$ (to 3 d.p.).

(b) If the true mean, μ, is 38.4, then the distribution of $\bar{X}$ is $N\left(38.4, \frac{0.250}{20}\right)$.
The probability of failing to reject H_0 is $P(37.78 < \bar{X} < 38.22) = 0.053$ (to 3 d.p.).

Exercises 15e

1. It is given that $X \sim N(\mu, 16)$. It is desired to test the null hypothesis $\mu = 12$ against the alternative hypothesis $\mu > 12$, with the probability of a Type I error being 1%. A random sample of 15 observations of X is taken and the sample mean $\bar{X}$ is taken to be the test statistic.
   (a) Find the acceptance and rejection regions.
   (b) For the case $\mu = 15$, find the probability of a Type II error and the power of the test.
2. A coin, believed to be fair, is tossed 100 times. The hypothesis that the coin is fair will be accepted if the number of heads obtained lies between 40 and 60, inclusive. It is given that $P(X \leq 60|p = 0.5) = 0.9824$, and $P(X \leq 39|p = 0.5) = 0.0176$,
   while $P(X \leq 60|p = 0.6) = 0.5379$, and $P(X \leq 39|p = 0.6) \approx 0$,
   where X is a binomial random variable with success probability p. Determine the probability of:
   (a) a Type I error,
   (b) the probability of a Type II error for the case where the probability of a head is 0.6.

15.11.2 The power curve

We have seen that the probability of a Type II error is a function of the actual parameter value. Since the power of a test is simply the probability of not making a Type II error, it too is a function of the actual parameter value. This dependence is referred to as the **power function** of the hypothesis test, while a plot of the power function (on the y-axis) against the parameter value (on the x-axis) is known as the **power curve**.

Example 15.7 (cont.)

We already have two values for P(Type II error), namely 0.855 when $\mu = 38.1$ and 0.053 when $\mu = 38.4$. These values correspond to values for power of 0.145 and 0.95, respectively. Also, by construction, since this is a 5% test, the power when $\mu = 38.0$ is 0.05. For any value of μ the power can be determined and the results are illustrated in the following diagram:

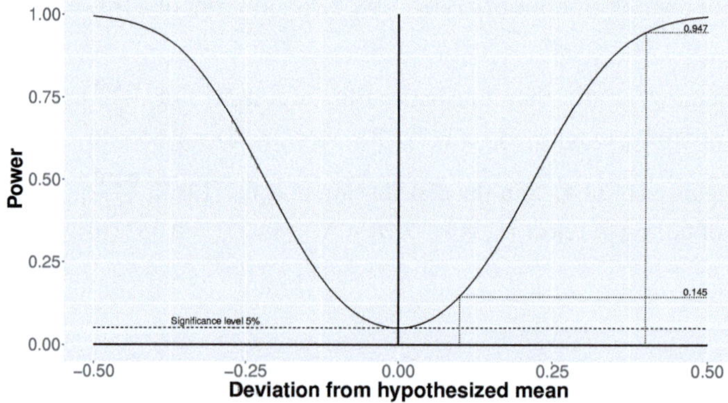

Similar calculations would apply to any hypothesis test and there are computer routines that automate the process.[4]

[4] In the case of a t-test, in order to determine the Type II error, the computer uses (not that we really need to know) the asymmetric **non-central t-distribution** (which has the symmetric t-distribution as a special case).

15.11.3 Power and sample size

In the previous example, a consequence of using a sample size of 20 was that, if the true mean differed from that specified by 0.1, then there was a high probability (0.875) of a Type II error. Suppose that we want to reduce that probability to just 5% (i.e. a power of 95%). How do we achieve that aim? The answer is to use a larger sample size. This gives us a more accurate estimate of the population mean (remember that the variance of the sample mean is proportional to $1/n$).

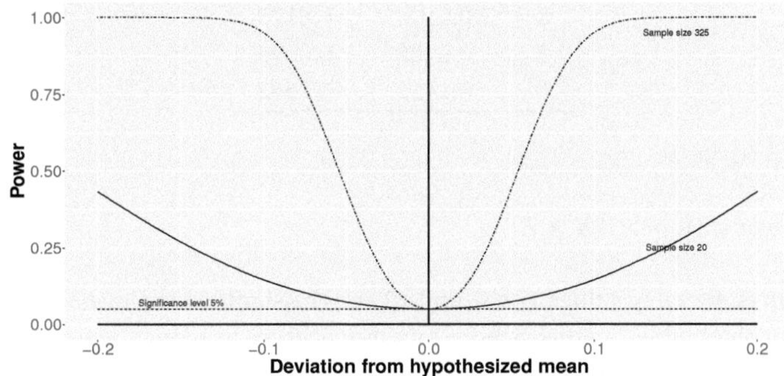

Figure 15.2 Increasing the sample size increases the power of a test.

Figure 15.2 compares the power curve for the previous situation with that which results from using a sample of $n = 325$ observations (the value suggested by the computer).[5]

15.12 Hypothesis tests for a proportion based on a small sample

The difficulties associated with this type of situation are best illustrated with an example. Consider the following problem:

> The standard treatment for a particular disease is successful on only 40% of occasions. A new treatment is introduced that is supposed to be better. Initially the treatment is given to just10 patients: the treatment is successful eight times. Does this provide significant evidence, at the 5% level, that the new treatment is significantly better than the standard treatment?

We will ignore the fact that the results of the new treatment might have appeared worse than the standard treatment, since there appears to be a strong expectation that the new treatment is at least as good as the old. So the test is one-sided and the hypotheses are

$$\begin{aligned} \mathrm{H_0}: &\quad p = 0.4, \\ \mathrm{H_1}: &\quad p > 0.4. \end{aligned}$$

Assuming $\mathrm{H_0}$, the distribution of the corresponding random variable, X, the number of successes, is

$$\mathrm{B}(10, 0.4).$$

[5] The computer reports that a power of 95% results from choosing $n = 324.9$, which we round up to 325.

The test is one-sided, since only if we see a large value of x are we going to reject H_0 in favour of H_1. We therefore need to look at the probabilities of large values of x occurring under the conditions specified by H_0:

r	$P(X = r)$	$P(X \geq r)$
10	0.0001	0.0001
9	0.0016	0.0017
8	0.0106	0.0123
7	0.0425	0.0548
6	0.1115	0.1662

The best choice appears to be

Reject H_0 in favour of H_1 only if $X \geq 7$.

The rejection region is the set of values 7, 8, 9, and 10 and the significance level is therefore 5.48% and not 5% (which is not obtainable). In this case we might refer to 5% as being the **nominal significance level**, with 5.48% being the **actual significance level** corresponding to the rejection region selected.

A conservative approach would be to define the critical region so that the actual significance level was not greater than the nominal significance level. In the present example this would imply using the following strategy:

Reject H_0 in favour of H_1 only if $X \geq 8$.

This strategy has an actual significance level of 1.23%.

There is no 'correct' solution: the pragmatic procedure is to examine the confidence interval associated with the observed outcome and accept the null hypothesis if the value that the hypothesis specifies lies in that interval.[a]

[a] But note that R provides 16 possible confidence intervals!

15.13 Hypothesis tests for a Poisson mean based on a small sample

The problems here are essentially the same as those encountered in Section 15.12. However, since Poisson distributions have infinite range, it is always sensible to work upwards from the outcome 0. The following example illustrates the procedure.

Example 15.8

A company uses a large number of floppy disks. At random intervals in time, disks develop faults: on average, 0.4% of black disks fail per month. The company also has blue disks. During a randomly chosen nine-month period, a random sample of 100 blue disks, develop a total of seven faults. Is there significant evidence, at the 5% level, that the failure rate of the blue disks is not 0.4% per month?

The test is two-sided. For convenience we will work with X, the total number of faults in 100 disks during a nine-month period. Since faults occur at random intervals in time, this has a Poisson distribution. The hypotheses are therefore

$$\begin{aligned} \mathrm{H}_0: &\quad \text{Rate} = 0.4\%, \\ \mathrm{H}_1: &\quad \text{Rate} \neq 0.4\%. \end{aligned}$$

Assuming H_0 is correct, the distribution of X is Poisson with mean $0.004 \times 100 \times 9 = 3.6$..

The test is two-sided, so we need to consider the cumulative probabilities associated with each tail of the distribution:

r	$P(X = r)$	$P(X \leq r)$	$P(X \geq r)$
0	0.0273	0.0273	1.0000
1	0.0984	0.1257	0.9727
2	0.1771		0.8743
3	0.2125		0.6973
4	0.1912		0.4848
5	0.1377		0.2936
6	0.0826		0.1559
7	0.0425		0.0733
8	0.0191		0.0308
9	0.0076		0.0117

In the lower tail, the nearest that we can get to 2.5% is 2.73%. In the upper tail, the nearest that we can get is 3.08%. We therefore propose the decision rule:

> Reject H_0 in favour of H_1 if the observed value of X is either 0, or at least 8. Otherwise, accept H_0.

The actual significance level is $2.73\% + 3.08\% = 5.81\%$.

The observed value of X was 7. Since 7 lies in the acceptance region $\{1, 2, 3, 4, 5, 6, 7\}$, we accept the null hypothesis (that the rate is 0.4%) using a significance level of 5.81%. The findings do not provide significant evidence that the blue disks have a different failure rate to the black disks.

Exercises 15f

1. A die is suspected of being biased towards the score of 6. It is thrown 10 times and the number n of sixes is observed. State suitable null and alternative hypotheses, and determine the acceptance and rejection regions for a test at the nominal 1% significance level. For a binomial distribution with $n = 10$ and $p = 1/6$, the probabilities of the various outcomes are summarized below:

Outcome	0	1	2	3	4	5	6	≥ 7
Probability	0.162	0.323	0.291	0.155	0.054	0.013	0.002	0.000

State the actual significance level.

2. I used to receive, on average, just two WhatsApp messages a day. Following the publication of a new book I suspect that the rate at which I get these messages will rise. Devise a suitable test using a 5% significance level. For a Poisson distribution with mean 2, the probabilities of the various outcomes are summarized below:

Outcome	0	1	2	3	4	5	6	≥ 7
Probability	0.135	0.271	0.271	0.180	0.090	0.036	0.012	0.004

State the actual significance level.

Key facts

- Two hypotheses:
  - **Null hypothesis**, H_0: Parameter has usual value or target value.
  - **Alternative hypothesis**, H_1: Parameter has changed, increased, decreased.
- **Critical region** = rejection region = range of values of the test statistic for which H_0 is rejected in favour of H_1. Remainder is acceptance region.
- **Significance level** = P(test statistic falls in rejection region given the situation described by H_0).
- **Test statistic**
  - **For a mean**, assuming a normal distribution or n large, symmetric distribution:

$$\frac{\bar{x}-\mu}{\sigma/\sqrt{n}} \sim \mathrm{N}(0,1).$$

  - **For a mean**, assuming a normal distribution with σ unknown:

$$\frac{\bar{x}-\mu}{s/\sqrt{n}} \sim t_{n-1}.$$

  - **For a Poisson parameter** (approximate test):

$$\frac{\bar{x}-\mu}{\sqrt{\mu}} \sim \mathrm{N}(0,1).$$

  - **For a proportion** (approximate test):

$$\frac{|\hat{p}-p|-0.5}{\sqrt{p(1-p)/n}} \sim \mathrm{N}(0,1).$$

  - **For a normal variance**:

$$(n-1)s^2/\sigma_0^2 \sim \chi^2_{n-1}.$$

R

All tests default to the 5% level and two-sided. The user can add `alternative="less"` or `alternative="greater"`, and also, for instance, `conf.level=0.99`, to override the defaults. The tests all report confidence intervals and *p*-values.

- **Test of normal mean, variance known**
  - Using a sample vector, `x`, first load the relevant library: `library(DescTools)`. Then issue the command: `ZTest(x,mu=pmu,sd_pop=psd)` where `pmu` is the hypothesized mean and `psd` is the known standard deviation.
  - Using a reported sample mean, create the vector `x` using `x=c(rep(xbar,times=n))`, then proceed as above.
  - A typical two-sided Type II error calculation has the form `pnorm(U,M,SD)-pnorm(L,M,SD)` where `M` is the true mean, `SD` is the known sd of the sample mean, and `U` and `L` are the upper and lower limits of the acceptance region. These can be calculated using (for the 95% case) `qnorm(c(0.025,0.975),`μ_0`,SD)`.
- **Test of normal mean, variance unknown**
  Use `t.test(x,pmu)` with `x` and `pmu` as defined above.
- **Test of Poisson mean**
  For a single value $\bar{x}$ from a distribution with hypothesized mean, μ, use `poisson.test(`$\bar{x}$`,1,`μ`)`. Alternatively, for a vector `x` of n values, use `poisson.test(x,n,`μ`)`.
- **Test of binomial proportion**
  With hypothesized value p and observed x successes in n trials, use: `binom.test(x,n,p)`.
- **Test for normal variance**
  - Using a reported sample variance, s^2, use `pchisq(y,(n-1))`, where $y = (n-1)s^2/\sigma_0^2$, n is the sample size, and σ_0^2 is the hypothesized variance.
  - Using a sample vector, `x`, load the `EnvStats` library, then use: `varTest(x,`σ_0^2`)`. Default is 5% test and two-sided.
- **Required tail percentage** Use e.g. `qbinom(0.01,n,p)`.
- **Shapiro–Wilk test for normality** With the data in a vector `x`, the command `shapiro.test(x)` returns the value of W and the corresponding p-value.

- **Power calculations**
  - First install the pwr library, then invoke it with the library(pwr) command.
  - For testing a hypothesis about the mean of a normal distribution, let d be the difference between the hypothesized and true means, s be the known standard deviation, n be the sample size, power be the power (or required power), and alternative be one of "two.sided", "less" or "greater". If the values of exactly three of d, n, power, and sig. level are provided, then the routine provides the value of the fourth.
    - With known variance use pwr.norm.test(n, d/s, sig.level, power, alternative).
    - With unknown variance use pwr.t.test(n, d/s, sig.level, power, type = "one.sample", alternative).
  - For a test of a hypothesized binomial proportion, p_0, against the alternative p_1, use pwr.p.test(h, n, sig.level, power, alternative) where

$$h = 2 * \arcsin(\sqrt{p_0}) - 2 * \arcsin(\sqrt{p_1}).$$

    If the values of exactly three of h, n, power, and sig. level are provided, then the routine provides the value of the fourth.

16

Two samples & paired samples

Here are some examples of the types of question with which this chapter is concerned.

- Is it true that, on a motorway, male drivers drive faster than female drivers?
  - We need to collect independent random samples of the speeds of male drivers and of female drivers. For each sex we calculate the mean speed. We then *compare the means* of the two samples.
- Is it true that male drivers are more likely to be prosecuted for speeding?
  - We need to collect independent random samples of male drivers and of female drivers. For each sex we calculate the proportion prosecuted for speeding. We then *compare the proportions* in the two samples.
- Is it true that, for right-handed people, the span of their right hand is greater than the span of their left hand?
  - This time each randomly chosen individual, who may range in size from Tarzan to Tom Thumb, provides two observations—one for the right-hand data set and one for the left. The variation in the sizes of the individuals does not matter—what matters is the difference *within* each of these pairs of observations: a *paired test* is appropriate.

16.1 The comparison of two means

We wish to test the hypothesis that two populations have the same mean. The general approach is the same as that set out in Section 15.3 for a single sample. The form of the test depends on what is known about the populations and (to an extent) whether the samples are small or large.

16.1.1 Normal distributions with known variances

This rather unlikely situation is included because it demonstrates the manipulation of normal random variables. The random variable X has unknown mean μ_x and known variance σ_x^2. The independent random variable Y has unknown mean μ_y and known variance σ_y^2.

The null hypothesis is

$$\mathrm{H_0}: \mu_x = \mu_y, \text{ or equivalently, } \mu_x - \mu_y = 0.$$

The alternative hypothesis may be two-sided,

$$\mathrm{H_1}: \mu_x \neq \mu_y,$$

or one-sided, e.g.

$$\mathrm{H_1}: \mu_x > \mu_y.$$

Since the hypotheses concern the population means, the test statistic will involve the sample means $\bar{x}$ and $\bar{y}$. Suppose that the samples have sizes n_x and n_y, then, since X and Y have normal distributions, the same will be true for $\bar{X}$ and $\bar{Y}$:

$$\bar{X} \sim \mathrm{N}\left(\mu_x, \sigma_x^2/n_x\right),$$

$$\bar{Y} \sim \mathrm{N}\left(\mu_y, \sigma_y^2/n_y\right).$$

Therefore $\bar{X} - \bar{Y}$ has a normal distribution with mean $\mu_x - \mu_y$ and variance $\sigma_x^2/n_x + \sigma_y^2/n_y$. Standardizing, we could write that

$$\frac{(\bar{X} - \bar{Y}) - (\mu_x - \mu_y)}{\sqrt{\sigma_x^2/n_x + \sigma_y^2/n_y}} \sim \mathrm{N}(0, 1).$$

Assuming $\mathrm{H_0}$, so that $\mu_x - \mu_y = 0$, we would then calculate:

$$z = \frac{\bar{x} - \bar{y}}{\sqrt{\sigma_x^2/n_x + \sigma_y^2/n_y}},$$

and proceed as usual.

In essence, this is what the computer will do behind the scenes.

16.1.1.1 Confidence interval for the common mean

If the null hypothesis $\mu_x = \mu_y$ is accepted, then all n_x observations on X $(x_1, x_2, \ldots, x_{n_x})$ and n_y observations on Y $(y_1, y_2, \ldots, y_{n_y})$ come from populations having the same mean. A natural **pooled estimate of the population mean**, denoted by $\hat{\mu}$, is therefore given by

$$\hat{\mu} = \frac{\sum_{i=1}^{n_x} x_i + \sum_{j=1}^{n_y} y_j}{n_x + n_y} = \frac{n_x\bar{x} + n_y\bar{y}}{n_x + n_y}. \tag{16.1}$$

The distribution of the corresponding random variable is

$$\mathrm{N}\left(\mu, \frac{n_x\sigma_x^2 + n_y\sigma_y^2}{(n_x + n_y)^2}\right).$$

Following the usual arguments, the corresponding 95% confidence interval for μ is

$$\left(\hat{\mu} - 1.96\frac{\sqrt{n_x\sigma_x^2 + n_y\sigma_y^2}}{(n_x + n_y)},\ \hat{\mu} + 1.96\frac{\sqrt{n_x\sigma_x^2 + n_y\sigma_y^2}}{(n_x + n_y)}\right).$$

If $\sigma_x^2 = \sigma_y^2$ $(= \sigma^2$, say), then, writing n for $n_x + n_y$, the confidence interval simplifies to become

$$\left(\hat{\mu} - 1.96\frac{\sigma}{\sqrt{n}},\quad \hat{\mu} + 1.96\frac{\sigma}{\sqrt{n}}\right).$$

16.1.2 Non-normal distributions, variances known, large samples

If the distributions are reasonably symmetric, then, because of the central limit theorem (Section 6.4), we can take the distributions of the sample means to be normal, and proceed as in Section 16.1.1.

16.1.3 Normal distributions with unknown common variance

Consider the hypotheses

$$\begin{aligned} &H_0: \mu_x = \mu_y,\\ &H_1: \mu_x > \mu_y. \end{aligned}$$

We are assuming that $\sigma_x^2 = \sigma_y^2$ $(= \sigma^2$, say). The samples have n_x and n_y observations, with sample means $\bar{x}$ and $\bar{y}$, respectively. Assuming H_0, $\bar{X} - \bar{Y}$ has mean 0 and variance $\sigma^2/n_x + \sigma^2/n_y$. However, σ^2 is unknown, so an estimate will be needed before progress can be made.

The sample information about variability is contained in the squared deviations of the observations from their respective means. The unbiased estimate of σ^2 based on the sample of x-values alone is given by s_x^2, where

$$s_x^2 = \frac{1}{n_x - 1}\sum(x_i - \bar{x})^2.$$

It follows that

$$\sum(x_i - \bar{x})^2 \text{ is an unbiased estimate of } (n_x - 1)\sigma^2.$$

Similarly, using

$$s_y^2 = \frac{1}{n_y - 1}\sum(y_j - \bar{y})^2,$$

we can state that

$$\sum(y_j - \bar{y})^2 \text{ is an unbiased estimate of } (n_y - 1)\sigma^2.$$

Adding these quantities together we find that

$$\sum(x_i - \bar{x})^2 + \sum(y_j - \bar{y})^2 \text{ is an unbiased estimate of } (n_x + n_y - 2)\sigma^2.$$

The so-called **pooled estimate of variance**, s_p^2, which is an unbiased estimate of σ^2, is therefore defined by

$$s_p^2 = \frac{\sum(x_i - \bar{x})^2 + \sum(y_i - \bar{y})^2}{n_x + n_y - 2}. \tag{16.2}$$

With normal distributions and an estimated variance, we once again use the t-distribution, with the test statistic being

$$t_p = \frac{\bar{x} - \bar{y}}{\sqrt{s_p^2(1/n_x + 1/n_y)}}. \tag{16.3}$$

In this case the reference distribution is a t-distribution with $n_x + n_y - 2$ degrees of freedom.

As a consequence of the central limit theorem, the same test would apply for large samples from non-normal distributions unless they were very skewed.

Put theory into practice: *This project relies on the observer being able to deduce the approximate age of a car using the car's number plate.*

Are the cars in the train station car park newer than those in the supermarket car park? This could be the case if the 'bread-winner' uses the newest car to drive to the station (and thence to work), while the bread-winner's partner takes an older car to do the shopping. Choose random samples of 50 cars in each situation, record the car approximate ages and perform an appropriate two-sample test to determine whether there is significant evidence to reject the hypothesis that the populations have a common mean.

16.1.4 Normal distributions with unknown variances

If the samples come from normal distributions that may have both different means and different variances, then the natural test statistic is

$$t_w = (\bar{x} - \bar{y}) \Bigg/ \sqrt{\frac{s_x^2}{n_x} + \frac{s_y^2}{n_y}}\,. \tag{16.4}$$

This is variously known as **Welch's** ***t*****-test**, the **Satterthwaite test**, or the **Satterthwaite–Welch test.**[1] The reference distribution is again a t-distribution. However, in this case the number of degrees of freedom, d_w, is not a simple integer. Instead it is given by

$$d_w = \left(\frac{s_x^2}{n_x} + \frac{s_y^2}{n_y}\right)^2 \Bigg/ \left(\frac{s_x^4}{n_x^2(n_x - 1)} + \frac{s_y^4}{n_y^2(n_y - 1)}\right).$$

If using tables, it would suffice to round to the nearest integer, but it is preferable to let the computer do the hard work.

Group project: How long does it take to thread a needle? Are females better at this task than males? Are left-handers better than right-handers? Are those wearing glasses better than those not wearing glasses? The answers to these questions can be quickly discovered! Record the times taken to the nearest second, obtain the class estimate of the common variance, and perform a two-sample test for a difference between the means. Decide beforehand what significance level to use and also whether you feel that a one-tailed or a two-tailed test is appropriate.

[1] The English statistician Bernard Welch first addressed this problem in a paper published in 1938. The American statistician Franklin Satterthwaite developed Welch's ideas in a 1941 paper. A 1947 paper by Welch further developed the theory and popularized the test.

16.2 Confidence interval for the difference between two normal means

16.2.1 Known variances

When σ_x^2 and σ_y^2 are known, we can obtain a confidence interval for the difference between the two population means by using an argument that by now has become rather familiar. We know that

$$\frac{(\bar{X} - \bar{Y}) - (\mu_x - \mu_y)}{\sqrt{\sigma_x^2/n_x + \sigma_y^2/n_y}} \sim \mathrm{N}(0, 1).$$

A 95% confidence interval for the difference is therefore given by

$$\left(\bar{x} - \bar{y} - 1.96\sqrt{\frac{\sigma_x^2}{n_x} + \frac{\sigma_y^2}{n_y}}, \quad \bar{x} - \bar{y} + 1.96\sqrt{\frac{\sigma_x^2}{n_x} + \frac{\sigma_y^2}{n_y}}\right).$$

The arguments given above, apply exactly when X and Y have normal distributions, and approximately (because of the central limit theorem) in other cases in which the sample sizes are large.

16.2.2 Unknown common variance

In this case we use s^2, the pooled estimate of the common variance, together with the appropriate t-distribution:

$$\frac{(\bar{X} - \bar{Y}) - (\mu_x - \mu_y)}{\sqrt{s^2(1/n_x + 1/n_y)}} \sim t_{n_x+n_y-2}.$$

The confidence interval becomes

$$\left(\bar{x} - \bar{y} - ks\sqrt{\frac{1}{n_x} + \frac{1}{n_y}}, \quad \bar{x} - \bar{y} + ks\sqrt{\frac{1}{n_x} + \frac{1}{n_y}}\right),$$

where k is the relevant percentage point from the t distribution with $(n_x + n_y - 2)$ degrees of freedom.

The above argument only applies exactly if X and Y have normal distributions.

16.2.3 Large samples

If the samples are large, but the variances are not known and cannot be assumed to be equal to one another, then only approximate confidence intervals can be calculated. The approximate 95% confidence interval based on the normal distribution and the central limit theorem is

$$\left(\bar{x} - \bar{y} - 1.96\sqrt{\frac{s_x^2}{n_x} + \frac{s_y^2}{n_y}}, \quad \bar{x} - \bar{y} + 1.96\sqrt{\frac{s_x^2}{n_x} + \frac{s_y^2}{n_y}}\right).$$

16.3 Paired samples

Before dealing with the testing of paired samples, we give some idea of the general motivation.

16.3.1 Experimental design

It is widely thought that reaction times are shorter in the morning, and generally increase as the day goes on. One way of testing reaction times uses a buzzer and a light. The light is programmed to flash at random intervals and the experimental subject has to press the buzzer as soon after as possible. The linked computer records the delay between the two actions. Some people (particularly sportsmen and sportswomen) have amazingly fast reactions.

How might we test the idea in the context of a group of students? Here is a suggestion:

> **Experiment 1**
> Two random samples of 40 students are selected. One of these samples, chosen at random, uses the apparatus during the first period of the day, while the second sample uses the apparatus during the last period of the day. The means of the two samples are then compared.

This will require a standard two-sample comparison of means, assuming a common variance and using the test described previously. There is nothing actually *wrong* with the procedure, but we could be misled. Suppose that, all the bookworms were in the first sample and all the athletes were in the second—we might well conclude that reaction times improve with time of day!

There is a second, more subtle, difficulty with Experiment 1. We have noted that reaction times may vary greatly from student to student. These variations between students may be much greater than any changes over the time of day. The latter may then pass unnoticed. Here is an improved experiment:

> **Experiment 2**
> A random sample of 40 students are selected. Each student in the sample is tested in the first period of the day, and again in the final period of the day. The differences in reaction times between the two periods for each student are calculated. The mean difference is compared with 0.

The problems with Experiment 1 have vanished. The variability *between* students plays no part. All that matters is the variability of the changes *within* each student's readings.

The two-sample test uses a pooled estimate of variance. The variance being estimated is $\sigma_S^2 + \sigma_e^2$, where σ_S^2 represents the variability of the students themselves and σ_e^2 represents the usual random errors.

The paired-sample test uses the estimated variance of the differences. Here the student effect cancels out, leaving the quantity being estimated as $2\sigma_e^2$. The factor 2 occurs because 2 measurements are involved. The paired-sample test is therefore more efficient and preferable whenever (as is usually the case) σ_S^2 noticeably exceeds σ_e^2.

Experiment 2 is a simple example of a **paired sample**, which itself is a very simple example of the application of **experimental design** in statistics. Before we continue with the analysis of a paired sample, here is an example of a more complicated design:

Experiment 3

The student group is divided into two populations: athletes and non-athletes. These sub-populations are themselves divided into two parts: older students and younger students. From each of the resulting four groups of students a random sample of five boys and five girls is chosen. Testing then proceeds as in Experiment 2.

This experiment also uses the paired-sample idea, and the values to be analysed will be differences between late and early reaction times. However, with this experiment we can also examine differences between athletes and non-athletes, between boys and girls and between older and younger students. Without taking any more readings than in the earlier experiments, we can answer questions about four separate possible effects all at once. Experimental design is a powerful idea!

16.3.2 The paired-sample comparison of means

Writing μ_d for the mean of the distribution of differences between the paired values, the hypothesis becomes

$$H_0 : \mu_d = 0,$$

with a one-sided or two-sided alternative as appropriate.

We have a single set of n pairs of values and are interested in the *differences* $d_1, d_2, \dots, d_n$, which, assuming H_0, are a random sample from a population with mean 0. An unbiased estimate of the unknown variance of this population, σ_d^2, is provided by s_d^2, defined in the usual way by

$$s_d^2 = \frac{1}{n-1}\sum(d_i - \bar{d})^2 = \frac{1}{n-1}\left\{\sum d_i^2 - \frac{1}{n}\left(\sum d_i\right)^2\right\},$$

where $\bar{d}$ is the mean of the sample differences.

Although the data arise from two sets of measurements, by working with differences we have effectively created a single-sample situation, so that the methods of the previous chapters apply. The test statistic is

$$t_d = \bar{d}/\sqrt{s^2/n}.$$

Example 16.1

Suppose that Experiment 2 on the reaction times is carried out, with the following results (in units of 0.001 seconds):

Subject	*1*	*2*	*3*	*4*	*5*	*6*	*7*	*8*	*9*	*10*
First period	*23*	*50*	*31*	*44*	*92*	*70*	*33*	*44*	*58*	*39*
Last period	*29*	*71*	*50*	*50*	*68*	*52*	*55*	*38*	*53*	*61*
Difference	*6*	*21*	*19*	*6*	*−24*	*−18*	*22*	*−6*	*−5*	*22*

Subject	*11*	*12*	*13*	*14*	*15*	*16*	*17*	*18*	*19*	*20*
First period	*44*	*42*	*60*	*61*	*77*	*33*	*31*	*22*	*25*	*82*
Last period	*68*	*66*	*82*	*59*	*68*	*63*	*40*	*49*	*51*	*35*
Difference	*22*	*40*	*−1*	*7*	*−14*	*7*	*18*	*29*	*10*	*−14*

Subject	*21*	*22*	*23*	*24*	*25*	*26*	*27*	*28*	*29*	*30*
First period	*44*	*69*	*38*	*55*	*29*	*70*	*88*	*81*	*43*	*55*
Last period	*68*	*68*	*45*	*57*	*43*	*75*	*99*	*62*	*50*	*61*
Difference	*24*	*−1*	*7*	*2*	*14*	*5*	*11*	*−19*	*7*	*6*

Subject	*31*	*32*	*33*	*34*	*35*	*36*	*37*	*38*	*39*	*40*
First period	*61*	*69*	*20*	*31*	*27*	*73*	*81*	*48*	*59*	*61*
Last period	*60*	*73*	*36*	*40*	*29*	*64*	*84*	*58*	*57*	*92*
Difference	*−1*	*4*	*16*	*9*	*2*	*−9*	*3*	*10*	*−2*	*31*

Analyse these data to determine whether there is significant evidence, at the 1% level, of an increase in reaction times,

***(a)** efficiently, using a paired-sample test,*

***(b)** inefficiently, using a two-sample test.*

Comment on the results.

(a) *Efficient analysis, using paired-sample test.*

The summary statistics are $\bar{d} = 6.65$ and $s_d^2 = 263.6$. There are 39 degrees of freedom. The hypotheses being compared are

$$\text{H}_0: \mu_d = 0,$$
$$\text{H}_1: \mu_d > 0.$$

The test statistic is $t = 2.59$, for which the corresponding p-value is approximately 0.7%, and so the hypothesis of no change in mean reaction time can be confidently rejected in favour of the alternative hypothesis that, by the final period, reaction times have increased. The analysis is efficient because it uses the knowledge that two observations have been generated by each individual.

(b) *Inefficient analysis, using two-sample test.*

The summary statistics are $\bar{x} = 51.575$, $\bar{y} = 58.225$,and the difference in the means is 6.65, as before. The pooled variance is 328.9 so that the test statistic is $t = 1.64$. There are 78 degrees of freedom, so the p-value is now about 5.3% and the increase in average time would be judged non-significant.

Comment:

The large variations in the individual reaction times, which range from 0.020 to 0.092 seconds during the first period, have obscured the relatively small mean increase (0.00665 seconds) between the two periods.

Exercises 16a

1. A machine assesses the life of a ballpoint pen by measuring the length of a continuous line drawn using the pen. A random sample of 80 pens of brand A have a total writing length of 96.84 km. A random sample of 80 pens of brand B have a total writing length of 95.75 km. It is known that, for both brands, the standard deviation of the writing length of a single pen is 0.15 km. The aim is to test whether the writing lengths of the two brands differ significantly.

   Without performing any calculations, state whether the test is one-sided or two-sided and whether the test statistic is z, t_p, t_w, or t_d.

2. A random sample of 10 yellow grapefruit are weighed and found to have mean 201 g and variance 234 g^2. The corresponding figures for 8 pink grapefruit are 222 g and 282 g^2. We wish to determine whether pink grapefruit are significantly heavier than yellow grapefruit. The masses of grapefruit may be assumed to be normally distributed.

   Without performing any calculations, state whether the test is one-sided or two-sided and whether the test statistic is z, t_p, t_w, or t_d.

3. An experiment to discover the movement of antibiotics in a certain variety of broad bean plants was made by treating 10 cut shoots and 10 rooted plants for 18 hours with a solution of chlorampherical. At the end of the experiment the concentrations of chlorampherical are recorded. Assuming normal distributions, we wish to determine whether there is a significant difference between the concentrations in the cut shoots and rooted plants.

   State whether the test is one-sided or two-sided and whether the test statistic is z, t_p, t_w, or t_d. Would your calculations be different if you were told that 10 separate containers were used, each containing one cut shoot and one rooted plant?

4. The quantities of beer in a random sample of 7 pints bought at each of pubs A and B are measured. It may be assumed that the variance of the quantities provided in pub A is the same as in pub B. The question of interest is whether the average quantities in each pub differ significantly.

   State whether the test is one-sided or two-sided and whether the test statistic is z, t_p, t_w, or t_d.

Group project: Does being right-handed imply that your right hand is more flexible? In order to find out, use a ruler to measure R, the span of your right hand from outstretched thumb tip to the end of your little finger. Now find L, the corresponding distance for your left hand. Calculate d, which is equal to $R-L$ for right-handers and $L-R$ for left-handers. The null hypothesis is that there is no difference between hand spans ($\mu_d = 0$) and the alternative is that the favoured hand has the larger span ($\mu_d > 0$). Is it true for you? How about for the group in general? Perform a paired-sample test.

16.4 The comparison of the variances of two normal distributions

We now consider hypotheses such as

$$\begin{aligned} \mathrm{H}_0&: \quad \sigma_x^2 = \sigma_y^2, \\ \mathrm{H}_1&: \quad \sigma_x^2 \neq \sigma_y^2, \end{aligned}$$

where σ_x^2 and σ_y^2 are the variances of two different populations. This test would be a natural preamble to a comparison of two means as in Section 16.1.3

Suppose we have two independent random samples of sizes n_x and n_y, with variances s_x^2 and s_y^2. The corresponding random variables are S_x^2 and S_y^2. If the distributions sampled are normal, with common variance σ^2, then

$$\frac{n_x - 1}{\sigma^2} S_x^2 \sim \chi^2_{n_x-1} \qquad \text{and} \qquad \frac{n_y - 1}{\sigma^2} S_y^2 \sim \chi^2_{n_y-1}.$$

From the definition of an F-distribution (Section 7.3), taking the scaled ratio of χ^2 random variables, we get

$$\frac{\frac{n_x-1}{\sigma^2} S_x^2}{n_x - 1} \Bigg/ \frac{\frac{n_y-1}{\sigma^2} S_y^2}{n_y - 1} \sim F_{n_x-1, n_y-1},$$

which simplifies greatly to give

$$S_x^2 / S_y^2 \sim F_{n_x-1, n_y-1}, \tag{16.5}$$

while also

$$S_y^2 / S_x^2 \sim F_{n_y-1, n_x-1}. \tag{16.6}$$

If the null hypothesis is accepted, then a pooled estimate of the common variance is given by s_p^2, given by Equation (16.2).

Example 16.2

An experiment is performed to investigate the effects of two alternative fertilizers on the growth of spinach plants. The plants were grown in controlled conditions, with 12 randomly selected plants being given fertilizer A, and a further 12 being given fertilizer B. Before the end of the experiment some plants were attacked by a fungus, and they were removed from the experiment. The final masses (x g) are summarized in the table below:

Fertilizer	*Sample size*	Σx	Σx^2
A	*11*	*1,098*	*175,644*
B	*10*	*1,083*	*145,350*

(a) *Show that, at the 5% significance level, the hypothesis that the two populations have equal variances is accepted.*

(b) *Obtain an estimate of the common variance.*

(c) *Determine a two-sided 99% symmetric confidence interval for the common value.*

(a) For fertilizers A and B, the unbiased estimates of the population variance are, respectively, $s_A^2 = 6604.36$ and $s_B^2 = 3117.90$. We therefore compare the ratio s_A^2/s_B^2 with tables of the upper-tail of an $F_{10,9}$ distribution.

The test is two-tailed, so we need the upper 2.5% point of the $F_{10,9}$ distribution. This is 3.96. Since the ratio (2.12) is less than this, there is no need to reject the null hypothesis that the variances are the same.

(b) The pooled estimate of the variance is given by

$$s^2 = \frac{1}{19}\{(175644 - 1098^2/11) + (145350 - 1083^2/10)\} = 4952.88.$$

(c) For the two-sided 99% confidence interval, we need the upper and lower 0.5% points of χ^2_{19} distribution which are, respectively, 38.58 and 6.844. The confidence interval is therefore

$$\left(\frac{19 \times 4952.88}{38.58}, \frac{19 \times 4952.88}{6.844}\right),$$

which simplifies to (2,440, 13,750).

16.5 Confidence interval for a variance ratio

Suppose that the variances σ_x^2 and σ_y^2 are not the same. We can repeat the arguments of the previous section, again assuming normal distributions, and see where they lead. We now have

$$\frac{n_x - 1}{\sigma_x^2} S_x^2 \sim \chi^2_{n_x - 1} \qquad \text{and} \qquad \frac{n_y - 1}{\sigma_y^2} S_y^2 \sim \chi^2_{n_y - 1}.$$

Once again taking the scaled ratio of χ^2 random variables, we get

$$\frac{S_x^2}{S_y^2} \frac{\sigma_y^2}{\sigma_x^2} \sim F_{n_x - 1, n_y - 1}. \tag{16.7}$$

For a symmetric 95% confidence interval, we need the upper and lower 2.5% points, which we denote by U and L, where U is the upper 2.5% point of an F_{n_x-1,n_y-1}-distribution, and L is the reciprocal of the upper 2.5% point of an F_{n_y-1,n_x-1}-distribution. We can then write

$$\mathrm{P}\left(L < \frac{S_x^2}{S_y^2} \frac{\sigma_y^2}{\sigma_x^2} < U\right) = 0.95.$$

Rearranging, we have

$$\mathrm{P}\left(\frac{S_y^2}{S_x^2} L < \frac{\sigma_y^2}{\sigma_x^2} < \frac{S_y^2}{S_x^2} U\right) = 0.95.$$

Consequently, using the observed s_x^2 and s_y^2, the 95% confidence interval for the ratio σ_y^2/σ_x^2 is

$$\left(\frac{s_y^2}{s_x^2} L, \ \frac{s_y^2}{s_x^2} U\right).$$

Exercises 16b

1. The result of a variance test (using R) is shown below:

```
data: x and y
F = 0.11266, num df = 7, denom df = 11, p-value = 0.008074
alternative hypothesis: true ratio of variances is not equal to 1
95 percent confidence interval:
0.02997375 0.53057102
sample estimates:
ratio of variances
0.1126605
```

   (a) What sizes were the samples?

   (b) Is the test one-sided or two-sided?

   (c) What is the outcome of the test?

Key facts

- **The comparison of two normal means:**
  $H_0: \mu_x = \mu_y$,
  $H_1: \mu_x \neq \mu_y$ or $H_1: \mu_x > \mu_y$, or $H_1: \mu_x < \mu_y$.

  - **Variances known:**

$$z = \frac{\bar{x} - \bar{y}}{\sqrt{\sigma_x^2/n_x + \sigma_y^2/n_y}}$$

    Assuming H_0, z is an observation from N(0,1).

  - **Common unknown variance**:

$$t_p = \frac{\bar{x} - \bar{y}}{\sqrt{s^2(1/n_x + 1/n_y)}},$$

    where

$$s_p^2 = \frac{\sum(x_i - \bar{x})^2 + \sum(y_j - \bar{y})^2}{n_x + n_y - 2}.$$

    Assuming H_0, t_p is an observation from a $t_{n_x+n_y-2}$ distribution.

  - **Unknown variances**:

$$t_w = (\bar{x} - \bar{y}) \Bigg/ \sqrt{\frac{s_x^2}{n_x} + \frac{s_y^2}{n_y}}$$

    Welch's t-test with a complicated formula for degrees of freedom.

  - **Paired samples:**

$$t = \frac{\bar{d}}{\sqrt{s_d^2/n}},$$

    where

$$s_d^2 = \frac{\sum d_i^2 - (\sum d_i)^2/n}{n-1}.$$

    If the observations come from normal distributions, then, assuming H_0, t is an observation from a t_{n-1} distribution.

- **Comparing two normal variances:**
  With s_x^2 and s_y^2 denoting the sample variances for samples of sizes n_x and n_y, respectively:

  - s_x^2/s_y^2 is an observation from a F_{n_x-1,n_y-1} distribution.
  - The confidence interval for the ratio σ_y^2/σ_x^2 is $(Ls_y^2/s_x^2,\ \ Us_y^2/s_x^2)$ where L is the reciprocal of the upper 2.5% point of an F_{n_y-1,n_x-1} distribution and U is the upper 2.5% point of an F_{n_x-1,n_y-1} distribution.

R

All tests default to the 5% level and two-sided. The default is unpaired samples. The user can add `alternative="less"` or `alternative="greater"`, `paired=TRUE`, and can add, for instance, `conf.level=0.99`, to override the defaults. The tests all report confidence intervals and *p*-values.

With data in the two vectors `x` and `y`:

- **Tests for equality of means**
  - The *z*-test with known variances requires:
    `ZTest(x, y, alternative)`
  - The two-sample *t*-test with variances equal but unknown requires:
    `t.test(x, y, var.equal=TRUE)`.
  - Otherwise, using the Welch *t*-test:
    `t.test(x, y)`
  - For paired samples:
    `t.test(x, y, paired=TRUE)`
- **Test for equality of normal variances**

  `var.test(x, y)`

17

Goodness of fit

Previous chapters have assumed that a particular type of distribution is appropriate, and they have focused on estimating and testing hypotheses about the parameter(s) of that distribution. In this chapter, the focus switches to the distribution itself, and we ask the question "Does the data support the assumption that this particular type of distribution is appropriate?"

17.1 The chi-squared test

Suppose, for example, that we roll an apparently normal six-sided die 60 times and obtain the following results:

Outcome	1	2	3	4	5	6
Frequency	4	7	16	8	8	17

In this sample there seem to be a rather large number of 3s and 6s—is this die fair, or is it biased? With a fair die the probability of each outcome is $1/6$. With 60 tosses the expected frequencies would each be $60 \times 1/6 = 10$:

Outcome	1	2	3	4	5	6
Frequency	10	10	10	10	10	10

The question of interest is whether the observed frequencies, O, and the expected frequencies, E, are reasonably close or unreasonably different:

Observed frequency, O	4	7	16	8	8	17
Expected frequency, E	10	10	10	10	10	10
Difference, $O - E$	−6	−3	6	−2	−2	7

The larger the magnitude (ignoring the sign) of the differences, the more the observed data differs from that expected according to our model (that the die was fair).

Suppose we now roll a second die 660 times, and obtain the following results:

Observed frequency, O	104	107	116	108	108	117
Expected frequency, E	110	110	110	110	110	110
Difference, $O - E$	−6	−3	6	−2	−2	7

This time the observed and expected frequencies seem remarkably close, yet the $O - E$ values are the same as before. This tells us that it is not simply the size of $O - E$ that matters, but also its *relative* size, $\frac{O-E}{E}$.

Combining these ideas might suggest using the product $(O - E) \times \frac{O-E}{E}$. so that the goodness of fit for outcome i is measured using $r_i^2 = \frac{(O_i-E_i)^2}{E_i}$. The smaller this quantity is, the better is the fit. The quantity r_i is called the **Pearson residual**:

$$r_i = \frac{O_i - E_i}{\sqrt{E_i}}. \tag{17.1}$$

An aggregate measure of **goodness of fit** of the model is then provided by X^2, defined by:

$$X^2 = \sum_{i=1}^{m} r_i^2 = \sum_{i=1}^{m} \frac{(O_i - E_i)^2}{E_i}, \tag{17.2}$$

where m is the number of different categories (six, in the case of a die). Significantly large values of X^2 suggest lack of fit.

Different samples (e.g. different sets of 60 rolls of the die) will give different sets of observed frequencies and hence different values for X^2. Since X^2 is a function of counts, there is only a finite number of possible values that it can take. The distribution of X^2 is therefore discrete. Nevertheless, in most circumstances, the number of possible values of X^2 is so large that its distribution can be well approximated by a continuous distribution. It was Karl Pearson[1] who showed that, when the probabilities of the various outcomes are correctly specified by the null hypothesis, X^2 is (approximately) an observation from a chi-squared distribution (Section 7.2) with $m - 1$ degrees of freedom.

Example 17.1

According to a genetic theory, when sweet peas having red flowers are crossed with sweet peas having blue flowers, the next generation of sweet peas have red, blue, and purple flowers in the proportions 1/4, 1/4, and 1/2, respectively. The outcomes in an actual experiment are as follows: 84 with red flowers, 92 with blue flowers, and 157 with purple flowers. Determine whether these results support the theory.

[1] For a brief biography of Karl Pearson see footnote in Section 14.2.1. Pearson introduced the X^2 test in 1900, but it was Fisher (see footnote in Section 7.3) that added the correct number of degrees of freedom in articles published in 1922 and 1924.

The null hypothesis is that the three proportions are 1/4, 1/4, and 1/2, with the alternative being simply that the null hypothesis is false. We set out the calculations in the following table[a]:

Type	O_i	E_i	$O_i - E_i$	r_i^2
Red	84	83.25	0.75	0.007
Blue	92	83.25	8.75	0.920
Purple	157	166.50	−9.50	0.542
Total	333	333	0	$X^2 = 1.469$

There are $3 - 1 = 2$ degrees of freedom. The computer reports a p-value of 0.48, which implies that the observed results are typical of those to be expected according to the theory.

[a] Note that $\sum(O_i - E_i) = 0$; this must always be the case.

Exercises 17a

1. Four coins are tossed 100 times. The number of heads obtained on each toss is recorded. The results are summarized below:

Number of heads	0	1	2	3	4
Frequency	5	23	39	19	14

   (a) Determine the theoretical probabilities of the five possible outcomes.

   (b) Use X^2 to test the hypothesis that the four coins are fair. Percentage points for χ^2-distributions were given in Table 7.2.

Put theory into practice: In the subject of statistics, we frequently use phrases such as 'randomly chosen' or 'at random'. The object of this project is to determine whether people can really choose things 'at random'. If they cannot, then tables of random numbers are needed!

Write the letters A B C D E in a horizontal line on a sheet of paper. Then ask people to 'choose a letter at random'. Record their choice. After you have recorded the choices of at least 25 people, test the hypothesis that all letters are chosen with equal probability.

Most research suggests that people are biased towards the ends of lists, and particularly toward the left of a horizontal list. You might like to repeat the experiment with a vertical list.

17.2 Small expected frequencies

The distribution of X^2 is discrete—the χ^2 distribution is continuous and is simply a convenient approximation which becomes less accurate as the expected frequencies become smaller. A rule that is often used to decide whether the approximation may be used is

'All expected frequencies must be equal to at least 5'.[2]

If the original categories chosen lead to many expected frequencies less than 5, or to any expected frequency less than 1, then it will be necessary to combine categories together. This combination may be done on any sensible grounds, but should be done *without reference to the observed frequencies* so as to avoid biasing the results. With numerical data it is natural to combine adjacent categories: for example, we might replace the three categories '7', '8', and '9' by the single category '7–9'.

Example 17.2

A test of a random number generator is provided by studying the lengths of 'runs' of digits. The probability of a run of length k (i.e. that a randomly chosen digit is followed by exactly $k-1$ similar digits) is $0.9\times0.1^{k-1}$. This is a geometric distribution (see Section 3.9). A sequence of supposedly random numbers are generated, and the following results are obtained:

Length of run	*1*	*2*	*3*	*4*	*5*	*6 or more*
Frequency	*8083*	*825*	*75*	*9*	*1*	*0*

Determine whether these results suggest that there is anything wrong with the random number generator.

The null hypothesis specifies that a run of length k has probability $0.9 \times 0.1^{k-1}$, with the alternative hypothesis stating that the null hypothesis is incorrect.

Run length	O_i	Probability	E_i	$O_i - E_i$	r_i^2
1	8083	0.900	8093.700	−10.700	0.014
2	825	0.090	809.370	15.630	0.302
3	75	0.009	80.937	−5.937	0.435
4 }	9 }	}	8.094 }		
5 } 4+	1 } 10	} 0.001	0.809 } 8.993	1.007	0.113
6+ }	0 }	}	0.090 }		
Total	8993	1.000	8993.000	0	0.864

[2] This rule errs on the safe side: many researchers happily permit a small proportion of expected frequencies to be less than 5.

The expected frequencies are shown in the table. That for run lengths of 6 or more was obtained by subtraction of the remainder from the known total of 8993. The last two expected frequencies are very small and we combine these with the previous category to form a category '4+'.

After combining categories together there are four groups and hence the χ_3^2 distribution is relevant. The reported p-value corresponding to $X^2 = 0.864$ is 0.834: the observed outcome is typical of what one would expect.

17.2.1 An alternative approach: Monte Carlo simulation

In 1963, Barnard[3] suggested a procedure, often referred to as a **Monte Carlo test**, for the general problem of deciding whether observed data are consistent with the theory under test. The procedure involves generating data (**simulating**) using the probabilities specified by the theory under test. Because of random variation, the simulated outcomes are unlikely to exactly match the the theoretical expected frequencies. The question is whether the simulated data consistently matches the theoretical values more closely than did the observed data. For each simulated set of data (each containing the same number off 'observations' as the observed data) we calculate the value of X^2. The hypothesis is rejected if the observed value of X^2 is more extreme than most of the values resulting from the simulations.

Example 17.2 (cont.)

We require the computer to generate the values 1, 2, 3, ..., with respective probabilities 0.9, 0.09, 0.009, ...[a] The first five sets of simulated data (each containing 8993 observations) are shown in the following table, together with the resulting values of X^2:

1	2	3	4	5	6+	X^2
8062	839	84	6	2	0	3.71
8087	815	77	10	4	0	13.35
8137	756	95	4	1	0	8.40
8079	830	74	9	1	0	1.38
8108	796	83	6	0	0	1.74

All five X^2 values are greater than that observed (0.864), which certainly suggests that the random number generator is working correctly. However, five simulations is nowhere near sufficient. We need at least 999, and, preferably 9999, or more, to be sure.[b]. To be super confident, I ran a total of 9999 simulations. Of these 9511 exceeded 0.864: far from being unreasonably high, the observed value of X^2 was actually in the lower 5% of values to be expected by chance when the random number generator is working correctly.

[a] In R the command is `1+rgeom(5993,0.9)`.
[b] Combined with the outcome for the observed data, we then have convenient 'round numbers'.

[3] George Alfred Barnard (1915–2002) was an English logician with a special interest in statistical inference. He was the founding Professor of Statistics at Essex University.

17.3 Goodness of fit to prescribed distribution type

We now turn to cases where the null hypothesis states that the data 'has a particular named distribution', but does *not* specify all the parameters of the distribution. A typical example would be the hypothesis

H_0: the number of emails received between 09:00 and 10:00 is an observation from a Poisson distribution.

The hypothesis does not specify *which* Poisson distribution, so we choose the most plausible one—this is the one having the same mean as the observed data. Because of this deliberate matching we are imposing constraints on the expected frequencies which reduces the value of d, the degrees of freedom of the approximating χ^2 distribution. The general rule is:

$$d = m - 1 - k, \tag{17.3}$$

where m is the number of different outcomes (after amalgamations to eliminate small expected frequencies) and k is the number of parameters estimated from the data.

Example 17.3

Eggs are packed in cartons of six. On arrival at a supermarket each pack is checked to make sure that no eggs are broken. Fred, the egg-checker, attempts to relieve his boredom by recording the numbers of broken eggs in a pack. After examining 5000 packs his results look like this:

Number of broken eggs	*0*	*1*	*2*	*3*	*4*	*5*	*6*
Number of packs	*4704*	*273*	*22*	*0*	*0*	*1*	*0*

Determine whether these results are consistent with the null hypothesis that egg breakages are independent of one another, with each of the six eggs in a pack being equally likely to break.

According to the null hypothesis, all six eggs are equally likely to break, with the breakages being independent of one another. If this is the case, then the number of broken eggs in a pack is an observation from a binomial distribution with $n = 6$. Fred examined a total of 30,000 eggs, of which $273 + (22 \times 2) + 5 = 322$ were found to be broken. The sample estimate of p is therefore 322/30,000. The X^2 calculations now proceed as usual, with, in this case, the last five categories being combined.

After combining categories 2 to 6, $m = 3$. One parameter (p) was estimated from the data and consequently $d = 3 - 1 - 1 = 1$. The observed value of X^2 is 28.84 which is a very large value given that a χ_1^2-distribution has mean 1. In fact the computer reports a p-value of 8×10^{-8}: we can confidently reject the null hypothesis. Examining the data we see that there were too many packs containing two (or more) broken eggs—it seems likely that egg breakages are not independent events, but are caused by packs being dropped and other accidents.

Example 17.4

A typist creates a 50-page typescript and gives it to the author for checking. She notes the numbers of errors on each page. The results are summarized below:

Number of errors	*0*	*1*	*2*	*3*	*4*	*5*	*6*	*7*	*8*	*9*	*10+*
Number of pages	*2*	*5*	*16*	*11*	*6*	*3*	*1*	*2*	*3*	*1*	*0*

Test, at the 5% significance level, the null hypothesis that the errors are randomly distributed through the typescript.

The null hypothesis states that the errors are distributed at random. If this is true then the counts should be observations from a Poisson distribution. In order to test the hypothesis we need first to estimate the mean of the Poisson distribution. In total there are 162 errors and hence the mean number per page is 3.24. The probability of a page containing r errors is therefore estimated as being P_r, where:

$$P_r = \frac{3.24^r e^{-3.24}}{r!},$$

and the corresponding expected frequency is $50P_r$. The calculations are set out below:

Errors		O_i		Probability		E_i		$O_i - E_i$	r_i^2
0	}	2	7	0.0392	0.1661	1.9582	8.3027	−1.3027	0.204
1		5		0.1269		6.3446			
2		16		0.2056		10.2782		5.7218	3.185
3		11		0.2220		11.1004		−0.1004	0.001
4		6		0.1798		8.9913		−2.9913	0.995
5		3		0.1165		5.8264		−2.8264	1.371
6	6+	1	7	0.1100		5.5009		1.5000	0.409
7		2							
8		3							
9		1							
10+		0							
Total		50		1.0000		50.0000		0	6.165

Notice that, in this case, categories were combined in each tail of the distribution. After these combinations, $m = 6$, and hence $d = 6 - 1 - 1 = 4$, since one parameter was estimated from the data. The computer reports that, for a χ_4^2 distribution, the probability of a value of greater than 6.165 is about 19%: we can therefore accept the hypothesis that the typist's errors occurred at random.[a]

[a] Without combining the categories, ten simulation runs (each of 2000 simulations) produced tail probabilities ranging from 0.049 to 0.070, giving mild evidence of non-randomness. This is very different to the result using the chi-squared approximation, suggesting that it is unwise to combine too many categories together when there is a long tail to the distribution.

Exercises 17b

1. The table below summarizes 200 observations on X:

x	0	1	2	3	4
Frequency	46	77	69	7	1

In each of the following three cases, state the number of degrees of freedom available to test the hypothesis that X has a binomial distribution. If you have access to a computer, determine the value of X^2 and the outcome of the test in each case.

(a) It is hypothesized that X has a binomial distribution with $n = 4$ and $p = 0.4$.

(b) It is hypothesized that X has a binomial distribution with $n = 4$. The chi-squared test uses all five categories.

(c) It is hypothesized that X has a binomial distribution with $n = 4$. The categories $x = 3$ and $x = 4$ are pooled.

The chi-squared test is often referred to as an **omnibus test** because it can be used in a wide variety of situations. This versatility hints at the fact that it is unlikely to be the best test (in the sense of being the most powerful test) for detecting departures from the specified distribution. In the case of the normal distribution we have already seen a more powerful test: the Shapiro test (Section 15.8.2).

17.4 Comparing distribution functions

A problem with data from continuous distributions is that, in order to use the X^2 test, we need to group the data. Using different groupings with the same data we could reach different conclusions. An alternative is to compare the theoretical distribution function, $\mathrm{F}(x)$, with its sample approximation. A large difference between these quantities would suggest that the theoretical distribution was incorrect. There are several tests that use this type of approach (which is related to the Q–Q plot that was briefly introduced in Section 15.8.1). Probably, the most commonly used of these tests is the Kolmogorov–Smirnov test.[4]

[4] Kolmogorov (see the brief biography in Section 9.4) suggested the test in 1933 and Nikolai Visil'yevich Smirnov (1900–66), a Russian mathematical statistician who studied in Moscow, developed the underlying theory in a paper published in 1939.

17.4.1 Kolmogorov–Smirnov test

Let $x_{(i)}$ be the ith largest of the n observations in a sample. Then the test statistic, D, is given by:

$$D = \max_i \left(F(x_{(i)}) - \frac{i-1}{n}, \quad \frac{i}{n} - F(x_{(i)}) \right). \tag{17.4}$$

In plain English, this is simply the largest difference between the theoretical distribution function and the sample distribution function.

Example 17.5

In this example the theoretical distribution is a normal distribution with mean zero and unit variance.

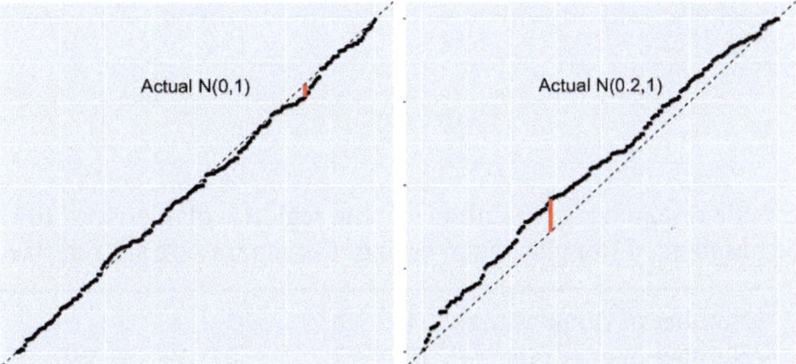

Figure 17.1 ***Q–Q*** **plots comparing the theoretical distribution (a standard normal distribution) with two sets of sample data.**

Figure 17.1 shows two Q–Q plots resulting from random samples of 200 observations taken from normal distributions with unit variance. In one case the distribution has mean 0 as hypothesized, but in the other case the mean is slightly larger (0.2). The difference in the D-values is marked. For the correct case $D = 0.04$, which is a value exceeded by chance on 90% of occasions. For the incorrect case D has risen to 0.10, with a p-value of just 2.8%.

17.5 The dispersion test

This is a specialized goodness-of-fit test for testing the hypothesis that a sample of observations has been taken from a population having a Poisson distribution with unknown mean. It is a more powerful alternative to the X^2 test.

If a random variable does have a Poisson distribution, then it will have mean equal to variance. If the ratio of the sample variance to the sample mean is very different from 1, then this would provide evidence against the Poisson hypothesis.

The test statistic, often known as **the index of dispersion**, is

$$I = \frac{(n-1)s^2}{\bar{x}},$$

where n is the sample size and $\bar{x}$ and s^2 are the sample mean and variance.

If the data come from a Poisson distribution, then I is an observation from a χ^2_{n-1} distribution. The test is usually two-sided, with unusually small or unusually large values leading to rejection of the null hypothesis of a Poisson distribution in favour of, respectively, a regular distribution, or a clustered distribution..

Example 17.5 (cont.)

In this case $\bar{x} = 3.24$ *and* $s^2 = 4.51$ *so that* $I = 49 \times 4.51/3.24 = 68.25$. *The reference distribution has* $50 - 1 = 49$ *degrees of freedom. The probability of a value greater than 68.25 is reported to be 3.8%. Whilst this is quite small, the dispersion test is two-sided, so that, for 'significance at the 5% level', the observed value must lie in an extreme 2.5% tail. As with the previous chi-squared test, we therefore accept the null hypothesis.*

Exercises 17c

1. In one corner of a research plot in California, the region is divided into 100 square quadrats of side 4m. The numbers of Douglas firs in each of the squares are summarized below.

Number of Douglas firs	0	1	2	3	4	5	6	7	8
Number of quadrats	63	27	1	5	2	1	0	0	1

(a) Verify that the standard deviation of the number of trees per quadrat is 1.250 (to 3 d.p.).

(b) Determine whether there is convincing evidence that the trees are not randomly distributed.

17.6 Contingency tables

Often data are collected on several variables at a time. A table that gives the frequencies for two or more variables simultaneously is called a **contingency table**. Here is an example table:

	Conservative	Liberal	Labour	
Male	313	124	391	828
Female	344	158	388	890
	657	282	779	1718

Sample data of this type are collected in order to answer interesting questions about the behaviour of the population. such as 'Are there differences in the way that males and females vote?'. If there are differences, then the variables vote and gender are said to be **associated**, whereas if there are no differences, then the variables are said to be **independent**.

The null hypothesis is that the variables are independent. If this is true then, in the population, the proportion of Conservative supporters who are male will be equal to the proportion of Liberal supporters who are male, and to the proportion of Labour supporters who are male. Furthermore, the proportion of males who support the Conservative Party, will be equal to the proportion of females who support the Conservative Party, and so on.

If the null hypothesis of independence is true, then the best estimate of the population proportion voting for the Conservatives is 657/1718 (= 0.3824). The expected number of males voting Conservative would therefore be:

$$828 \times 657/1718 = 316.64,$$

and the number of females would be:

$$890 \times 657/1718 = 340.36.$$

These expected values are easily calculated using the formula:

$$\frac{\text{Row total} \times \text{Column total}}{\text{Grand total}}. \tag{17.5}$$

Observed frequencies		
313	124	391
344	158	388

Expected frequencies		
316.64	135.91	375.44
340.36	146.09	403.56

The value of X^2 is calculated as before:

$$X^2 = \frac{(313 - 316.64)^2}{316.64} + \cdots + \frac{(388 - 403.56)^2}{403.56} = 3.34.$$

With large expected frequencies, the distribution of X^2 is closely approximated by a χ^2 distribution with d degrees of freedom, where, for a contingency table with r rows and c columns,

$$d = (r - 1)(c - 1). \tag{17.6}$$

The reason for this value for d can be seen by looking again at the expected frequencies:

316.64	135.91	?	828
?	?	?	890
657	282	779	1718

After calculating the $(r - 1)(c - 1)$ (= 2) expected frequencies that are shown, the remainder are not 'free'—their values are fixed by the need for them to sum to the known row and column totals.

For the voting data, since (according to the computer) the probability of a value from a χ^2_2 distribution being greater than 3.34 is about 19%, we would conclude that the voting patterns of the males and females do not differ significantly.

Example 17.6

The following data refer to visits to patients in a mental hospital:

	Length of stay in hospital (years)			
	2 to 10	*10 to 20*	*More than 20*	*Total*
Visited regularly	*43*	*16*	*3*	*62*
Visited sometimes	*6*	*11*	*10*	*27*
Never visited	*9*	*18*	*16*	*43*
Total	*58*	*45*	*29*	*132*

Verify that the association between length of stay and the frequency with which a patient is visited is significant at the 0.1% level.

The null hypothesis states that length of stay and frequency of visit are independent of one another, with the alternative being that the two are associated.

Since there are three rows and three columns, the relevant chi-squared distribution has $(3-1) \times (3-1) = 4$ degrees of freedom. If the variables were independent, then the expected frequencies could be calculated using Equation (17.5). For example, the expected frequency for regular visits to recent patients would be $(62 \times 58)/132 = 27.24$. The calculations find that $X^2 = 35.17$ which is a value that greatly exceeds the upper 0.1% point of a χ_4^2 distribution (18.47): it is clear that there is a strong association between the classifying variables.

It is often useful to set out the individual cell contributions to X^2 in a table like that of the data, since this can help to show up any pattern in the lack of fit.

	r_i^2	
9.11	1.25	8.28
2.90	0.35	2.79
5.18	0.76	4.55

For this data set, focusing on the cells with large observed frequencies, the major lack of fit arises from the corner cells—the NW and SE corners have much larger frequencies than would have been expected if there had been no association. These cells correspond to the regular visiting of recently admitted patients and the infrequent visits to the very long-stay patients.

Exercises 17d

1. A voter survey obtains information on the political party supported and the highest academic qualification held by the interviewee. The results are summarized in the following table:

	Conservative	Labour	Other party
GCSE	62	111	38
A level	57	53	25
Degree	24	15	20

(a) State the number of degrees of freedom for a chi-squared test.

(b) Verify that the value of X^2 is 19.3 (to 1 d.p.)

(c) Demonstrate that there is significant evidence of an association between academic achievement and political affiliation.

(d) State which cell has the largest magnitude residual.

17.7 The 2×2 table: the comparison of two proportions

We denote the proportions of successes in two populations by p_1 and p_2, with the failure proportions being denoted by q_1 and q_2. The methods used for comparing the hypotheses,

$$\begin{aligned} H_0&: p_1 = p_2, \\ H_1&: p_1 \neq p_2, \end{aligned}$$

have attracted a great deal of discussion because the simple appearance of the situation is misleading.

Suppose that, from population 1, we have a sample of m observations containing a successes, while from population 2 we have a sample of n observations that contains b successes. Let the respective numbers of failures be $c\ (= m - a)$ and $d\ (= n - b)$. We can set the results out as a 2×2 table:

	Success	Failure	Total
Sample 1	a	c	m
Sample 2	b	d	n
Total	r	s	N

Using Equation (17.5), the corresponding expected frequencies are given by

	Success	Failure	Total
Sample 1	mr/N	ms/N	m
Sample 2	nr/N	ns/N	n
Total	r	s	N

Now

$$a - \frac{mr}{N} = \frac{1}{N}\{a \times (a+b+c+d) - (a+c) \times (a+b)\} = \frac{1}{N}(ad - bc)$$

with corresponding expressions for the other differences. Collecting these together reveals that, for a general 2×2 table, X^2 can be calculated using:

$$X^2 = \frac{N(ad - bc)^2}{mnrs}. \tag{17.7}$$

As usual, the discrete distribution of X^2 values is quite well approximated by a chi-squared distribution when the cell frequencies are reasonably large. In this case there is just one degree of freedom.

Example 17.7

In the United States a **double blind**[a] *experiment was conducted to investigate whether a daily dose of aspirin would alter the risk of having a heart attack. In this experiment, the doctors were also the patients: each doctor was sent a large box of pills and was instructed to take one pill a day. Half the boxes contained aspirins, with the other half containing a* **placebo.**[b] *The outcome of the trial was as follows. Of 11,037 doctors given aspirin, 104 had heart attacks during the period of the trial, whereas of 11,034 doctors given the placebo, 189 had heart attacks. Investigate the significance of this result.*

[a] A 'double blind' experiment is one in which neither the person administering the drug, nor the patient, know what drug is being administered. This eliminates possible biases.
[b] A dummy pill containing no medicine.

The hypotheses being compared are:

$$\mathrm{H}_0\colon p_1 = p_2,$$
$$\mathrm{H}_1\colon p_1 \neq p_2.$$

Set out as a 2×2 table, the information in the question is:

	Success	Failure	Total
Aspirin	10,933	104	11,037
Placebo	10,845	189	11,034
Total	21,778	293	22,071

The value of X^2 is given by;

$$X^2 = \frac{22,071(10,933 \times 189 - 10,845 \times 104)^2}{11,037 \times 11,034 \times 21,778 \times 293} = 25.01.$$

which corresponds to a tail probability of approximately 5.7×10^{-7}. There can be no doubt of the benefit of taking a daily aspirin in order to reduce the risk of a heart attack.

17.7.1 *The equivalent test using the normal distribution*

We noted in Section 7.2.1 that, if Z has a standard normal distribution, then Z^2 has a χ_1^2-distribution. Of course, the reverse is true, so we can write

$$X = (ad - bc)\sqrt{\frac{N}{mnrs}} \sim N(0, 1).$$

17.7.2 *Fisher's exact test*

Before continuing, we introduce some related terms, beginning with **odds**. This is simply the ratio of the proportion of successes to the proportion of failures. For the first sample, this would be p_1/q_1, where $q_1 = 1 - p_1$. According to H_0,

$$\frac{p_1}{q_1} = \frac{p_2}{q_2},$$

which implies that

$$\theta = \frac{p_1/q_1}{p_2/q_2} = \frac{p_1 q_2}{p_2 q_1} = 1. \tag{17.8}$$

The quantity θ is called the **odds ratio**. So the hypothesis of equality of proportions is equivalent to a hypothesis that the odds-ratio is 1.

The test introduced by Fisher[5] in 1934 uses no approximations. It is often referred to simply as the **exact test**, although, of course, there are many other tests that are exact. To understand Fisher's approach, consider the following situation:

> To test the effectiveness of two drugs, a group of N patients are randomly divided into two groups, with m being given one drug, and n given the other drug. If the drugs are equally effective, then the precise assignment of the drugs to the patients will have no bearing on the outcome: there would have been r successes and s failures under any allocation.

The sample sizes are evidently fixed, but so are the numbers of successes and failures. The only difference is that the latter numbers are unknown.

With all four marginal totals fixed, we need only consider the value of one cell. For example, concentrating on a, we can write $b = m - a$, $c = r - a$, and $d = n - r + a$:

	Success	Failure	Total
Sample 1	a	$m - a$	m
Sample 2	$r - a$	$n - r + a$	n
Total	r	s	N

[5] See footnote in Section 7.3.

Given the fixed margins, with A denoting the random variable describing the number of successes in sample 1, we have:

$$\mathrm{P}(A = a) = \frac{m!n!r!s!}{N!a!b!c!d!}. \tag{17.9}$$

This is the hypergeometric distribution of Section 3.15.

In deriving Equation (17.9), Fisher was considering the distribution of θ, the odds-ratio given by Equation (17.8). In terms of θ, the hypotheses become:

$$\mathrm{H}_0 : \theta = 1,$$
$$\mathrm{H}_1 : \theta \neq 1.$$

If the alternative hypothesis is that the proportion of successes in sample 1 is greater than that in sample 2, we must now add to $\mathrm{P}(A = a)$ the probabilities of any possible outcomes where, given the observed margins, θ takes a greater value than that observed. If the alternative hypothesis is simply that the two success probabilities are not the same, then the usual procedure is to combine $\mathrm{P}(A = a)$ with the probabilities of all values of A that are less likely than the value observed.

Example 17.8

A young surgeon performs four complex operations and only one patient survives. An experienced surgeon performs 14 such operations of which 10 are successful.

(a) *Is there significant evidence, at the 5% level, of a difference between the success rates of the two surgeons?*

(b) *Is there significant evidence, at the 5% level, that the success rate for the young surgeon is less than that of the experienced surgeon?*

With the rows corresponding to the allocation to the surgeons and the columns corresponding to the outcome of the operation, the outcome is summarized in the following table:

1	3
10	4

The null hypothesis that the two surgeons have the same success rates implies that, for each patient, the operation is as likely to be a success with the young surgeon as with the experienced surgeon: the outcome of the operation has not been affected by the choice of surgeon. Since the treatment was successful for 11 of the patients, one of the following outcome tables was sure to occur:

0	4
11	3

1	3
10	4

2	2
9	5

3	1
8	6

4	0
7	7

Using Equation (17.9) the corresponding probabilities are:

$$\frac{49}{4284} \qquad \frac{539}{4284} \qquad \frac{1617}{4284} \qquad \frac{1617}{4284} \qquad \frac{462}{4284}.$$

(a) For a two-sided test we must add to the probability of the observed outcome, 539/4284, the smaller probabilities from *both* tails: 49/4284 and 462/4284. The total is 1050/4284 = 0.245, which is much greater than 5%. So, at the 5% level, there is no significant evidence of a difference in the success rates of the surgeons.

(b) In this case we need only consider the single tail. Since (49 + 539)/4284 = 0.137, we again find no significant difference. Examining the probabilities, we find that the only situation that would have led to a significant difference, at the 5% level, would have been if the inexperienced surgeon had not performed any successful operations.

At the time that Fisher introduced his test, and for decades afterwards, the test could only feasibly be used for 2×2 tables with small cell frequencies. However, the logic of the test is equally applicable to *any two-way table*. Any restriction on its use due to the sizes of the cell frequencies will depend on the power of the computer.

17.7.3 Derivation of Equation (17.9)

For convenience, we repeat the outcome table here:

	Success	Failure	Total
Sample 1	a	c	m
Sample 2	b	d	n
Total	r	s	N

The null hypothesis is that the probabilities of success are the same for the two samples. Denoting the common probability by p, the probability of a successes out of the m observations in sample 1 is:

$$\binom{m}{a} p^a (1-p)^{m-a},$$

with the corresponding probability for sample 2 being:

$$\binom{n}{r-a} p^{r-a} (1-p)^{(n-r+a)}.$$

So, in the absence of other information, the probability of the observed outcome is:

$$\binom{m}{a} p^a (1-p)^{m-a} \times \binom{n}{r-a} p^{r-a} (1-p)^{(n-r+a)} = \binom{m}{a}\binom{n}{r-a} p^r (1-p)^s.$$

However, we know that there is a total of r successes, for which the probability is:

$$\sum_a \binom{m}{a}\binom{n}{r-a} p^r (1-p)^s = p^r (1-p)^s \sum_a \binom{m}{a}\binom{n}{r-a}.$$

The summation here runs from $a = 0$ to $a = min(m, r)$.

Comparing the coefficients of x^r in $(1+x)^m(1+x)^n$ and in $(1+x)^{m+n}$ we find that:

$$\sum_a \binom{m}{a}\binom{n}{r-a} = \binom{m+n}{r}.$$

Hence, the probability of the observed outcome *conditional* on the two sets of marginal totals is:

$$\binom{m}{a}\binom{n}{r-a} p^r(1-p)^s \Big/ p^r(1-p)^s \binom{m+n}{r} = \frac{m!n!r!s!}{a!b!c!d!N!}.$$

17.7.4 The Yates correction

Returning to the application of the X^2-test, Yates[6] suggested that, for the special case of a 2×2 table, the χ^2 approximation is improved by making the small adjustment of reducing the difference between the observed and expected cell frequencies by 0.5. We denote the resulting test statistic by X_c^2, given by:

$$X_c^2 = \sum_{i=1}^{4} \frac{(|O_i - E_i| - 0.5)^2}{E_i}.$$

There are a number of ways of simplifying this expression. For example, if we denote the four cells by a, b, c, and d, as before, then

$$X_c^2 = \frac{N(|ad - bc| - N/2)^2}{mnrs}. \tag{17.10}$$

The correction suggested by Yates is a disguised form of the usual continuity correction used when approximating a binomial distribution by a normal distribution.

A century after its introduction the correction continues to be a bone of contention. It is included here for completeness, since it appears in older introductory texts. Nowadays, it will rarely be required since, in most cases, Fisher's exact test can be used.

[6] Frank Yates (1902–94), an English statistician, became Head of Statistics at Rothamsted, the agricultural research institute in Hertfordshire. I once gave a talk concerning the 2×2 table, with Yates sitting in the front row: it was unnerving!

Example 17.9

The following data comes from a study concerning a possible cure for the common cold. A random sample of 279 French skiers were divided into two groups. Both groups took a pill each day. The pill taken by one group contained the possible cure, whereas the other group took an identical-looking pill that contained only sugar. Of the 139 skiers taking the possible cure, 17 caught a cold. Of the 140 skiers taking the sugar pill, 31 caught a cold. Does this provide significant evidence, at the 5% level, that the possible cure has worked?

The null hypothesis is that the outcome (cold or not) is independent of the treatment (sugar or cure), with the alternative being that there is an association.

A summary of the observed and expected frequencies is given below:

	Observed				*Expected*		
	Cold	No cold			Cold	No cold	
Sugar	31	109	140	Sugar	24.09	115.91	140
Cure	17	122	139	Cure	23.91	115.09	139
	48	231	279		48	231	279

Using the exact test we obtain a tail probability of 3.8%, suggesting that there is evidence of a difference between the success rates of the cure and the sugar pill. Examining the expected frequencies we can see that the difference is as expected: the cure is apparently curing.

For the record, $X^2 = 4.81$ and $X_c^2 = 4.14$, corresponding to tail probabilities of 2.8 and 4.2%, respectively.

It is always worth examining the expected frequencies (and the r_i values) to see the extent and location of differences between the observed and expected values.

Exercises 17e

1. A survey of the traffic passing along a particular road concentrates on the age of the car and the gender of the driver. The results are as follows:

	New car	Old car
Male	117	63
Female	52	48

Test whether there is significant evidence of an association between car age and the gender of the driver:

(a) Using the uncorrected chi-squared test.
(b) Using the Yates correction.
(c) Using Fisher's exact test.

17.8 *Multi-way contingency tables

In practice it is often the case that there are more than two relevant variables. In social science applications there are usually several key variables. Here are some: age, gender, social class, religion, location, political allegiance, occupation, parental occupation.

Frequently therefore, the analyst must consider several variables simultaneously. The following example gives an indication of the difficulties that arise.

Example 17.10

In June 1970, a representative sample of poorly educated white American teenagers were asked how many children they hoped to have when they were married. It was expected that their responses might vary according to their age and gender. The results are summarized in the following table:

	Males			Females		
No. of children	≤ 2	3	≥ 4	≤ 2	3	≥ 4
Age 12–15	24	18	20	27	17	23
Age 16–17	7	11	10	21	6	23

So here we have three classifying variables: age (A), gender (B), and number of children (C). We could form two-way tables from any pair of variables. For example:

No. of children	≤ 2	3	≥ 4
Male	31	29	30
Female	48	23	46

No. of children	≤ 2	3	≥ 4
Age 12–15	51	35	43
Age 16–17	28	17	33

Denoting any association (i.e. lack of independence) between a respondent's age and the number of children hoped for, as the AC association, it is evident that one might also study the AB and BC associations. However, these are not the only possibilities, since the association between a respondent's age and the number of children hoped for, might differ according to the respondent's gender: this would be denoted as the ABC interaction.

When further variables are included, the number of possible multi-variable associations rapidly increases. The strategy for analysing this type of data uses so-called **log-linear models**, which are outside the scope of this introductory volume.

Key facts

- **The chi-squared test of goodness of fit**

$$X^2 = \sum_{i=1}^{m} \frac{(O_i - E_i)^2}{E_i}.$$

  compares

  H_0: the results are a random sample from the supposed distribution,
  H_1: H_0 is incorrect.

  - For the χ^2 approximation to be valid, E_i-values must be reasonably large (say, ≥ 5): smaller expected frequencies should be eliminated by combining categories.
  - When there are m categories (after combination), d, the number of degrees of freedom of the distribution, is given by either:
    * $d = m - 1$, for the case of probabilities completely prescribed by H_0,
    * $d = m - 1 - k$, for the case where k parameters are estimated from the data,
  - An alternative to the χ^2-approximation is to use a Monte Carlo test.

- **Independence in an $r \times c$ contingency table**

  For a table having (after combination) r rows and c columns:

  - H_0: the variables are independent,
    H_1: the variables are associated.
  - X^2 has approximately a χ^2 distribution with $(r-1)(c-1)$ degrees of freedom.
  - For 2×2 tables, the approximation is improved using the **Yates correction**:

$$X_c^2 = \sum_{i=1}^{m} \frac{(|O_i - E_i| - 0.5)^2}{E_i}.$$

  - **Fisher's exact test**: Samples of sizes m and $N - m = n$, give totals of r successes and s failures, with individual cell counts a, b, c, and d. The probability of this outcome is:

$$\frac{m!n!r!s!}{N!a!b!c!d!}.$$

- **Kolmogorov–Smirnov test** Compares the sample distribution function with the distribution function of the distribution specified by H_0. Test statistic, D is given by

$$D = \max_i \left(F(x_{(i)} - \frac{i-1}{n}, \ \frac{i}{n} - F(x_{(i)}) \right).$$

- **Dispersion test** Used to test whether data are consistent with a Poisson distribution. The test statistic is:

$$I = \frac{(n-1)s^2}{\bar{x}},$$

where n is the sample size. If the data come from a Poisson distribution, then I is an observation from a χ^2_{n-1} distribution.

R

- To compare an observed vector of counts, `x` with a set of theoretical probabilities `prob` use `chisq.test(x,p=prob)`. But check that the degrees of freedom are correct. You may need to ignore the quoted p-value, and, with d denoting the true number of degrees of freedom, use `csq=as.numeric(chisq.test(x,p=prob)[1]);` `1-pchisq(csq,d)`. Alternatively, using simulation, the option `simulate.p.value=TRUE` should be added to the `chisq.test` command.

- Expected frequencies are given by `chisq.test(x)$expected`

- Residuals are given by `chisq.test(x,p=prob)$residuals`.

- **Kolmogorov–Smirnov test**: With the data in a vector `x` use `ks.test(x, pType, Parameters)` where `pType` specifies the distribution function for the supposed type of distribution and `Parameters` specifies the parameters of that distribution. For example: `ks.test(y,"pnorm",mean=0,sd=1)`.

- **Large sample test for equality of proportions**: `prop.test(A)` where `A` is a 2×2 array—use `A<-array(c(a,b,c,d), c(2,2))` where a, b, c, d are the four counts.

- **Fisher's exact test**: `fisher.test(A, alternative)` The test returns the p-value, the maximum likelihood estimate of the odds ratio and a confidence interval for that odds ratio.

18

Correlation

Previous chapters have concentrated principally on developing methods and models for a *single* random variable, but many data sets provide information about *several* variables, with the question of interest being whether there are connections between these variables. In this chapter, and in Chapter 19, we concentrate on methods suitable for use with two quantitative variables, usually denoted by x and y.

Often all the data are collected at more or less the same time, for example:

x	y
Take-off speed of ski-jumper	Distance jumped
No. of red blood cells in sample of blood	No. of white blood cells in sample
Hand span	Foot length
Size of house	Value of house
Depth of soil sample from lake bottom	Amount of water content in sample

However, sometimes data are collected on one variable later than the other variable, though the link (the same individual, same plot of land, same family, etc.) is clear:

x	y
Mark in mock exam	Mark in real exam (three months later)
Amount of fertilizer	Amount of growth
Height of father	Height of son when 18

In some of the above cases, while the left-hand variable, x, may affect the right-hand variable, y, the reverse cannot be true—these are cases in which the regression methods introduced in Chapter 19 will be particularly suitable.

In other cases (such as the counts of red and white blood cells) both variables are influenced by some unmeasured variable (e.g. the condition of the patient). Cases like these are the subject of the current chapter, which deals with the sample equivalent of population correlation (which was briefly introduced in Section 10.2.2).

If changes in the values of one variable are reflected in changes in the other variable, then the two variables are said to display **correlation**. In cases where increasing values of one variable, x, are accompanied by increasing values of the other variable, y, the variables are said to display **positive**

correlation. The opposite situation (**negative correlation**) is one where increasing values of one variable are associated with decreasing values of the other variable.

We now introduce some simplifying notation:

$$S_{xx} = \sum(x_i - \bar{x})x_i = \sum(x_i - \bar{x})^2, \tag{18.1}$$

$$S_{yy} = \sum(y_i - \bar{y})y_i = \sum(y_i - \bar{y})^2, \tag{18.2}$$

$$S_{xy} = \sum(x_i - \bar{x})y_i = \sum x_i(y_i - \bar{y}) = \sum(x_i - \bar{x})(y_i - \bar{y}). \tag{18.3}$$

The quantities S_{xx} and S_{yy} are the sample variances of the x-values and the y-values multiplied by $(n-1)$. In the same way S_{xy} is the product of $(n-1)$ and the **sample covariance.**[1]

Example 18.1

The marks obtained by 10 students on test papers in mathematics and physics are given in the following table:

Student	1	2	3	4	5	6	7	8	9	10
Maths	65	45	40	55	60	50	80	30	70	65
physics	60	60	55	70	80	40	85	50	70	80

Display these data using an appropriate diagram.

A scatter diagram is appropriate:

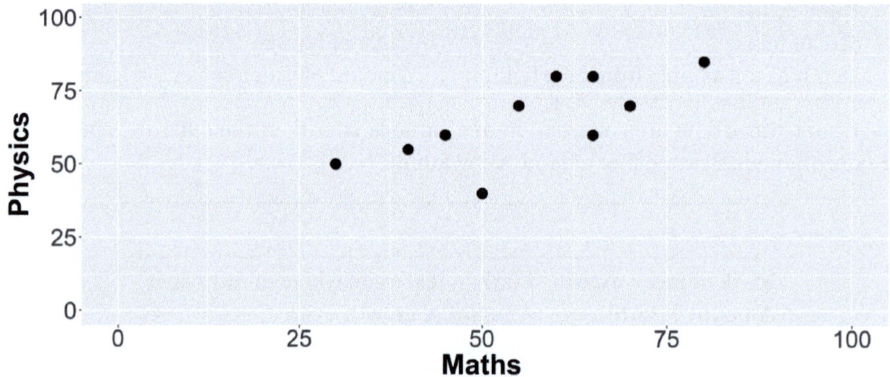

We can see that, on the whole, higher marks on one subject are reflected by higher marks for the other subject. The two sets of marks are positively correlated.

[1] The *population* covariance was defined in Section 10.2.2.

18.1 The product-moment correlation coefficient

The fact that students who get high marks in physics tend to be the ones who get high marks in mathematics, and vice versa is not unexpected. This seems clear enough — but what do we really mean by a 'high mark'? We do not have any particular number in mind. A high mark is simply 'a mark above the average mark'. If, for a random sample of students, the individual marks in mathematics are denoted by $x_1, x_2, ..., x_n$, with a mean $\bar{x}$, then we are interested in the values of $x_1 - \bar{x}$, $x_2 - \bar{x}$, etc.

Suppose we denote the corresponding physics marks by $y_1, y_2, ..., y_n$, with mean $\bar{y}$. The suggestion is that if $x_i - \bar{x}$ is positive (corresponding to a 'high' mark in mathematics by the ith student), then $y_i - \bar{y}$ is also likely to be positive.

Of course, if high marks go together, then so do low marks, so we anticipate that a negative value for $x_i - \bar{x}$ will usually correspond to a negative value for $y_i - \bar{y}$.

We see that positive correlation results in most products $(x_i - \bar{x})(y_i - \bar{y})$ being positive. and this will be reflected in a positive value for their sum, S_{xy}.

Example 18.1 (cont.)

Student i	*Maths* x_i	*physics* y_i	$x_i - \bar{x}$	$y_i - \bar{y}$	$(x_i - \bar{x})(y_i - \bar{y})$
1	65	60	9	−5	−45
2	45	60	−11	−5	55
3	40	55	−16	−10	160
4	55	70	−1	5	−5
5	60	80	4	15	60
6	50	40	−6	−25	150
7	80	85	24	20	480
8	30	50	−26	−15	390
9	70	70	14	5	70
10	65	80	9	15	135

The mean mathematics mark is 56. Of the five students who get less than this, four also get less than the average mark (65) in physics. There is a corresponding agreement for the marks that are above average. The consequence is that eight of the ten $(x_i - \bar{x})(y_i - \bar{y})$ products are positive, as is their sum, $S_{xy}(= 1450)$.

The quantity S_{xy} is affected by changes in the scale. For example, if in the previous example, the marks had been out of 1000, instead of out of 100, with both the x-values and y-values multiplied by 10, then the S_{xy} values would have been multiplied by 100. For this reason, Galton[2] suggested calculating a quantity, r, that is *unaffected by scale changes*. This quantity, which Galton termed the co-relation, is now called the (sample) **product-moment correlation coefficient**[3] (or, simply, the **correlation**)

[2] Sir Francis Galton (1822–1911) had a varied career: he could reasonably be described as a doctor, explorer (in Africa), meteorologist (he coined the word 'anticyclone'), biometrician, and statistician.

[3] It is also referred to as **Pearson's correlation coefficient** in acknowledgement of the work done by Karl Pearson (see footnote in Section 14.2.1) on the underlying theory.

and is calculated using:

$$r = \frac{S_{xy}}{\sqrt{S_{xx}S_{yy}}}. \tag{18.4}$$

It is the value of S_{xy} that determines whether the correlation coefficient is negative, zero, or positive. The limiting values -1 and 1 correspond to cases where the points are collinear with, respectively, negative and positive slopes.

The value of r is unaffected by changes in the units of measurements.

Example 18.1 (cont.)

In this case $S_{xx} = 2040$ and $S_{yy} = 1900$ so that $r = 1450/\sqrt{2040 \times 1900} = 0.7365$. There is a sizeable positive correlation between the two sets of marks.

18.1.1 Demonstration that collinearity implies that $r = \pm 1$

If the points are perfectly collinear then $y_i = a + bx_i$ for all i. This means that

$$S_{xy} = \sum(x_i - \bar{x})y_i = \sum(x_i - \bar{x})a + \sum(x_i - \bar{x})bx_i = 0 + bS_{xx},$$

since $\sum(x_i - \bar{x}) = 0$.

Similarly

$$S_{yy} = \sum(y_i - \bar{y})y_i = \sum(y_i - \bar{y})a + \sum(y_i - \bar{y})bx_i = bS_{xy} = b^2S_{xx},$$

using the previous result.

Substituting into the fraction $S_{xy}/\sqrt{S_{xx}S_{yy}}$ we see that $r = b/\sqrt{b^2}$, which is equal to 1 or -1, depending upon whether b is positive or negative. Hence, *collinear points imply* $r = \pm 1$. The converse is also true.

18.1.2 Testing the hypothesis that $\rho = 0$

The sample quantity, r, is an estimate of the population correlation, ρ. If the underlying variables X and Y are uncorrelated, then $\rho = 0$. Any appreciable departure from 0 will be of interest, so we concentrate on the hypotheses H_0: $\rho = 0$ and H_1: $\rho \neq 0$. We will reject H_0, in favour of H_1, if $|r|$ is unusually close to 1.

The distribution of r depends on the distributions of X and Y. It is often reasonable to assume that the distributions are normal, in which case the critical values for $|r|$ are given in the following table:

n	5%	1%	n	5%	1%	n	5%	1%	n	5%	1%
4	.950	.990	7	.754	.874	10	.632	.765	13	.553	.684
5	.878	.959	8	.707	.834	11	.602	.735	14	.532	.661
6	.811	.917	9	.666	.798	12	.576	.708	15	.514	.641

Example 18.1 (cont.)

For the marks data we had $n = 10$ and $r = 0.7365$. When $n = 10$, the 5% critical value is 0.632, while the 1% point critical value is 0.765. The observed value is therefore significant at the 5% level but not (quite) at the 1% level. We would conclude that there is rather strong evidence to reject the hypothesis that X and Y are uncorrelated.

Assuming that X and Y have normal distributions and are uncorrelated, $r\sqrt{(n-2)/(1-r^2)}$ will be an observation from a t_{n-2}-distribution.

18.1.3 Correlation versus relation

Figure 18.1 illustrates three cases where $r = 0$:

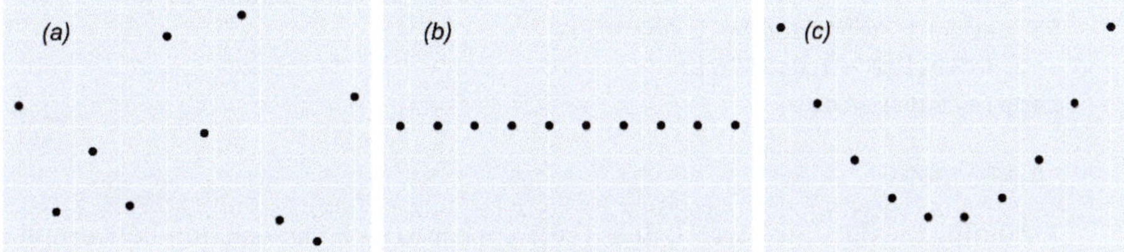

Figure 18.1 Three examples where $r = 0$.

Figure 18.1 (a) illustrates a case where the y-values vary independently of the value of x. In case (b) the y-value is fixed and independent of the x-value. The most interesting case is case (c) where y and x are certainly related, but the relationship is quadratic and not linear. This demonstrates that r is a measure of the extent to which the variables are *linearly* related. It is *not* an indicator of whether the variables are independent of one another.

Independence implies zero correlation, but zero correlation does *not* imply independence.

Exercises 18a

1. The following data refer to x, the average temperature (in degrees Fahrenheit) and y, the average butterfat content of a group of cows (expressed as a percentage of the milk).

Temp	64	65	65	64	61	55	39	41	46	59
Butterfat	4.65	4.58	4.67	4.60	4.83	4.55	5.14	4.71	4.69	4.65
Temp	56	56	62	37	37	45	57	58	60	55
Butterfat	4.36	4.82	4.65	4.66	4.95	4.60	4.68	4.65	4.60	4.46

It is given that $\sum x = 1,082$, $\sum y = 93.5$, $\sum x^2 = 60,304$, $\sum xy = 5,044.5$, and $\sum y^2 = 437.6406$.

(a) Plot the data.

(b) A computer program reports the following:

```
t = -2.1575, df = 18, p-value = 0.04473
alternative hypothesis: true correlation is not equal to 0
95 percent confidence interval:
 -0.74614139 -0.01347013
sample estimates:
       cor
-0.4532889
```

Assuming the data are correct, does there appear to be an association between butterfat content and temperature?

(c) It is found that the eleventh data item (56, 4.36) is incorrect. Ignoring that data item, determine a revised value for r.

2. State, with a reason, the effect on the value of the product-moment correlation coefficient, r, of:

(a) multiplying every x-value by 10,

(b) subtracting 10 from every y-value.

18.2 Nonsense correlation: storks and gooseberry bushes

Previously we observed that a lack of correlation between two variables does not necessarily imply that the variables are unrelated. The reverse is also true: two variables may appear highly correlated without there being any direct relationship. Suppose that the value of the variable Z has a direct bearing on the values of both variable X and variable Y, then in this case it will appear that the variables X and Y are inter-related.

In more innocent times, when children asked where babies came from, the answer was sometimes that they were found under gooseberry bushes. If they then asked how did they get there, the answer might be that they were brought there by storks. If data were available for those parts of Europe where storks nest on rooftops, then it is very probable that positive correlations would be found between the number of storks, the number of gooseberry bushes, and the number of babies. The reason is that there is a so-called **latent variable** affecting all three of storks, bushes, and babies: namely, population size.

The rise in the world's population is the latent variable that underlies many spurious correlations. For example, the decrease in the world's whale population is strongly negatively correlated with the increase in the number of cars on the road, yet there is no direct relation between the two.

Example 18.2

In 1997, when he was at the University of Hong Kong, Chris Lloyd (subsequently Professor of Statistics at the University of Melbourne), was consulted about an apparently amazing correlation between two measurements made on young babies.[a] The two variables ($x1$ and $x2$) are plotted against one another in Figure 18.2.

[a] The data are currently (2025) available at https://gksmyth.github.io/ozdasl/general/babies.html.

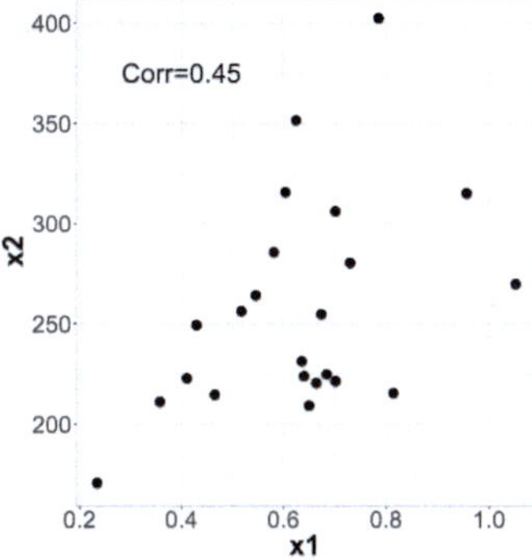

Figure 18.2 The apparent relation between two measurements of features of babies.

The correlation of 0.45 is substantial and apparently significant ($p = 0.03$). However, on questioning, Professor Lloyd found out that there were actually three groups of babies, nominally aged either 3, 12, or 24 months old. Figure 18.3 shows the resulting plot.

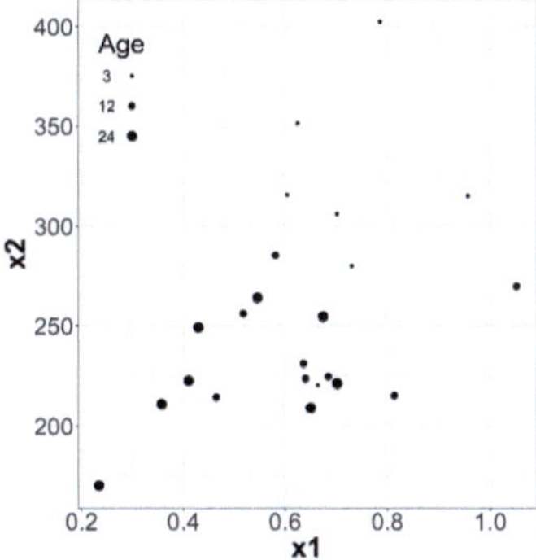

Figure 18.3 The relation between the measurements showing the age groups to which the babies belong.

Considering each age group separately tells a different story, with correlations of 0.19 ($p = 0.71$) for the 3-month group, 0.19 ($p = 0.69$) for the 12-month group, and 0.53 ($p = 0.22$) for the 24-month group. Even here the correlations are probably inflated since, for example, the 24-month group includes babies aged between 18 months and 30 months.

18.3 The ecological fallacy: immigration and illiteracy

When we study groups of people, we may find that two characteristics, X and Y, are such that if a group has, on average, a high value for X, then it is the case that the same groups have, on average, a high value for Y: if we plot the average values for X and Y, then we might see a clear trend. An ecological fallacy would result if we then argued that, because an individual is a member of a group with a high value of X, then, necessarily, that individual will have a high value of Y.

An example is provided if an onlooker, who observes a member of a group associated with a particularly high level of crime, then assumes that, because of that membership, the individual is likely to be a criminal. The potential for false conclusions was made apparent by a 1935[4] paper that studied the relationship between immigration and illiteracy (amongst those aged at least 15) in the United States. Using the data provided by the 1930 census (available online), there were at that time more than 4 million people in the United States who were described as illiterate. The division between native-born and foreigners is set out in the following table:

	Foreign-born	Native	Total
Illiterate	1.30	2.98	4.28
Literate	11.92	82.52	94.44
Total	13.22	85.50	98.72

We see that at that time about 3.5% (2.98/85.50) of native Americans were illiterate, whereas the figure for the foreign-born inhabitants was unsurprisingly much higher (roughly 10%).

The census reports not only the national figures, but also the figures at state and division levels. Figure 18.4 shows how the percentage deemed to be illiterate varies with the percentage of immigrants in the nine divisions of the United States. Looking only at this figure, one might conclude that it is the immigrants who are the more literate, whereas, in fact, the reverse is the case.

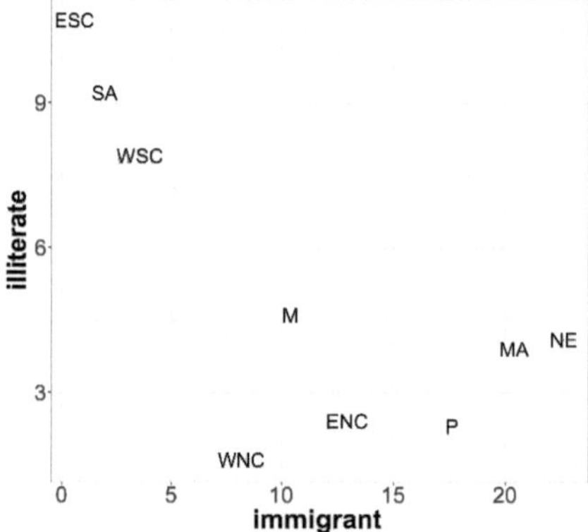

Figure 18.4 The divisions of the United States show lower levels of literacy where there are higher proportions of immigrants.

[4] W. S. Robinson, Ecological correlations and the behavior of individuals' *American Sociological Review*, 351–357.

The explanation is that in the 1930s most immigrants settled in the well-populated eastern states where the inhabitants were highly educated, whereas relatively few settled in the southern states where literacy levels were comparatively low. Since foreign-born inhabitants accounted for just 11% of the 1930 population, the illiteracy levels reported largely reflect the illiteracy levels of the native population.

18.4 Simpson's paradox: amputation or injection?

This paradox takes its name from a paper published in 1951,[5] though the situation had been discussed in papers at the start of the twentieth century. Here is an example.

Example 18.3

A patient has a particular disease for which there are two possible treatments, amputation, or injection. The patient has a phone consultation with the doctor. To decide which treatment to use, the doctor examines the past records and finds the following results:

Treatment	Success	Failure	Success rate
Amputation	8	3	73%
Injection	7	7	50%

The doctor concludes that amputation is required. The worried patient suggests that the success rates might depend upon a patient's gender. The patient has a deep voice so the doctor looks at the records for males and concludes that the patient was correct, since the male records suggest that injection is preferable to amputation. Here are the male records:

Treatment	Success	Failure	Success rate
Amputation	7	1	88%
Injection	2	0	100%

When the patient meets the doctor to receive the injection it is apparent that she has an unusually deep voice: the doctor should have consulted the female records which were as follows:

Treatment	Success	Failure	Success rate
Amputation	1	2	33%
Injection	5	7	42%

Fortunately, the decision to give the injection is seen to be correct. Each sub-table favours injection, but the combined table paradoxically favours amputation.

The key to Simpson's paradox is the presence of a third (latent) variable that is missing from the original combined table. In this example it is the patient's gender: the success of either treatment is greatly affected by whether the patient is male or female.

[5] E. H. Simpson, The interpretation of interaction in contingency tables, *Journal of the Royal Statistical Society, Series B (Methodological)*, **13 (2)**, 238–241.

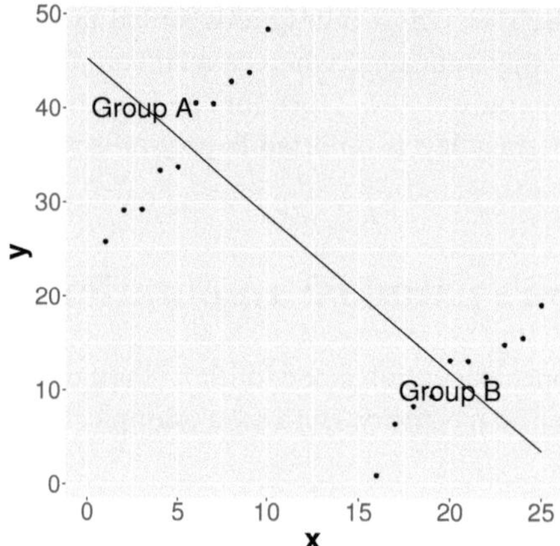

Figure 18.5 Although the two groups each show a positive slope for the relation between x and y, taking the data as a single entity suggests a line with a negative slope.

Figure 18.5 shows a graphical analogue of the situation. In the figure it is obvious that there are two distinct groups with very different values for the two variables: the two groups should be considered separately. In the same way, in the example, the fictitious doctor should have been aware that the disease affected males and females in a different fashion.

18.5 Rank correlation

Assuming that the two sets of data, x and y, are numerical, it is always possible to determine the value of the product-moment correlation coefficient, r. However, the standard assessment of the significance of the observed value (using tables, the t-distribution, or the computer) relies on the assumption that both data sets come from normal distributions.

We now relax the assumption of normality: indeed, we relax the requirement that the values are numerical. All that is required is that the data items can be placed in order of magnitude.; for example, 'small', 'a bit larger', 'a bit larger still'. Several measures have been suggested, with the two commonly used being those due to Spearman and to Kendall.

18.5.1 Spearman's rank correlation coefficient, r_s

Spearman[6] introduced his measure in 1904. Suppose that there are n pairs of observations, with the values of each variable being replaced by the ranks 1 to n. Let d_i be the difference in the ranks assigned to the items in the ith pair. Then r_s is given by

$$r_s = 1 - \frac{6\Sigma d_i^2}{n(n^2 - 1)}. \quad (18.5)$$

[6] Charles Spearman (1863–1945) was an Englishman who, after a distinguished career in the army, turned to the study of psychology, introducing statistical methods to that discipline.

It does not matter whether rank 1 refers to the largest, or to the smallest, provided it is the same for both variables.

Example 18.4

Ten pairs of values of x *and* y *are as follows:*

x	8	18	23	29	46	54	60	71	143	297
y	7	1	12	16	21	37	46	19	55	122

Plot these values on a scatter diagram. Does there appear to be a relation between x *and* y*?*

The scatter diagram shows a clear relationship—as x increases, so (on the whole) y increases. The variables are certainly not normally distributed, so a measure of rank correlation is appropriate.

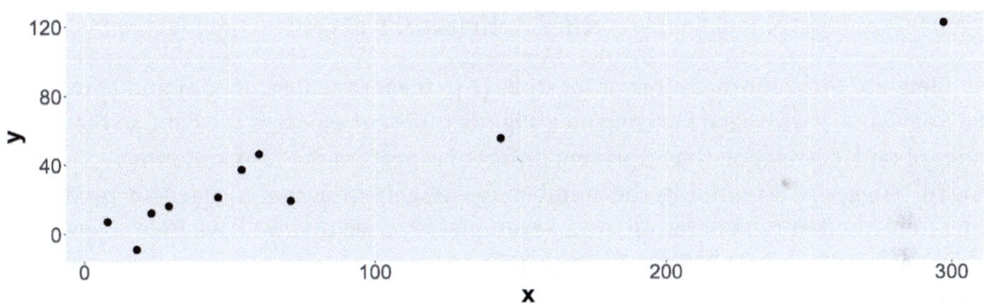

The two sets of ranks and their differences are as follows:

Rank of x	1	2	3	4	5	6	7	8	9	10
Rank of y	2	1	3	4	6	7	8	5	9	10
Absolute difference, $\|d\|$	1	1	0	0	1	1	1	3	0	0

In this case $\Sigma d_i^2 = 14$. Thus:

$$r_s = 1 - \frac{6 \times 14}{10 \times 99} = 0.915.$$

This is obviously a very positive correlation: the null hypothesis that the ranks are unrelated is rejected since the computer reports a p-value of about 0.0005.

If two or more items are ranked equally, then it is conventional to award the average of the corresponding ranks that could have been awarded (the so-called **tied rank**).

18.5.1.1 Using r_S for non-linear relationships

The product-moment correlation is concerned with the extent to which X and Y are *linearly* related. Suppose, however, that we had the following data:

x	1	2	3	4	5	6
y	1	8	27	64	125	216

The values *exactly* satisfy the simple relation $y = x^3$, but the product-moment correlation coefficient, r, is not equal to 1 because the relation is not linear (in fact, $r = 0.938$). By contrast, $r_S = 1$ in all cases where y increases as x increases and $r_S = -1$ if y decreases as x increases.

Rather tedious algebra demonstrates that r_S is, in fact, the product-moment correlation coefficient for ranks.

18.5.2 Kendall's τ

Just as there are different measures of location (e.g. mean, median, mode) and spread (e.g. range, standard deviation), with each focusing on a slightly different aspect of the data, so there are different measures of rank correlation that represent different aspects of the rank orderings.

Kendall[7] suggested examining the number of 'neighbour-swaps' needed to produce one rank ordering from another rank ordering. As an example, suppose that we have four objects, and the two orderings (A,B,D,C) and (D,B,A,C). To convert the second ordering into the first using neighbour-swaps we could proceed as follows:

(D,B,A,C)	The second ordering
(D,A,B,C)	Swap A and B
(A,D,B,C)	Swap A and D
(A,B,D,C)	Swap B and D to give the first ordering

Denoting the minimum number of neighbour-swaps needed by Q, Kendall proposed using the measure τ, defined by:

$$\tau = 1 - \frac{4Q}{n(n-1)}.$$

[7] Sir Maurice Kendall (1907–83) was Professor of Statistics at the London School of Economics from 1949 to 1961. Subsequently he was director of the World Fertility Study. His initial paper on rank correlation was published in 1938, though 'Kendall's τ' only became widely used following the publication, in 1948, of his book *Rank Correlation Methods*.

The symbol τ is the Greek letter 'tau' which is pronounced either like the first syllable of the word 'towel' or like the word 'tor'.

The swapping idea that underlies Kendall's τ is difficult to use with larger values of n. Fortunately, there is an easier way of counting the minimum number of neighbour-swaps! The procedure is as follows:

1. Rearrange the items in the order specified by the first ranking.
2. Write down the ranks assigned to these items by the second ranking to give a reordered second ranking.
3. Consider each rank in the reordered second ranking in turn, and count how many rank *of those to the right* are smaller than the rank being considered.
4. Sum these numbers. This sum is Q, the required minimum number of neighbour-swaps.

Example 18.4 (cont.)

Ranked in order by x	1	2	3	4	5	6	7	8	9	10
Rank of y	2	1	3	4	6	7	8	5	9	10
Smaller numbers of y to the right	1	0	0	0	1	1	1	0	0	0

Thus $Q = 4$ and $\tau = 1 - 16/90 = 0.822$. The null hypothesis that the the ranks are unrelated is rejected since the computer reports that the associated p-value is 0.0004, confirming that x and y are highly correlated.

The value given by Kendall's tau is usually less than that given by Spearman's statistic: the two values should not be compared.

Exercises 18b

1. An expert is asked to arrange seven plates in order of their date of manufacture. The results are as follows:

Actual date	1690	1710	1780	1810	1857	1896	1920
Expert's rank	1	2	5	4	3	7	6

The computer reports the following:

```
Spearman's rank correlation rho
S = 10, p-value = 0.03413
alternative hypothesis: true rho is not equal to 0
sample estimates:
      rho
0.8214286

Kendall's rank correlation tau
T = 17, p-value = 0.06905
alternative hypothesis: true tau is not equal to 0
sample estimates:
      tau
0.6190476
```

(a) Using the notation of this chapter, to what quantities do S and T correspond?

(b) State your conclusions concerning the ability of the expert.

2. The yield of a crop, y, is believed to be dependent on the monthly rainfall, x. measurements at nine locations are as follows;

y	8.3	10.1	15.2	6.4	11.8	12.2	13.4	11.9	9.9
x	14.7	10.4	18.8	13.1	14.9	13.8	16.8	11.8	12.2

(a) Determine the value of Spearman's r_S.

(b) Determine the value of Kendall's τ.

(c) State your conclusions.

3. The moisture content, M, of core samples of mud from a lake is measured as a percentage. The depth, d metres, at which the core is collected is also recorded. The results are as follows:

d	0	5	10	15	20	25	30	35
m	90	82	56	42	30	21	20	18

(a) Plot the data.

(b) Without performing any calculations, state, with a reason, whether the product moment correlation coefficient, r, will have the same value as Spearman"s r_S.

Key facts

- With:

$$
\begin{aligned}
S_{xx} &= \sum(x_i - \bar{x})x_i = \sum(x_i - \bar{x})^2, \\
S_{yy} &= \sum(y_i - \bar{y})y_i = \sum(y_i - \bar{y})^2, \\
S_{xy} &= \sum(x_i - \bar{x})y_i = \sum x_i(y_i - \bar{y}) = \sum(x_i - \bar{x})(y_i - \bar{y}).
\end{aligned}
$$

  Product-moment correlation coefficient, $r = S_{xy}/\sqrt{S_{xx}S_{yy}}$.

- With n pairs of values, replacing x and y by their ranks r_x and r_y, and writing $d = r_x - r_y$, **Spearman's rank correlation coefficient** is given by $r_s = 1 - 6\sum d^2/n(n^2 - 1)$.

- Setting the pairs in the order from 1 to n for one variable, consider each rank in the reordered second ranking in turn, and count how many ranks of those to the right are smaller than the number being considered. Let Q denote the sum of these values. **Kendall's rank correlation coefficient** is given by $\tau = 1 - 4Q/n(n - 1)$.

R

- With data in the vectors `x` and `y`, the command `cor(x,y)` returns the value of r.

- The command `cor.test(x,y)` provides the value of r together with a confidence interval based on the t-distribution and also a p-value for testing the independence hypothesis.

- The commands `cor.test(x,y, method="spearman")` and `cor.test(x,y, method="kendall")` produce the output illustrated in Exercises 17b.

19

Regression

Correlation was principally concerned with measuring the *strength* of a linear relationship between two variables. In this chapter, continuing that theme, we attempt to determine the *form* of that linear relationship. We therefore begin by reviewing the mathematical description of a straight line.

19.1 The equation of a straight line

When studying statistics the standard notation for the equation of a straight line is

$$y = a + bx,$$

where a is a constant known as the **intercept** and b is a constant known as the **slope**. xa

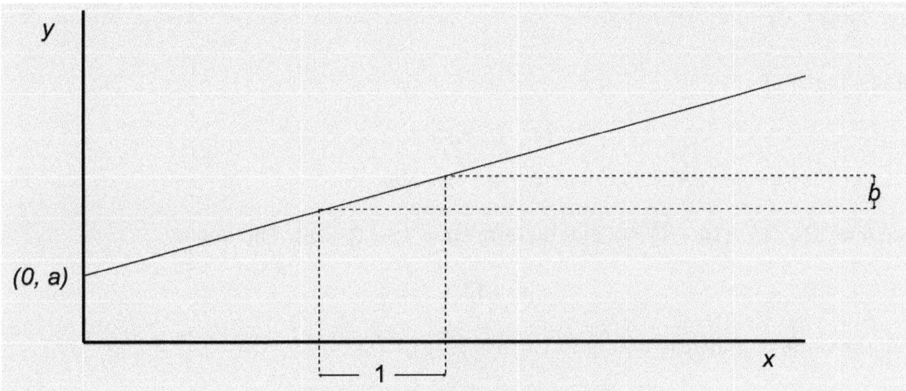

Figure 19.1 The line $y = a + bx$.

If we increase x to $(x + 1)$, the value of y changes from $\{a + bx\}$ to $\{a + b(x + 1)\}$, which is a change of b; thus b measures the amount of change in y for unit change in x. The quantity a represents the value of y when x is equal to 0, and therefore prescribes where the line crosses the y-axis (see Figure 19.1).

In the context of regression, the values of x are fixed ('controlled') and the values of y are subject to random variation. In general x is known as the **independent variable** and Y is the **dependent variable** (which is a random variable and hence is denoted with a capital letter).

19.1.1 Determining the equation

Suppose that we have drawn a line on a scatter diagram. How do we determine the equation of that line? The answer is quite simple. We first determine the coordinates of two points lying on the line. These can be any points, though it is a good idea to choose points near the edges of the diagram. Denote the points by (x_1, y_1) and (x_2, y_2). Then

$$\begin{aligned} y_1 &= a + bx_1, \\ y_2 &= a + bx_2. \end{aligned}$$

Subtracting, we get

$$y_1 - y_2 = b(x_1 - x_2)$$

and hence

$$b = (y_1 - y_2)/(x_1 - x_2). \tag{19.1}$$

To find the value of a it is easiest to substitute our value for b into either of the original equations:

$$a = (y_1 - bx_1) = (y_2 - bx_2).$$

Example 19.1

> *A line goes through the points (1,15) and (10, 33). The point (5,y) also lies on the line. Determine the value of y.*

We know that

$$\begin{aligned} 15 &= a + b, \\ 33 &= a + 10b. \end{aligned}$$

Hence $b = (33 - 15)/(10 - 1) = 2$ and therefore $a = 15 - 2 = 13$. The line is

$$y = 13 + 2x,$$

so that the value of y corresponding to $x = 5$ is $13 + (2 \times 5) = 23$.

Exercises 19a

1. A line goes through the points (4,6) and (12, 22). Find the equation of the line.
2. The regression line is $y = 4x + 2$. The observed value of y was 12. Determine the value of x corresponding to that value of y.

19.2 Why 'regression'?

It is the result of a paper entitled 'Regression towards Mediocrity in Hereditary Stature' that appeared in *The Journal of the Anthropological Institute of Great Britain and Ireland* in 1886. This was the work of Sir Francis Galton,[1] who had collected data referring to the heights of fathers and their adult sons. Some of Galton's data are summarized below and are shown in the diagram together with a dotted line corresponding to equality of the heights of the two generations.

Mean height of fathers (inches)	72.5	70.5	68.5	66.5	64.5
Mean height of their adult sons (inches)	72.25	69.5	68.25	67.25	65.75

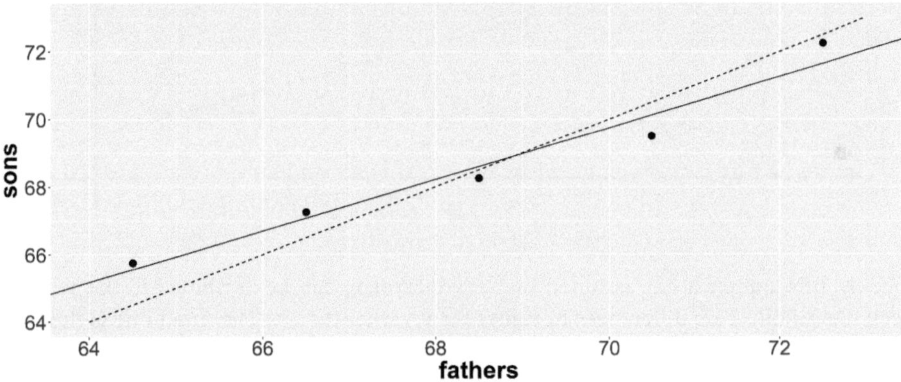

Galton plotted the data and noticed three things:

1. On average, the heights of the adult children of tall parents were greater than the heights of the adult children of short parents.
2. The two averages appeared to be (more or less) *linearly* related.
3. On the whole, the children of tall parents are shorter than their parents, (72.25 < 72.5, etc.), whereas the children of short parents are taller than their parents, (65.75 > 64.5, etc.): the values *regress* towards the mean.

These findings led Galton to refer to his summary line as being a **regression line**, and this name is now used to describe quite general relationships.

19.3 The method of least squares

Suppose that we have the n observations: $(x_1, y_1), (x_2, y_2), ..., (x_n, y_n)$. Our aim is that the values that we choose for a and b, should be such that the line $y = a + bx$, fits these data as well as possible. The discrepancy between the ith observation, and the y-value suggested by the fitted line is the **residual**, denoted by e_i and given by

$$e_i = y_i - (a + bx_i). \tag{19.2}$$

[1] See footnote in Section 18.1.

If the values of a and b are well chosen, then all of the residuals e_1, e_2, ..., e_n will be small in magnitude. Some of the residuals will be negative and some will be positive. Handling absolute values causes mathematical difficulties, so we work with their squares. A line that fits the data well will give a relatively small value for $\sum e_i^2$. The **least-squares line** is the line that actually minimizes this quantity and it is sometimes called the **line of best fit**.

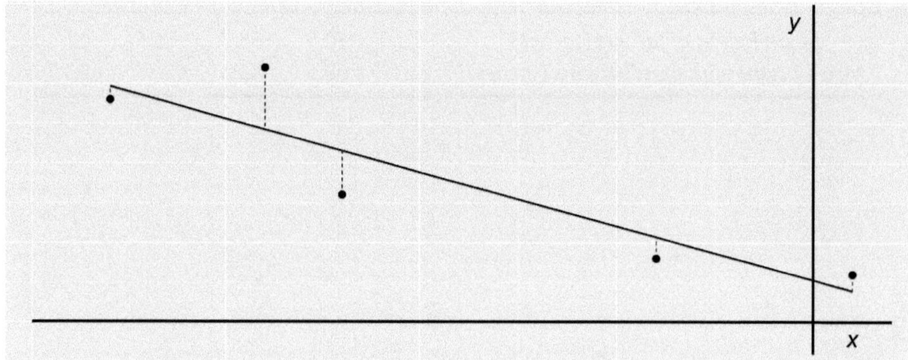

Implicitly, this approach has assumed that any lack of fit can be attributed to the y-value being incorrect: the residuals are differences in values for y, rather than x. The fitted line is the regression line of y upon x.

We could instead have assumed that any lack of fit to a straight line was a result of incorrect x-values. We would then be looking at different residuals (to a slightly different fitted line—the regression line of x upon y):

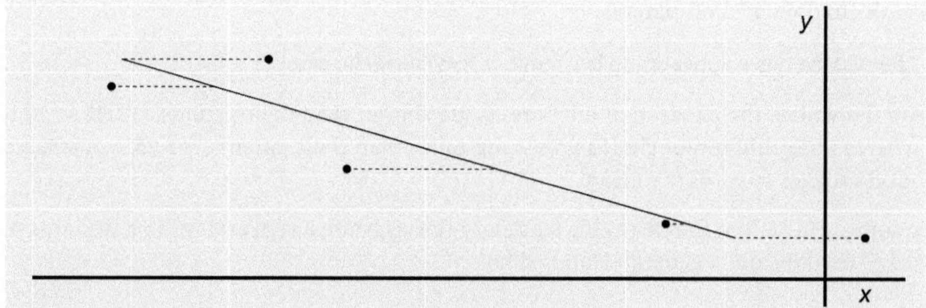

The choice between the two regression lines depends upon which variable we decide is the more prone to random variation or mis-measurement. Traditionally this is the variable that we denote as Y, where we are using the capital letter to state that it is a random variable. Often it will be obvious which is the Y-variable:

x	Y
Number of bricks in pile	Weight of pile of bricks
The price of a commodity	The number sold

As noted earlier, Y is often described as the **dependent variable** having an unpredictable value, whereas x is described as the **independent variable** whose value has been fixed. If there is an underlying linear relationship, then this relationship connects x not to an individual y-value, but to the mean of the y-values *for that particular value of* x. The mean of Y, given the particular value x, is denoted by

$$\mathrm{E}(Y|x),$$

which is more formally called the **conditional expectation** of Y. The formal linear regression model is

$$\mathrm{E}(Y|x) = a + bx. \tag{19.3}$$

Implicitly we have been paying equal attention to every pair of observations. With Y being the random variable, this implies that the Y-values should have equal variance:

$$\mathrm{Var}(Y|x) = \sigma^2. \tag{19.4}$$

19.3.1 *The least-squares estimates for* a *and* b

Henceforth, Y is the random variable, and the value of x is taken to be error free. The residual sum of squares, which is to be minimized, is therefore given by

$$\sum e_i^2 = \sum\{y_i - (a + bx_i)\}^2 = \sum(y_i - a - bx_i)^2. \tag{19.5}$$

To find the least-squares estimates, $\hat{a}$ and $\hat{b}$, the most straightforward approach uses partial differentiation. Differentiating with respect to a gives

$$-2\sum(y_i - \hat{a} - \hat{b}x_i) = 0,$$

so that

$$\hat{a} = \bar{y} - \hat{b}\bar{x}. \tag{19.6}$$

Differentiating the sum of squares with respect to b gives

$$-2\sum x_i(y_i - \hat{a} - \hat{b}x_i) = 0,$$

so that

$$\sum x_i y_i - \hat{a}\sum x_i - \hat{b}\sum x_i^2 = 0.$$

Substituting for $\hat{a}$ gives

$$\sum x_i y_i - \sum x_i \bar{y} + \hat{b}\sum x_i \bar{x} - \hat{b}\sum x_i^2 = 0.$$

Rearranging gives

$$\sum x_i(y_i - \bar{y}) - \hat{b}\sum x_i(x_i - \bar{x}) = 0.$$

Using the simplifying notation of Chapter 18,

$$S_{xy} - \hat{b}S_{xx} = 0.$$

Hence

$$\hat{b} = S_{xy}/S_{xx}. \tag{19.7}$$

Substituting the least-squares estimate into Equation (19.2) gives

$$\hat{e}_i = y_i - (\hat{a} + \hat{b}x_i).$$

The estimated residuals always sum to 0,

$$\begin{aligned} \sum \hat{e}_i &= \sum y_i - \sum(\hat{a} + \hat{b}x_i) \\ &= \sum y_i - n(\bar{y} - \hat{b}\bar{x}) - \hat{b}\sum x_i \\ &= \left(\sum y_i - n\bar{y}\right) - \hat{b}\left(\sum x_i - n\bar{x}\right). \end{aligned}$$

Since both bracketed terms are equal to 0.

$$\sum \hat{e}_i = 0. \tag{19.8}$$

Substituting the least-squares estimates given by Equations (19.6) and (19.7) into Equation (19.5) gives the **deviance**,[2] D:

$$D = \sum \hat{e}_i^2 = S_{yy} - S_{xy}^2/S_{xx}. \tag{19.9}$$

Example 19.2

The electricity consumption of an electric vehicle depends upon the speed at which it is driven. The approximate results from a study of freeway travel collected during a five-month study in California, are summarized in the following table:

mph	5	10	15	20	25	30	35
kW	1	2.5	4	5.5	6	9	7
mph	40	45	50	55	60	65	70
kW	12	13	14	13	16	19	22

Investigate these data.

With Y denoting kW and x denoting mph, we find that $\hat{a} = -0.7912$. and $\hat{b} = 0.2954$. As the figure demonstrates, the least-squares line is indeed a good fit to the data.

[2] An alternative description is the **residual sum of squares**.

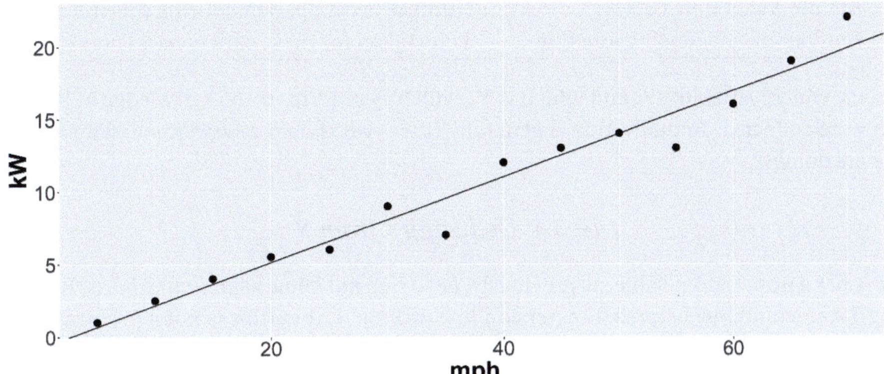

The predicted average kW for a car travelling at a steady 42 mph is therefore given by $-0.7912 + 42 \times 0.2954 \approx 11.6$.

19.3.2 *Linear regression and correlation*

In Chapter 18 we saw that the correlation coefficient, r, was a measure of the extent to which the pairs of values of x and y are collinear. There must therefore be a connection to a linear regression line. In describing correlation, we treated the two variables in the same way, whereas with linear regression we have an asymmetric approach, in which one variable depends on the other. For the regression line $\mathrm{E}(Y|x) = a + bx$, we have found that the slope, b, is estimated by

$$\hat{b} = S_{xy}/S_{xx}.$$

Now suppose that we reverse the roles of the two variables, with X becoming the random variable being dependent on the value of y.. Denote the resulting regression line by $\mathrm{E}(X|y) = c + dy$. The estimated value for d will be

$$\hat{d} = S_{yx}/S_{yy}.$$

Comparing these estimates with the formula for r, the correlation coefficient, given by Equation (18.4), we see that (since $S_{xy} = S_{yx}$)

$$\hat{b}\hat{d} = r^2.$$

19.3.3 *Distinguishing* x *and* Y

Here are two pairs of examples of x and Y-variables. In each case the x-variable has a nonrandom value set by the person carrying out the investigation, while the Y-variable has an unpredictable (random) value.

x	Y
Length of chemical reaction (mins)	Amount of compound produced (g)
Amount of chemical compound (g)	Time taken to produce this amount (mins)
An interval of time (hrs)	Number of cars passing during this interval
Number of cars passing junction	Time taken for these cars to pass (hrs)

To decide which variable is x, and which is Y, evidently requires some knowledge of how and why the data were collected. Actually, this is generally true—we should always know why we are doing what we are doing!

19.3.4 *Deducing* x *from* Y

Suppose x has a nonrandom value, as previously, *but we do not know what that value is*. If we have the resulting Y-value and the estimated regression line of Y on x, then this is not a problem! We simply use the line 'backwards':

$$x = \frac{y - a}{b}.$$

Example 19.3

An experiment is conducted to determine the effects of varying amounts of fertiliser (x, measured in grams per square metre) on the average crop of potatoes (y, measured in kg per plant). The results were as follows:

x	3.25	1.5	0.75	0.0	2.5	8.0	4.0	*
y	2.4	2.2	2.3	2.0	2.3	3.6	3.0	2.9

*One x-value, indicated by a * has been mislaid. Estimate the value of the missing value, using a least-squares procedure.*

Omitting the eighth pair of values, the fitted line is $y = 1.9700 + 0.2005x$. The estimated value of x corresponding to $y = 2.8$ is therefore given by

$$x = \frac{2.9 - 1.97}{0.2005} = 4.63.$$

Since it appears that the chosen values of x are multiples of 0.25, we can reasonably deduce that the missing value was either 4.5 or 4.75 grams per square metre.

Group project: Use a tape measure to measure (in mm) the circumference of your right wrist and the length of your right foot. Pool these results with others so as to obtain around 20 observations. Plot these on a scatter diagram. For the benefit of future would-be foot measurers, determine the regression line of foot length on wrist circumference.

19.4 Transformations, extrapolation, and outliers

Not all relationships are linear. However, there are quite a few nonlinear relations which can easily be turned into the linear form. Here are some examples:

$y = ax^b$	Take logarithms	$\log(y) = \log(a) + b\log(x)$
$y = ae^{bx}$	Take natural logarithms	$\ln(y) = \ln(a) + bx$
$y = (a + bx)^k$	Take kth root	$y^{1/k} = a + bx.$

For many relations no transformation is needed because, over the restricted range of the data, the relation does appear to be linear. As an example, consider the following fictitious data:

x Amount of fertilizer per m^2	y Yield of tomatoes per plant
10 g	1.4 kg
20 g	1.6 kg
30 g	1.8 kg

In this tiny data set there is an exact linear relation between the yield and the amount of fertilizer, namely $y = 0.02x + 1.2$. How we use that relation will vary with the situation.
Here are some examples:

1. We can reasonably guess that, for example, if we had applied 25 g of fertilizer, then we would have got a yield of about 1.7 kg. This is a sensible guess, because 25 g is a value similar to those in the original data.

2. We can expect that 35 g of fertilizer would give a yield of about 1.9 kg. This is reasonable because the original data involved a range from 10 30 g of fertilizer and 35 g is only a relatively small increase beyond the end of that range.

3. We can expect that 60 g of fertilizer might lead to a yield in excess of 2 kg, as predicted by the formula. However, this is little more than a guess, since the original range of investigation (10 g to 30 g) is very different from the 60 g that we are now considering.

4. If we use 600 g of fertilizer, then the formula predicts over 13 kg of tomatoes. This is obviously nonsense! In practice the yield would probably be 0 because the poor plants would be smothered in fertilizer!

   Our linear relation cannot possibly hold for *all* values of the variables, however well it appears to describe the relation in the given data.

The above shows that the least-squares regression line is *not* a substitute for common sense! Care is required, since thoughtless extrapolation can lead to stupid statements. If a cricketer were to make successive scores of 1, 10, and 100, we would be unwise to predict a score of 1000 for his next effort!

The third topic in this section is 'outliers'. An **outlier** is an observation that has values that are very different from the values possessed by the rest of the data. The calculations for regression involve quantities such as $(x_i - \bar{x})$ and $(y_i - \bar{y})$. If one (or both) of these is large—because observation i is an outlier—then observation i is likely to dictate the values of the parameter estimates. Figure 19.2 illustrates a case in which the precise location of a single outlier essentially dictates the equation of the regression line.

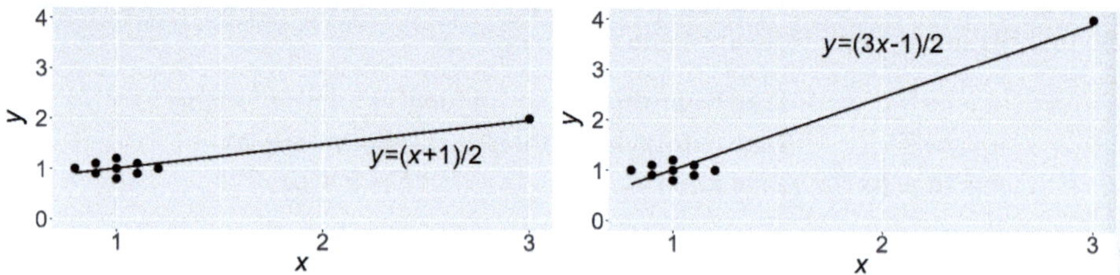

Figure 19.2 A single outlier can have a notable effect on the fitted line.

Nine of the points have the same values in both cases. These values show no special relation between Y and x. The slope of the regression line is essentially decided by the location of the outlying point.

Example 19.4

The following observations on x *and* Y *have been reported.*

x	132	246	188	343	512	442	377	413	421	334
y	116	188	136	215	300	266	239	253	180	218

Plot the data using a scatter diagram and verify that there is an outlier. Supposing that this is due to a typographical error, suggest a correction.

The diagram shows that all the observations bar one lie close to a straight line.

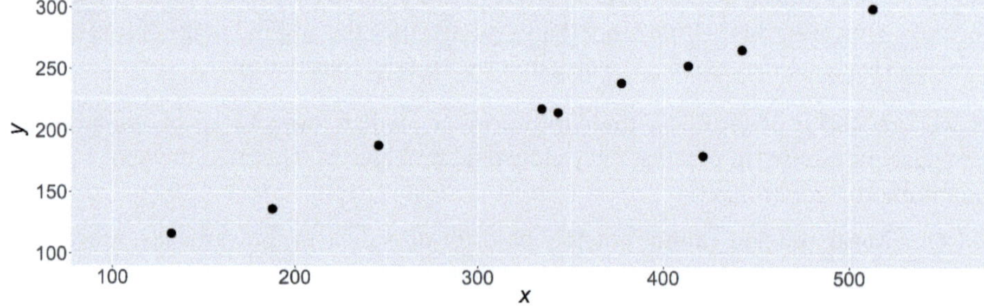

The exception is the point recorded as (421, 180). The most common typographical error involves the interchange of neighbouring digits. It is possible that the point should have been recorded as (241, 180).

Exercises 19b

1. Eight pairs of observations on the variables x and y are as follows:

x	1.2	0.5	0.8	0.1	2.3	1.1	1.8	2.2
y	8.1	4.3	7.1	3.5	12.8	8.4	9.9	11.5

(a) Plot these points on a scatter diagram.

(b) State whether b should be positive or negative.

(c) Determine the coordinates of one point through which the regression line will pass.

2. The radiation intensity I at time t, from a radioactive source, is given by the formula $I = I_0 e^{kt}$, where I_0 and k are constants. Explain how, given a set of values of t and I, linear regression could be used to obtain estimates of I_0 and k.

3. It has been proposed that linear regression could be used to provide a realistic estimated value for y, given that $x = 20$. For each of the following cases, you are asked to state, with a reason, whether this is a reasonable task.

(a)

x	2	5	8	11	17	25	30
y	0	15	22	235	52	80	95

(b)

x	2	5	8	10	7	5	3
y	0	15	22	35	32	20	5

(c)

x	2	5	8	11	17	25	30
y	0	15	22	35	52	80	95

(d)

x	12	15	18	21	27	35	40
y	100	75	50	15	52	80	95

19.5 Properties of the estimators

19.5.1 The slope estimator, $\widehat{B}$

Recall that the y-values are independent observations from distributions having the common variance σ^2. We first consider the slope estimator

$$\widehat{B} = \frac{S_{xY}}{S_{xx}} = \frac{1}{S_{xx}} \sum (x_1 - \bar{x}) Y_i.$$

$$
\begin{aligned}
\mathrm{E}\left(\widehat{B}\right) &= \frac{1}{S_{xx}}\sum(x_i-\bar{x})\mathrm{E}(Y_i)\\
&= \frac{1}{S_{xx}}\sum(x_i-\bar{x})(a+bx_i)\\
&= \frac{a}{S_{xx}}\sum(x_i-\bar{x})+\frac{b}{S_{xx}}\sum(x_i-\bar{x})x_i\\
&= 0+\frac{b}{S_{xx}}\times S_{xx}\\
&= b.
\end{aligned}
$$

The variance is given by

$$
\begin{aligned}
\mathrm{Var}(\widehat{B}) &= \left(\frac{x_1-\bar{x}}{S_{xx}}\right)^2\sigma^2+\left(\frac{x_2-\bar{x}}{S_{xx}}\right)^2\sigma^2+\cdots+\left(\frac{x_n-\bar{x}}{S_{xx}}\right)^2\sigma^2\\
&= \frac{(x_1-\bar{x})^2+(x_2-\bar{x})^2+\cdots+(x_n-\bar{x})^2}{S_{xx}^2}\sigma^2\\
&= \frac{S_{xx}}{S_{xx}^2}\sigma^2\\
&= \frac{\sigma^2}{S_{xx}}.
\end{aligned}
$$

The estimator $\widehat{B}$ is unbiased, with mean b and variance $\frac{\sigma^2}{S_{xx}}$.

Recalling that S_{xx} is a multiple of the variance of the x-values, we see that the greater the variability of the x-values, the more precisely do we know the slope of the line.

19.5.2 Confidence interval for the slope

Information about the common variance is provided by D, given by Equation (19.9). After making allowance for the estimation of the values of a and b, there are effectively $n-2$ independent pieces of information available and it can be shown that an unbiased estimate of σ^2 is provided by

$$\widehat{\sigma^2} = D/(n-2). \tag{19.10}$$

Writing $\hat{b}$ in the form

$$\hat{b} = \frac{x_1-\bar{x}}{S_{xx}}y_1+\frac{x_2-\bar{x}}{S_{xx}}y_2+\cdots+\frac{x_n-\bar{x}}{S_{xx}}y_n,$$

we see that it is a linear combination of the y-values. If we assume that the y-values are observations from normal distributions, then it follows that $\hat{b}$ is also an observation from a normal distribution (see Section 6.3).

If $Y_1, Y_2, ..., Y_n$ have normal distributions, then so does $\widehat{B}$.

Using $D/(n-2)$ as the estimate[3] of σ^2, because the true value is unknown, leads, as usual, to a move from a normal distribution to a t-distribution. In this case that distribution has $(n-2)$ degrees of freedom. Denoting the upper 2.5% point of a t_{n-2}-distribution by $t_{n-2}(.025)$, a symmetric two-sided 95% confidence interval for b is given by

$$\left(\hat{b} - t_{n-2}(.025)\sqrt{\frac{D/(n-2)}{S_{xx}}},\ \hat{b} + t_{n-2}(.025)\sqrt{\frac{D/(n-2)}{S_{xx}}}\right).$$

19.5.3 *Significance test for the slope*

This can be performed in the usual way. Alternatively, the result of such a test can be deduced by studying the corresponding confidence interval. For example, if the population value specified by the null hypothesis falls inside the confidence interval, then the hypothesis will be accepted at the corresponding level.

Example 19.5

The range of an electric car is dependent on many factors. One key factor is the battery size (measured in kWh). The table shows the estimated[a] ranges and battery sizes for randomly chosen models representing seven manufacturers.

Battery size (kWh)	*81.2*	*37.8*	*90.0*	*48.1*	*77.0*	*78.0*	*84.7*
Estimated range (Miles)	*265*	*140*	*215*	*185*	*280*	*250*	*235*

Source: ev-database.org

Examine the dependence of range on battery size.

[a] Estimation based on a hypothesized mix of urban and motorway driving.

We begin with a plot of the data (together with the least-squares line).

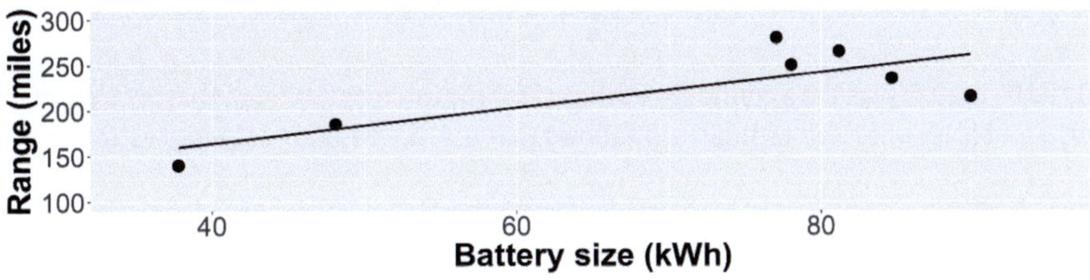

[3] The usual sample variance, $S_{yy}/(n-1)$, is not an appropriate estimate of σ^2 since some of the variation in the y-values may be explained by the varying x-values.

In practice, the results of the calculations would be reported by the computer with a single command, but here we show some of the underlying calculations which begin by determining the sums of the two variables, the sums of their squared values, and the sum of the cross-products:

$$\sum x = 496.8, \sum y = 1570, \sum x^2 = 37622.98, \sum y^2 = 366400, \sum xy = 116023.$$

We next calculate S_{xx}, etc,

$$S_{xx} = 2364.374, \quad S_{yy} = 14271.43, \quad S_{xy} = 4597.857.$$

The estimated slope is $\hat{b} = S_{xy}/S_{xx} = 1.945$, with the estimated intercept being $\hat{a} = \bar{y} - \hat{b}\bar{x} =$ 86.272. Thus the estimated relationship is roughly

$$\text{Range in miles} = 85 + 2 \times \text{Battery capacity in kWh}.$$

The estimate of the error variance is

$$\widehat{\sigma^2} = D/(n-2) = (S_{yy} - S_{xy}^2/S_{xx})/5 = 1066.1.$$

The upper 2.5% point of a t_5-distribution is 2.57, so the 95% confidence interval for the slope is

$$\left(1.945 \pm 2.57\sqrt{\frac{1066.1}{2364.374}}\right) = (0.22, 3.67).$$

Since the interval does not include 0 we can conclude that there is significant evidence, at the 5% level, of a dependence of the estimated range on the battery size.

19.5.4 *The intercept estimator,* $\widehat{\mathrm{A}}$

Since $\mathrm{E}(Y_i) = a + bx_i$, $\mathrm{E}(\bar{Y}) = a + b\bar{x}$ where $\bar{Y}$ is the random variable corresponding to the observed $\bar{y}$. Now $\hat{a} = \bar{y} - \hat{b}\bar{x}$, so, writing $\widehat{A}$ as the random variable corresponding to $\hat{a}$, we find that

$$\mathrm{E}(\widehat{A}) = (a + b\bar{x}) - \mathrm{E}(\widehat{B})\bar{x} = a.$$

So $\hat{a}$ is also an unbiased estimate.[4] Following calculations similar to those shown for the regression line (below), the variance is found to be

$$\sigma^2\left(\frac{1}{n} + \frac{\bar{x}^2}{S_{xx}}\right).$$

This could be be used to provide a confidence interval for the intercept, though that is rarely of interest. Since both $\hat{a}$ and $\hat{b}$ are unbiased estimates, it follows that $\hat{y} = \hat{a} + \hat{b}x$ is an unbiased estimate of $a + bx$.

If $Y_1, Y_2, \ldots, Y_n$ have normal distributions, then so does $\widehat{A}$.

[4] Proof of the fact that all least-squares estimators are unbiased is beyond the scope of this book.

19.5.5 *Confidence interval for a regression line*

Consider a specific x-value, x_k. For this value of x, the estimated value of y will be

$$\widehat{y_k} = \hat{a} + \hat{b}x_k = \bar{y} + \hat{b}(x_k - \bar{x}).$$

Substituting for the least-squares estimates,

$$\widehat{y_k} = \frac{1}{n}\sum y_i - \frac{x_k - \bar{x}}{S_{xx}}\sum(x_i - \bar{x})y_i.$$

To simplify the working denote $\frac{x_k - \bar{x}}{S_{xx}}$ by C. Notice that C is a constant. Hence,

$$\widehat{y_k} = \sum y_i\left\{\frac{1}{n} + C(x_i - \bar{x})\right\}.$$

We now find the variance of Y_k, the corresponding random variable:

$$\begin{aligned}
\text{Var}(\widehat{Y_k}) &= \sum \sigma^2\left\{\frac{1}{n^2} + \frac{2C}{n}(x_i - \bar{x}) + C^2(x_i - \bar{x})^2\right\} \\
&= \sigma^2\left\{\sum\frac{1}{n^2} + \frac{2C}{n}\sum(x_i - \bar{x}) + C^2\sum(x_i - \bar{x})^2\right\} \\
&= \sigma^2\left(n \times \frac{1}{n^2} + 0 + C^2 S_{xx}\right) \\
&= \sigma^2\left(\frac{1}{n} + \frac{(x_k - \bar{x})^2}{S_{xx}}\right).
\end{aligned}$$

Notice that the variance is not the same for all values of x: it is least for values close to $\bar{x}$ and progressively increases as $|x - \bar{x}|$ increases. The consequence is that the bounds for the line are curved.

Since both $\hat{a}$ and $\hat{b}$ are unbiased estimates, it follows that $\hat{y} = \hat{a} + \hat{b}x$ is an unbiased estimate of $a + bx$.

If $Y_1, Y_2, ..., Y_n$ have normal distributions, then so does $\widehat{Y_k}$.

Example 19.5 (cont.)

The figure shows the region within which we are 95% confident that the linear relationship between x and y lies:

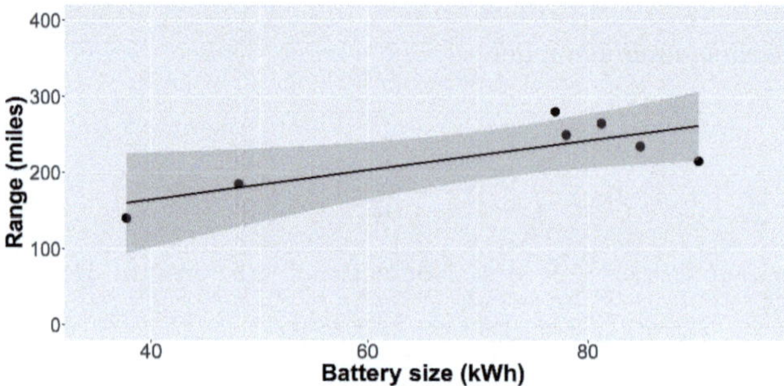

The true linear relationship connecting battery size and range could be represented by *any* line that stays within the (noticeably curved) shaded area.

19.5.6 Prediction interval for future observations

The confidence interval for the regression line gives us an idea of where the line sits: any line that stays within the shaded area might be the true line. Our fitted line represents our best guess. Wherever the true line sits, we cannot expect that future observations will lie directly on that line. Random variation will be liable to displace the observations above or below the line. Therefore, when we ask about the possible y-values of future observations, we must expect to get much greater uncertainty (an extra σ^2) than that concerning the line itself. The variance associated with a future observation at $x = x_f$ is

$$\sigma^2\left(1 + \frac{1}{n} + \frac{(x_f - \bar{x})^2}{S_{xx}}\right).$$

The first term, σ^2, usually dominates this expression, so that the curvature of the resulting interval (when moving across values of x_f) will be much less apparent than for the confidence bounds.

Example 19.5 (cont.)

Continuing with the electric vehicle data, we now add the prediction bounds, together with the results for the next 10 cars randomly chosen from the database:

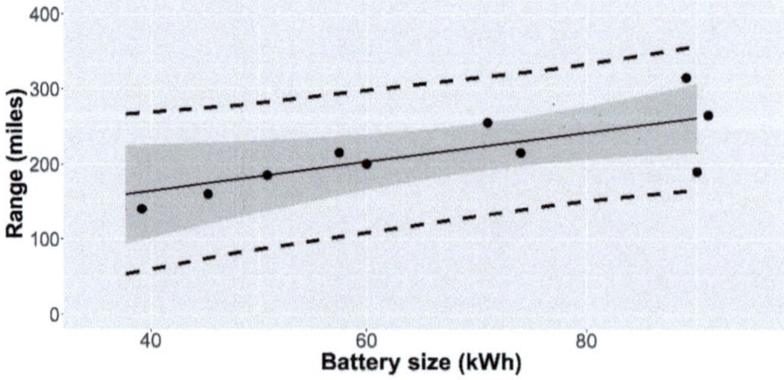

Most of the new data points lie within the confidence region for the fitted line, as one might expect. The two that lie outside that region nevertheless lie well within the prediction region (which is marked by the dashed lines).

Exercises 19c

1. Samples of water were collected from lakes in the vicinity of a Canadian copper smelting factory because it was believed that debris and dust from the factory were contaminating the neighbourhood. The pH values were determined for each sample (pure water has a pH value of 7). The results were

x: Distance (km)	3.9	6.5	13.5	41.9	47.7	52.3	61.3	75.5	90.3
y: pH	3.40	3.20	4.20	5.19	4.41	6.75	7.01	6.40	4.75

Summary statistics include: $S_{xx} = 7465.66$, $S_{xy} = 235.41$, and $S_{yy} = 15.94$.

(a) Fit an appropriate regression line of the form $\mathrm{E}(Y|x) = \alpha+\beta x$, giving the estimated values of α and β to the accuracy that you feel is appropriate.

(b) Taking the value of the upper 2.5% point of a t_7-distribution to be 2.365, give a 95% confidence interval for β.

(c) State, with a reason, whether there is significant evidence at the 5% level, that the pH of the samples is varying with distance from the smelter.

(d) Determine a 95% prediction interval for the pH value of the water from a lake at a distance of 80 km from the smelter.

2. Suppose that it is known that the relationship between two variables x and Y is linear, with the random variable Y having the value 0 when $x = 0$, so that the appropriate relationship between the variables is $\mathrm{E}(Y|x) = \beta x$. Determine an expression for the least-squares estimator of β. Illustrate the procedure using a scatter diagram that shows in what sense the estimator is 'least squares'.

Put theory into practice: Here is an obvious 'anatomical' practical. Collect the heights and weights of about 30 people. Plot the data on a scatter diagram. If the data refer to both males and females, then use different symbols for the two sexes and note whether there appear to be differences between the sexes. For the data referring to your own sex, determine the regression lines of height on weight and weight on height. Use these lines to estimate the average height of someone of your weight and the average weight of someone of your height.

Put theory into practice: How do people 'see' scatter diagrams? If the previous scatter diagram, or one showing a more pronounced relation, was presented to a 13-year-old with the instruction 'Using a ruler, draw a line through these points so as to show the relationship between x and y as clearly as possible', what line would they draw? Would it approximate the regression line of Y on x, or the regression line of X on y, or would it lie halfway between these? Would an 18-year-old have a different preference? Are people who have learnt about regression more likely to approximate the regression line of Y on x than the regression line of X on y?

19.6 Analysis of variance (ANOVA)

Consider the following observations on the random variable, Y:

y	12	18	22	25	30	36	39

It appears that Y is rather variable: the values are widely dispersed about the mean (which is 26), with $\sum(y - \bar{y}^2) = 562$. So the estimate of the variance is $562/6 \approx 93.7$.

Now, however, it is revealed that these seven y-values were not all collected under the same conditions: they do, in fact, correspond to the cases where $x = 1, \dots, 7$. The best fit line is (approximately) $y = 8.143 + 4.464x$. We now have the following information:

x	1	2	3	4	5	6	7
y	12	18	22	25	30	36	39
Estimate using $y = 8.143 + 4.464x$	12.6	17.1	21.53	26.0	30.5	34.9	39.4
Residual	-0.6	0.9	0.5	-1.0	-0.5	1.1	-0.4

The sum of the squared residuals is 3.96, so that the deviance is given by $D = 3.96/(7 - 2) = 0.793$. The majority of the apparent variability in the y-values is now explained, with the estimated variance of Y having been reduced from 9.37 to 0.793.

In this case it is obvious that x and Y are closely related. Before proceeding further, we need to consider what would happen if this were not the case. We know that, if Y did not depend on x, then $s^2 = S_{yy}/(n-1)$ would be an unbiased estimate of σ^2 so that then

$$S_{yy} \text{ would be an estimate of } (n-1)\sigma^2,$$

and S_{yy}/σ^2 would be an observation from a χ^2_{n-1}-distribution (see Section 15.9).

When the dependence of Y on x is taken into account, $D/(n-2)$ is an unbiased estimate of σ^2 and it can be shown that D/σ^2 has a chi-squared distribution with $(n-2)$ degrees of freedom. This implies that

$$D \text{ is an unbiased estimate of } (n-2)\sigma^2.$$

Subtracting these results,

$$\text{if } Y \text{ does not depend on } x, \text{ then } (S_{yy} - D) \text{ is an estimate of } \sigma^2.$$

If Y *does* depend on x, then $S_{yy} - D > \sigma^2$. Thus $S_{yy} - D$ provides the key information concerning any dependence of Y on x.

Since we have that $D/\sigma^2 \sim \chi^2_{n-2}$, while, if Y and x are unrelated, $S_{yy}/\sigma^2 \sim \chi^2_{n-1}$, it is plausible that

$$S_{yy} - D = S_{yy} - (S_{yy} - S_{xy}^2/S_{xx}) = S_{xy}^2/S_{xx} \sim \chi^2_1.$$

It can be shown that this is the case and that this chi-squared distribution is independent of the distribution of D. We can now put these results together in an ANOVA table:

Source of variation	D. f.	Sum of squares	Mean square	Distribution
Regression on x	1	$R = S_{xy}^2/S_{xx}$	R	$\sigma^2\chi^2_1$
Residual variation	$n-2$	D	$D/(n-2)$	$\sigma^2\chi^2_{n-2}$
Total	$(n-1)$	S_{yy}		

This leads (at last!) to a means of assessing whether the supposed relationship between Y and x is significant. The ratio of R divided by $D/(n-2)$ involves the ratio of two independent chi-squared distributions and that ratio will have an $F_{1,n-2}$-distribution (see Section 7.3). An observed value lying in the upper-tail of that distribution will be an indication of a dependence of Y on x.

The quantity $1-D/S_{yy}$ which measures how much of the variation in y-values is explained by the model is called the **coefficient of determination**.

It is usual for one column of the summary table of the analysis of variance to report the ratio of interest, so, for the example with which we started this section, we would have

Source of variation	D. f.	Sum of squares	Mean square	Variance ratio
Regression on x	1	558.04	558.04	703.83
Residual variation	5	3.96	0.79	
Total	6	562.00		

The computer reports that the probability of a variance ratio that large or larger, is about 1.4×10^{-6}: there can be no doubt concerning the dependence of Y on x.

Example 19.5 (cont.)

For the EV range data, the computer reports (using an unnecessary number of significant figures!) the following:

Source of variation	D. f.	Sum of squares	Mean square	F-value	Pr(>F)
Battery size	1	8941.2	8941.2	8.3872	0.03395
Residual variation	5	5330.3	1066.1		

The most informative value in this table is the tail probability reported for the F-distribution. This is 0.03395, which is a little less than 5% and certainly indicates that a car's range is dependent on the battery size. The fact that that tail probability is not much lower suggests that, as one would have anticipated, there are other factors that are relevant.

19.7 Multiple regression

As the electric car example has suggested, there are usually several explanatory variables that may affect the value of the dependent variable. The apparent relevance of a variable is liable to be affected by which other variables have been selected as part of the model. To see how this might occur, consider the following trivial example:

y : The weight of a person (kg, to the nearest kg)
x_1 : The height of that person (cm, to the nearest cm)
x_2 : The height of that person (mm, to the nearest mm)

Weight predictions based on either of the two explanatory variables might be expected to be reasonably accurate, but there will be no need for both variables. Taken on its own, x_1 may appear rather important as a predictor of weight. However, if x_2 is already in the model, then x_1 will provide no useful information.

When there are many possible explanatory variables, the techniques for efficient variable selection are beyond the scope of this book. However, when there are just two or three variables, it will often be possible to explore all the possibilities, using ANOVA tables, to arrive at a conclusion.

Example 19.5 (cont.)

We now repeat the electric car data, but with the addition of a candidate second explanatory variable, the maximum rate at which the battery can be charged. This is not obviously connected to the car's range, but a high maximum rate might indicate the use of advanced technology.

x_1	Battery size (kWh)	81.2	37.8	90.0	48.1	77.0	78.0	84.7
x_2	Maximum rate (Mph)	390	230	180	280	420	380	220
y	Estimated range (Miles)	265	140	215	185	280	250	235

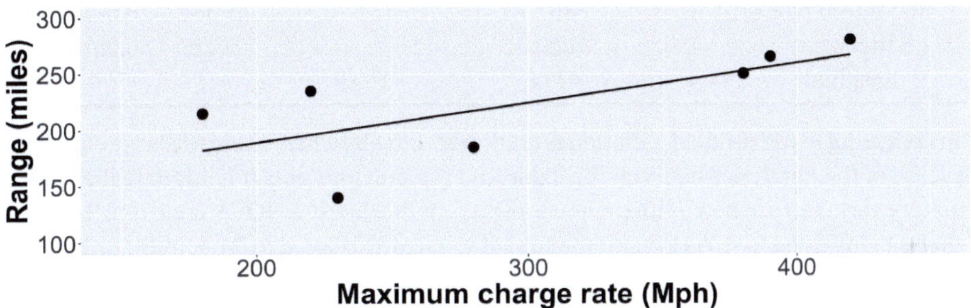

The plot of maximum charge rate against distance certainly indicates a connection, with the strength indicated by the corresponding ANOVA table:

Source of variation	D. f.	Sum of squares	Mean square	F-value	Pr(>F)
Maximum rate	1	6807.7		4.5605	0.0858
Residual variation	5	7463.7	1492.7		

The tail probability of greater than 8% implies that, on its own, the maximum charge rate is not a convincing predictor of range. We now consider the model in which the maximum range is dependent on both explanatory variables. The estimated relation is reported as

$$\text{Range} = 10.45 + 1.74 \times \text{Battery size} + 0.30 \times \text{Maximum charge rate}.$$

The fact that $0.30 < 1.74$ is irrelevant, since these values would change if we changed the units with which battery size and charge rate were measured. To get a proper idea of the importance of the variables we must study the ANOVA table. Fitting the variable 'Battery size' first we obtain the following:

Source of variation	D. f.	Sum of squares	Mean square	F-value	Pr(>F)
Battery size	1	8941.2		80.591	0.0008521
Maximum rate	1	4886.5		44.044	0.0026751
Residual	4	443.8	110.9		

There are several features to be noted here:

- The residual mean square, which is the estimate of the error variance, is greatly reduced from the previous estimates.

- For *both* variables, their significance is greatly increased. In particular, the maximum charge rate is now seen to be highly significant with a tail probability of less than 0.3%.

In case this seems straightforward, consider the following ANOVA table, which refers to the same data:

Source of variation	D. f.	Sum of squares	Mean square	F-value	Pr(>F)
Maximum rate	1	6807.7		61.361	0.001434
Battery size	1	7020.0		63.275	0.001353
Residual	4	443.8	110.9		

The line referring to the residual variation is unaltered, but elsewhere there are large changes. To make sense of the difference between this table and the previous one, it helps to realize that the explanatory variables are being fitted one-at-a-time. In this latest ANOVA table it is 'Maximum rate' that is being fitted first. As a result the sum of squares attributed to 'Maximum rate' (6807.7) is the sum of squares reported earlier when this was the only explanatory variable. The 7020.0 explained by 'Battery size' represents the major part of the previous 'Error' residual for that single-variable model.

Hopefully, the extended discussion required for this simple example will explain why an entire book can be assigned to a discussion of situations involving more than two variables.

Put theory into practice: Conduct your own investigation into how the prices of used cars depend upon age, mileage, etc., by transcribing data from some convenient online source such as *Parkers Car Guide* (UK) or the *Kelley Blue Book* (USA). Does the depreciation rate vary greatly according to the make, model, or original price of the car?

Key facts

Define S_{xx}, S_{yy}, and S_{xy} by, for example,

$$S_{xy} = \sum\{(x_i - \bar{x})(y_i - \bar{y})\} = \sum(x_i - \bar{x})y_i = \sum x_i y_i - \frac{1}{n}\sum x_i \sum y_i,$$

- **The least-squares line of best fit**, which is **the regression line of** y **on** x, is $y = \hat{a} + \hat{b}x$, where

$$\hat{b} = \frac{S_{xy}}{S_{xx}},$$
$$\hat{a} = \bar{y} - b\bar{x}.$$

This choice for a and b minimizes the sum of the squared residuals: $\sum\{y_i - (a + bx_i)\}^2$. The resulting minimum is the **deviance**, or **residual sum of squares**, D, where

$$D = S_{yy} - S_{xy}^2/S_{xx}.$$

- A **confidence interval for the slope**, b, is

$$\hat{b} \pm c\sqrt{\frac{D/(n-2)}{S_{xx}}},$$

where c is the relevant value from a t_{n-2} distribution.

- A **confidence interval for the line** at the point where $x = x_k$ is formed using

$$\hat{y} \pm c\sqrt{\frac{D}{(n-2)}\left(\frac{1}{n} + \frac{(x_k - \bar{x})^2}{S_{xx}}\right)}.$$

R

- With the data in the vectors x and y the **regression line** is found using the command lm(y~x). The 'lm' is short for linear model.
- To see the **fitted values** given by the line use lm(y~x)$fitted.
- To see the **residuals** use lm(y~x)$residuals.
- To obtain **confidence intervals** for the slope and intercept use confint(lm(y~x)).
- To obtain a confidence interval, or a prediction interval, for particular values of x, place the values in a data frame df, then use either predict(lm(y~x), df,interval = "confidence") or predict(lm(y~x), df,interval = "prediction")
- To display the confidence bounds and prediction interval using the given formulae with basic R commands is tedious (but possible!). The presentation used here required library(ggplot2) and then the use of the ggplot, geom_point and geom_smooth commands.
- With two variables use e.g. lm(y~x1+x2)

20

*The Bayesian approach

In this short chapter, we will briefly examine an approach to data analysis, that provides an alternative to the **frequentist** methods presented in Chapters 14 to 19. The term 'frequentist' refers to the interpretation of probability as being the limit of relative frequency, with, as a result, population quantities being directly estimated by their sample counterparts.

To give an idea of the difference in the two attitudes to data analysis, consider the simple situation where a coin is tossed. The coin has two sides: 'heads' and 'tails'. Assuming that the coin has no obvious peculiarity, the frequentist approach will be to hypothesize that the probability of a head is 0.5, toss the coin a few times, and then test the hypothesis. If the coin appears biased, with probability r/n, the frequentist will quote that ratio as an estimate, and then provide a confidence interval following the methods described earlier.

The Bayesian will argue that there is a zero possibility of any coin being *exactly* fair: every coin has scratches and other blemishes, while the different patterns on the two sides will contribute to bias. So the probability of a head will not be the 0.5 asserted by the frequentist. Nevertheless, the Bayesian will concede that the probability of a head *is* likely to be near 0.5, but without ruling out other possible values. After a few coin tosses, the Bayesian 's view concerning the probable value of p will have been modified, but without ruling out any possibility.

In essence, the Bayesian approach is to acknowledge that there is uncertainty about population parameters. This uncertainty is expressed in the form of a probability distribution (the so-called **prior distribution**) for the unknown parameter value. When data become available, they are used to update our ideas about the parameter value to give a revised probability distribution (the **posterior distribution**).

The updating relies upon Bayes' rule, which was introduced in Section 2.5. In its simplest form, this states that, for two events A and B,

$$\mathrm{P}(A|B) = \frac{\mathrm{P}(B|A)\mathrm{P}(A)}{\mathrm{P}(B)}.$$

In the current context, the roles of A and B are taken by the unknown parameter (θ, say), and the observed data:

$$\mathrm{P}(\theta)|\mathrm{data}) = \frac{\mathrm{P}(\mathrm{data}|\theta)\mathrm{P}(\theta)}{\mathrm{P}(\mathrm{data})}. \tag{20.1}$$

Here $\mathrm{P}(\theta)$ is the prior distribution and $\mathrm{P}(\theta|\mathrm{data})$ is the posterior distribution.

The quantity $\mathrm{P}(\mathrm{data}|\theta)$ is the **likelihood** that we first met in Section 14.2.2.

The quantity P(data) is calculated using the total probability theorem (Section 2.4):

$$\mathrm{P(data)} = \int_{\mathrm{All}\ \theta} \mathrm{P(data}|\theta)\mathrm{P}(\theta)\mathrm{d}\theta. \tag{20.2}$$

For a point estimate, there are several possibilities, all of which rely on the posterior distribution. Examples are the mean of the posterior distribution, the posterior mode, and the posterior median.

20.1 Conjugate priors

Equation (20.2) is easy to write down, but could be computationally troublesome, depending on the form of the prior distribution. Different prior distributions will lead to different posterior distributions. A conjugate prior is a prior that has been chosen from a family of distributions such that the posterior distribution is another member of that family. A good choice will avoid the need for any summation or integration.

20.1.1 Beta prior for a binomial distribution

Because of its limited range (from 0 to 1) and great flexibility, in the case of a proportion, a suitable choice of prior is a member of the family of beta distributions (Section 5.7). For some values of the parameters α and β, with θ now denoting the probability of a success, we have

$$\mathrm{P}(\theta) = \theta^{\alpha-1}(1-\theta)^{\beta-1}/\mathrm{B}(\alpha,\beta).$$

The situation is binomial, so, with r successes in n trials,

$$\mathrm{P(data}|\theta) = \binom{n}{r}\theta^r(1-\theta)^{n-r}.$$

In this case, therefore,

$$\begin{aligned}
\mathrm{P(data)} &= \int_0^1 \binom{n}{r}\theta^r(1-\theta)^{n-r} \times \frac{1}{\mathrm{B}(\alpha,\beta)}\theta^{\alpha-1}(1-\theta)^{\beta-1}\mathrm{d}\theta \\
&= \frac{1}{\mathrm{B}(\alpha,\beta)}\binom{n}{r}\int_0^1 \theta^{r+\alpha-1}(1-\theta)^{n-r+\beta-1}\mathrm{d}\theta \\
&= \binom{n}{r} \times \mathrm{B}(r+\alpha, n-r+\beta)\,/\mathrm{B}(\alpha,\beta)\,.
\end{aligned}$$

Putting this altogether we have

$$\begin{aligned}
\mathrm{P}(\theta)|\mathrm{data}) &= \frac{\left\{\binom{n}{r}\theta^r(1-\theta)^{n-r}\right\} \times \left\{\theta^{\alpha-1}(1-\theta)^{\beta-1}\,/\mathrm{B}(\alpha,\beta)\right\}}{\binom{n}{r} \times \mathrm{B}(r+\alpha, n-r+\beta)\,/\mathrm{B}(\alpha,\beta)} \\
&= \theta^{r+\alpha-1}(1-\theta)^{n-r+\beta-1}/\mathrm{B}(r+\alpha, n-r+\beta),
\end{aligned}$$

which is the pdf of a beta distribution with parameters $(r+\alpha)$ and $(n-r+\beta)$.

Thus the initial values of α and β can be interpreted as the notional historical numbers of successes and failures.

Example 20.1

A magician produces a coin from his pocket. We know nothing about this coin. It might be bent; it might be fair; it might be double-headed; it might be double-tailed. The magician tosses the coin 10 times and obtains heads on nine occasions. Plot the prior and posterior distributions.

Since we knew nothing about the coin, the only feasible choice for the prior distribution, $f(\theta)$, is the uniform distribution, which corresponds to a beta distribution with both parameters set equal to 1. With nine heads and one tail, the updated posterior distribution is a $B(10, 2)$-distribution.

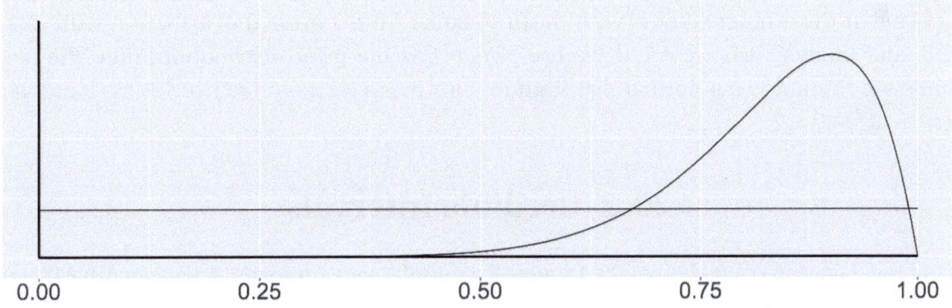

The posterior distribution has a mode at $(\alpha - 1)/(\alpha + \beta - 2) = 0.9$, a median of 0.85, and a mean of $\alpha/(\alpha + \beta) = 10/12 == 0.83$. A frequentist would be estimating the probability of a head as $r/n = 9/10 = 0.9$.

Beta distributions are also used as conjugate priors for the Bernoulli, geometric, and negative binomial distributions, since in each case the focus is on the value of a probability.

20.1.2 Gamma prior for a Poisson distribution

To simplify the presentation, the terms in the distributions that do not involve the unknown parameters will be expressed as the constants C_1 and C_2.

Suppose we have n observations $x_1, x_2, \dots, x_n$, from a Poisson distribution with parameter θ. Their joint likelihood is

$$P(\text{data})|\theta) = C_1 \theta^{n\bar{x}} e^{-n\theta},$$

where $\bar{x}$ is the sample mean.

With Poisson observations, a convenient choice of prior is the gamma distribution with parameters α and β, which is given by

$$P(\theta) = C_2 \theta^{\alpha-1} e^{-\beta\theta}.$$

The next step is to determine P(data), which is given by

$$\begin{aligned} \text{P(data)} &= C_1C_2\int_0^\infty \theta^{n\bar{x}}\text{e}^{-n\theta}\theta^{\alpha-1}\text{e}^{-\beta\theta}\text{d}\theta \\ &= C_1C_2\int_0^\infty \theta^{n\bar{x}+\alpha-1}\text{e}^{-(n+\beta)\theta}\text{d}\theta. \end{aligned}$$

Thus,

$$\text{P}(\theta|\text{data}) = \frac{C_1C_2\theta^{n\bar{x}+\alpha-1}\text{e}^{-(n+\beta)\theta}}{C_1C_2\int_0^\infty \theta^{n\bar{x}+\alpha-1}\text{e}^{-(n+\beta)\theta}.\text{d}\theta}.$$

The C_1C_2 terms cancel and we are left with a gamma distribution with parameters $(n\bar{x}+\alpha)$ and $(n+\beta)$.

20.1.3 Normal prior for a normal distribution

The normal distribution is its own conjugate distribution, though, with two parameters, the formulae are a little more tedious.

Suppose that the n observations (with mean $\bar{y}$) come from a normal distribution with unknown mean, θ, and known variance σ^2. If we use $\text{N}(\mu, \tau^2)$ as the prior distribution, then the posterior distribution is found to be a normal distribution with mean $(\sigma^2\mu + \tau^2 n\bar{x})/(\sigma^2 + n\tau^2)$ and variance $\sigma^2\tau^2/(\sigma^2 + n\tau^2)$.

20.2 Credible intervals

These are the Bayesian equivalent of the frequentist's confidence intervals. A 95% credible interval for θ would be bounded by the 2.5 and 97.5% points of the posterior distribution.

Example 20.1 (cont.)

Having observed nine successes in 10 trials, then, using a uniform prior to give a B(10, 2)-distribution posterior, the 95% credible interval would be (0.587, 0.977). This is nearly identical to the interval that would be obtained by a frequentist using Wilson's method (Section 14.4.2) which is (0.596, 0.982).

Key facts

- $$\mathrm{P}(\theta)|\mathrm{data}) = \mathrm{P}(\mathrm{data}|\theta)\mathrm{P}(\theta)\,/\mathrm{P}(\mathrm{data}),$$

  where $\mathrm{P}(\theta)$ is the **prior distribution**, $\mathrm{P}(\mathrm{data}|\theta)$ is the **likelihood**, $\mathrm{P}(\theta|\mathrm{data})$ is the **posterior distribution**, and given by

  $$\mathrm{P}(\mathrm{data}) = \int_{\mathrm{All}\ \theta} \mathrm{P}(\mathrm{data}|\theta)\mathrm{P}(\theta)d\theta.$$

- **Conjugate priors** are chosen so that the posterior distribution is a member of the same family.
  - Beta distribution for binomial,
  - Gamma distribution for Poisson,
  - Normal distribution for normal.
- **Credible intervals** use e.g. the 2.5 and 97.5% points of the posterior distribution.

R

- With X having a beta distribution with parameters α and β, the command `qbeta(δ, α, β)` reports the value x for which $\mathrm{P}(X < x) = \delta$.
- With X having a gamma distribution with parameters α and β, the command `qgamma(δ, α, β)` reports the value x for which $\mathrm{P}(X < x) = \delta$.
- With X having a normal distribution with parameters μ and σ, the command `qnorm(δ, μ, σ)` reports the value x for which $\mathrm{P}(X < x) = \delta$.

INDEX